Plant Genetic Engineering and Biotechnology

Plant Genetic Engineering and Biotechnology

Editors

Zanmin Hu
Han Xiao
Yi Ren
Chengming Fan

MDPI

Basel • Beijing • Wuhan • Barcelona • Belgrade • Novi Sad • Cluj • Manchester

Editors
Zanmin Hu
Chinese Academy of Sciences
Beijing
China

Han Xiao
Chinese Academy of Sciences
Shanghai
China

Yi Ren
Beijing Key Laboratory of
Vegetable Germplasm
Improvement
Beijing
China

Chengming Fan
Chinese Academy of Sciences
Beijing
China

Editorial Office
MDPI
St. Alban-Anlage 66
4052 Basel, Switzerland

This is a reprint of articles from the Special Issue published online in the open access journal *Plants* (ISSN 2223-7747) (available at: https://www.mdpi.com/journal/plants/special_issues/6E29F1YH3U).

For citation purposes, cite each article independently as indicated on the article page online and as indicated below:

Lastname, A.A.; Lastname, B.B. Article Title. *Journal Name* **Year**, *Volume Number*, Page Range.

ISBN 978-3-7258-1035-2 (Hbk)
ISBN 978-3-7258-1036-9 (PDF)
doi.org/10.3390/books978-3-7258-1036-9

Contents

plants

MDPI

Article

Genome-Wide Identification of the Cytochrome P450 Superfamily Genes and Targeted Editing of *BnCYP704B1* Confers Male Sterility in Rapeseed

Zhilai Wang [1,†], Yanfeng Zhang [2,†], Min Song [2,3,†], Xiuhua Tang [1], Shuhua Huang [2], Bin Linhu [3], Ping Jin [1], Weike Guo [1], Fang Li [1], Liwen Xing [1], Ran An [2], Xiaona Zhou [1], Wenfang Hao [1], Jianxin Mu [2,*] and Changgen Xie [1,*]

1 State Key Laboratory of Crop Stress Biology for Arid Areas and College of Life Sciences, Northwest A&F University, Yangling 712100, China
2 Hybrid Rapeseed Research Centre of Shaanxi Province, Yangling 712100, China
3 College of Agronomy, Northwest A&F University, Yangling 712100, China
* Correspondence: jxmsxyc@163.com (J.M.); changen.xie@nwafu.edu.cn (C.X.); Tel.: +86-29-87081418 (C.X.); Fax: 86-29-87080062 (C.X.)
† These authors contributed equally to this work.

Abstract: The cytochrome P450 (CYP450) monooxygenase superfamily, which is involved in the biosynthesis pathways of many primary and secondary metabolites, plays prominent roles in plant growth and development. However, systemic information about CYP450s in *Brassica napus* (BnCYP450) was previously undiscovered and their biological significance are far from understood. Members of clan 86 CYP450s, such as CYP704Bs, are essential for the formation of pollen exine in plant male reproduction, and the targeted mutagenesis of *CYP704B* genes has been used to create new male sterile lines in many crops. In the present study, a total of 687 *BnCYP450* genes were identified in *Brassica napus* cultivar "Zhongshuang 11" (ZS11), which has nearly 2.8-fold as many *CYP450* members as in *Arabidopsis thaliana*. It is rationally estimated since *Brassica napus* is a tetraploid oil plant with a larger genome compared with *Arabidopsis thaliana*. The *BnCYP450* genes were divided into 47 subfamilies and clustered into nine clans. Phylogenetic relationship analysis reveals that *CYP86* clan consists of four subfamilies and 109 *BnCYP450s*. Members of *CYP86* clan genes display specific expression profiles in different tissues and in response to ABA and abiotic stresses. Two *BnCYP450s* within the *CYP704* subfamily from *CYP86* clan, *BnCYP704B1a* and *BnCYP704B1b*, display high similarity to *MS26* (*Male Sterility 26*, also known as *CYP704B1*). These two *BnCYP704B1* genes were specifically expressed in young buds. We then simultaneously knocked-out these two *BnCYP704B1* genes through a clustered regularly interspaced short palindromic repeats/CRISPR-associated protein 9 (CRISPR/Cas9) genome engineering system. The edited plants displayed a pollenless, sterile phenotype in mature anthers, suggesting that we successfully reproduced genic male sterility (GMS, also known as nuclear male sterility) lines in *Brassica napus*. This study provides a systemic view of *BnCYP450s* and offers a strategy to facilitate the commercial utility of the CRISPR/Cas9 system for the rapid generation of GMS in rapeseed via knocking-out *GMS* controlling genes.

Keywords: cytochrome P450; genic male sterility; *MS26*; CRISPR/Cas9; genome engineering

Citation: Wang, Z.; Zhang, Y.; Song, M.; Tang, X.; Huang, S.; Linhu, B.; Jin, P.; Guo, W.; Li, F.; Xing, L.; et al. Genome-Wide Identification of the Cytochrome P450 Superfamily Genes and Targeted Editing of *BnCYP704B1* Confers Male Sterility in Rapeseed. *Plants* **2023**, *12*, 365. https://doi.org/10.3390/plants12020365

Academic Editors: Zanmin Hu, Han Xiao, Yi Ren and Chengming Fan

Received: 11 December 2022
Revised: 28 December 2022
Accepted: 4 January 2023
Published: 12 January 2023

1. Introduction

Rapeseed (*Brassica napus* L., also known as canola, genome AACC, 2n = 4x = 38) is one of the most important vegetable oil crops across the world and is an allotetraploid that developed through doubling of chromosomes after the hybridization between *Brassica rapa* (genome AA, 2n = 20) and *Brassica oleracea* (genome CC, 2n = 18) [1,2]. The widespread adoption of hybrid vigor has contributed significantly to rapid and widespread increases in rapeseed production over the past few decades. Since then, the use of several pollination control systems has been proposed to promote heterosis in rapeseed, including cytoplasmic male sterility (CMS), genic male sterility (GMS), self-incompatibility, and chemical-induced male sterility (CIMS) [3]. Compared with other approaches, the major advantage of GMS is that it brings about nearly complete male sterility to a hybrid breeding program.

There are several GMS control loci used in hybrid rapeseed production, such as the S45AB, 117AB, Oro, 7365AB, 9012AB, Rs1046AB, Yi3A, and 609A systems [4–8]. In some of these systems, the genes controlling the male sterility have been determined. For instance, the male sterility of S45A and 117A is controlled by two duplicate genes, *BnMs1* and *BnMs2* [7]. The male sterile phenotypes of 7365A and 9012A are coupled with the loss of the *BnMs3* gene [5,6,8]. However, the main obstacle to using GMS remains the cross-breeding of the male sterility into new rapeseed varieties and there are only a few GMS lines available for conventional genetic manipulation.

Recently, targeted mutagenesis of *GMS* control genes has been used to create new male sterile lines in many crops [9–18]. One of these *GMS* genes is *ZmMS26* (also known as *ZmCYP704B1*), which is a cytochrome *P450-like* gene first identified in maize [15,16,19]. *ZmMS26* encodes long-chain fatty acid omega-hydroxylase required for sporopollenin biosynthesis and essential for pollen exine formation [11,15]. Since then, protein sequences with high homology to the ZmMS26 were identified, and targeted mutagenesis was performed to create new male sterile germplasms in many plants, such as maize, rice, sorghum, and wheat [11,14–16,20]. However, the rapid deployment of genic male sterile lines of rapeseed at the commercial breeding level derived from artificial manipulation remains controversial.

The cytochrome P450 (CYP450) monooxygenase superfamily is one of the largest enzymatic protein families of heme-thiolate proteins [21,22]. The CYP450s catalyze a great number of NADPH- and/or O_2-dependent oxygenation/hydroxylation reactions in many organisms [21,23]. The *CYP450* genes are widely identified in most organisms, including bacteria, plants, animals, and humans [21–28]. In plants, the nomenclature of *CYP450* genes is generally classified into two groups: A type and non-A type [21,22]. The A type *CYP450s* are specific to plants; the non-A type is non-plant-specific *CYP450s* which display high sequence similarity to animals and fungal *CYP450s* [22,23]. According to the current evolutionary relationship, plant *CYP450* subfamilies are grouped into different clans including five single-subfamily clans (clan 51, clan 710, clan 711, clan 74 and clan 97) and four multiple-subfamily clans (clan 72, clan 85, clan 86 and clan 71) [21,22,24]. The clan 71 was categorized into A type *CYP450s*, and the other eight clans were categorized into non-A type *CYP450s* [21–23].

A great number of *CYP450* genes have crucial roles in the biosynthesis of metabolites with distinct biological significance [21,25]. Plenty of the clan 73, clan 84, and clan 98 *CYP450* genes are involved in the phenylpropanoid biosynthetic pathway, results in producing a mass of phenol related compounds, i.e., flavonoids, suberin, lignin, and polyphenols [21,25]. Members of the *CYP703*, *CYP704*, and *CYP86* subfamilies are responsible for the biosynthesis of pollen exine and anther cutin, which are known as fatty acid hydroxylase [7,29–32]. The *CYP701A* and *CYP88A* genes are necessary for gibberellic acid (GA) biosynthesis, which encode ent-kaurene oxidase and ent-kaurenoic acid oxidase respectively [24,33,34]. The CYP707A is involved in abscisic acid (ABA) 8′-hydroxylation, which play an important role in ABA catabolism [35,36]. The CYP450 enzymes of the CYP74 subfamily are essential for jasmonate (JA) synthesis, which function as allene oxide synthase (AOS) [24,37]. Members of the CYP90, CYP72, CYP92, CYP85, CYP734, and CYP724 subfamilies are necessary for brassinosteroid (BR) biosynthesis [21,38]. Besides, CYP450s also act in fighting against abiotic and biotic stresses [21].

In the current study, we systematically identified and analyzed the *BnCYP450* genes in the rapeseed genome. In addition, we report the introduction of GMS into rapeseed by simultaneously knocking-out two anther-specific *ZmMS26* homologs through gene editing using the CRISPR (clustered regularly interspaced short palindromic repeats)/Cas9 (CRISPR-associated protein 9) system.

2. Results

2.1. Genome-Wide Identification of BnCYP450s

There are 687 putative BnCYP450s with conserved P450 domains isolated from the genome of *Brassica napus* cultivar "Zhongshuang 11" (ZS11), which is around 2.8 folds of AtCYP450s. This is not surprising since *Brassica napus* is tetraploid and Arabidopsis is diploid with a very small genome size. We found 318 and 367 *BnCYP450* genes in the A and C subgenome respectively (Table 1). The *BnCYP450s* were renamed in terms of their sequences' similarity to *AtCYP450s* (Table S1). The protein length of BnCYP450s was between 80 and 1971 amino acids (aa) with predicted molecular weight (Mw) in the range of 8.67–224 kDa. The theoretical isoelectric points (pI) of BnCYP450s ranged from 4.1 to 11.52 (Table S1). The pI of 547 BnCYP450s (79.6%) showed values greater than 7. Other characteristics of BnCYP450s, such as gene position and accession number, are presented in Table S1.

Table 1. Enumeration of AtCYP450 and BnCYP450 gene superfamily.

Classifications		Family	*Arabidopsis thaliana*	*Brassica napus*	*Brassica napus* A Subgenome	*Brassica napus* C Subgenome
Different Types	**Different Clans**					
Non-A type	CYP51 clan	CYP51	2	9	3	6
	CYP72 clan	CYP709	3	10	4	6
		CYP714	2	5	2	3
		CYP715	1	5	2	3
		CYP72	9	25	12	13
		CYP721	1	2	1	1
		CYP734	1	2	1	1
		CYP735	2	4	2	2
	CYP710 clan	CYP710	4	6	2	4
	CYP711 clan	CYP711	1	2	1	1
	CYP74 clan	CYP74	2	4	2	2
	CYP85 clan	CYP702	6	7	3	4
		CYP707	4	17	8	9
		CYP708	4	9	6	3
		CYP716	3	0		
		CYP718	1	2	1	1
		CYP720	1	2	1	1
		CYP722	1	3	1	2
		CYP724	1	4	2	2
		CYP85	2	8	4	4
		CYP87	1	4	2	2
		CYP88	2	5	2	3
		CYP90	4	18	9	9
	CYP86 clan	CYP704	3	8	4	4
		CYP86	11	29	14	15
		CYP94	6	20	9	10
		CYP96	14	52	24	28
	CYP97 clan	CYP97	3	13	7	6
Total non-A type			95	275	129	145
A type	CYP71 clan	CYP701	1	2	1	1
		CYP703	1	4	2	2
		CYP705	25	47	19	28
		CYP706	8	16	8	8
		CYP71	50	138	72	66
		CYP712	2	4	2	2
		CYP73	1	7	3	4
		CYP75	1	2	1	1
		CYP76	9	29	12	17
		CYP77	5	16	5	11
		CYP78	6	23	11	12
		CYP79	10	25	11	14
		CYP81	18	54	23	30
		CYP82	5	13	4	9
		CYP83	2	6	3	3
		CYP84	2	9	4	5
		CYP89	7	7	4	3
		CYP93	1	2	0	2
		CYP98	3	8	4	4
Total A-type			157	412	189	222
Total			252	687	318	367

2.2. Classification and Phylogenetic Analyses of BnCYP450 Proteins

We divided BnCYP450 proteins into 46 subfamilies according to the P450 nomenclature (Table S1). To explore the evolutionary relationships between AtCYP450s and Bn-

CYP450s, an unrooted NJ tree with the 251 AtCYP450s and 687 BnCYP450s was constructed (Figure S1). Based on it, BnCYP450s were also categorized into nine clans. However, there were no members identified in CYP716 subfamily. The *BnCYP450s* were also classified into non-A type and A-type, which is consistent to previous evolutionary analysis [23]. The non-A type BnCYP450s consisted of eight clans with 27 subfamilies containing 275 members. The A-type BnCYP450s were all grouped into CYP71 clan with 19 subfamilies comprising 412 members (Table 1). The CYP721, CYP734, CYP718, CYP720, CYP701, CYP75, and CYP93 subfamilies all had only two members, while the largest subfamily CYP71 harbored 138 members.

2.3. Gene Duplication and Collinearity Analysis of the BnCYP450 Genes

To study the putative tandem and segmental duplication events, we determined the chromosomal distribution of *BnCYP450* family members. In total, the isolated 687 *BnCYP450* genes, except *BnCYP81F4a* and *BnCYP94B2a*, were widely distributed on 19 chromosomes (Table S1). *BnCYP81F4a* and *BnCYP94B2a* were anchored on HiCscaffold2150 and HiC-scaffold2213 respectively. The distributing density of *BnCYP450s* on chromosomes was extremely uneven. A relatively high density was observed on chromosomes A09, C03 and C07 (more than 50 genes) while a relatively low density on chromosomes A07, A10 and C06 (fewer than 20 genes). We identified a total of 32 tandem duplication events and 608 segmental duplications in the *Brassica napus* genome according to the BLAST and MCScanX results (Figure 1 and Tables S2 and S3). Moreover, only nine segmental duplication events were observed within the same chromosome and five of them were observed on chromosomes C04, whereas 599 segmental duplications were detected across chromosomes (Table S3). Further analysis revealed that there are 88 duplication events located on the AA subgenome and 91 events on the CC subgenome, whereas there are 429 events occurred between AA and CC subgenomes (Tables S4 and S5). In terms of the U triangle, allopolyploidization may have key roles in the expansion of the *BnCYP450s* in *Brassica napus*. To ascertain the evolution of the *BnCYP450* genes, we determined the synteny between *Brassica napus* and *Arabidopsis thaliana* at the whole genome level. A total of 458 collinear gene pairs were observed between the two genomes (Figure S2 and Table S6). Most (113 out of 142) *AtCYP450s* genes have multiple orthologous genes in *Brassica napus*. For instance, AtCYP706A4 has 11 collinear *BnCYP450* genes. However, there are 29 *AtCYP450* genes having only one collinear *BnCYP450* gene (Table S6).

2.4. Gene Structure Analyses of the Clan 86 BnCYP450 Genes

As above mentioned, members of clan 86 CYP450s, such as CYP704Bs, are essential for the formation of pollen exine in plant male reproduction and targeted mutagenesis of *CYP704B* genes has been used to create new male sterile lines in many crops [11,14,16,30]. Therefore, we are focused on analysis of clan 86 CYP450s. The overall exon/intron profile is an index to understand the phylogenetic relationships within a particular gene family [39]. So, we first determined the intron and exon organization of the clan 86 *BnCYP450* genes. The majority of *CYP96* subfamily genes (29 out of 52) contains no intron. Most members of the *CYP86* (12 out of 29) and *CYP94* (9 out of 20) subfamilies also have intronless genes (Figure 2). All members of the *CYP704* subfamily have at least three introns. There are two genes from *CYP94* subfamily (*BnCYP94B1a* and *BnCYP94B1f*) and one gene from *CYP96* subfamily (*BnCYP96A15c*) having at most 21 introns (Figure 2). In short, these gene architectures within a subfamily may also be attributed to the high quality of 'ZS11' genome assembly after comparing with the 'Darmor-bzh' genome assembly as previously described [1].

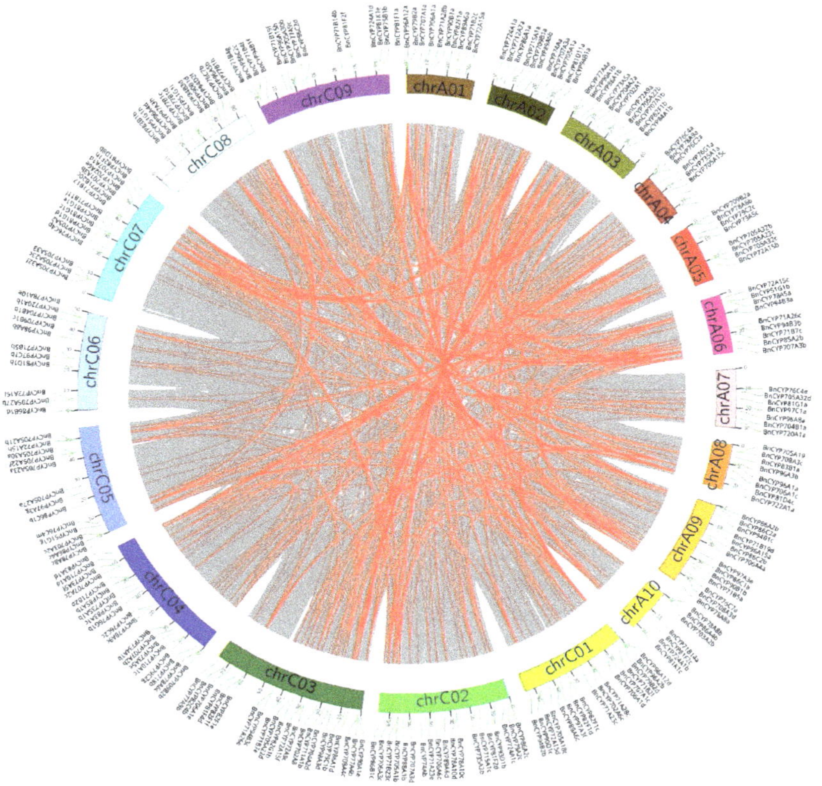

Figure 1. The synteny analysis of *BnCYP450* genes in *Brassica napus*. The gray lines display all synteny blocks, and the red lines indicate the segmental duplicate gene pairs in *Brassica napus* Cultivar ZS11 genome respectively.

2.5. Tissue-Specific Transcript Accumulation Patterns of the Clan 86 BnCYP450 Genes

In order to determine the transcript abundance of the clan 86 *BnCYP450* genes, we investigated their expression profiles in 12 different tissues (stem, sepal, pistil, stamen, ovule, pericarp, blossomy pistil, wilting pistil, root, flower, leaf and silique) using public available RNA-seq data of ZS11 as above mentioned [40,41]. Eighty-seven genes were observed to be expressed in at least one tissue, while 22 genes were not observed to be expressed in any of the analyzed tissues (Figure 3 and Table S7). Moreover, three genes were specifically expressed in one tissue, whereas 24 genes were expressed in all of the analyzed tissues (Figure 3 and Table S7). There are 13 genes highly expressed in flower. These results suggest that some members of the clan 86 *BnCYP450s* may have distinct roles in cell differentiation and tissue development, and the flower expressed members may play conserved roles in control of fertility.

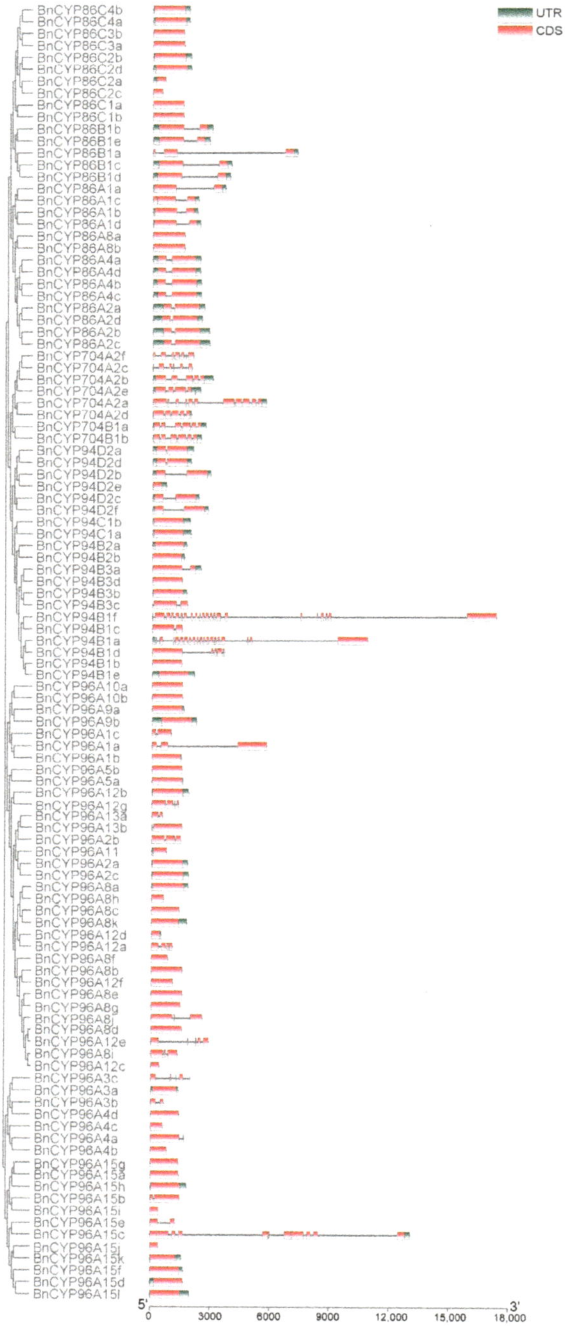

Figure 2. Schematic exon/intron structures of the clan 86 *BnCYP450* genes. The red boxes represent exons and black lines represent introns. The UTR region of are indicated in green boxes. The length of CDS can be estimated by the scale at the bottom.

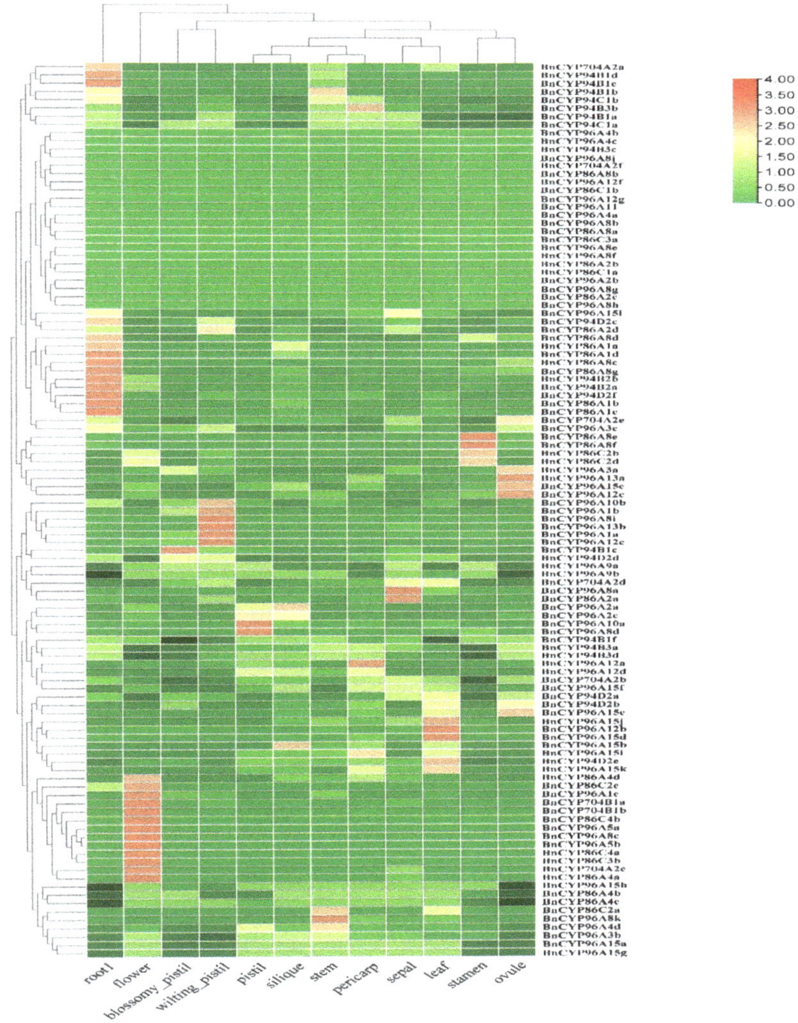

Figure 3. Expression profiles of the clan 86 *BnCYP450* genes in different tissues. The expression level is equal to the mean values and transforms $\log_2$ values for normalization. The color scale represents relative expression levels from low (green colored) to high (red colored).

2.6. Expression Profiles of the Clan 86 BnCYP450 Genes in Response to Abiotic Stresses

To examine the transcript abundance of the clan 86 *BnCYP450* genes in response to abiotic stresses, we investigated their expression profiles under dehydration, NaCl, ABA, and cold conditions with publicly available RNA-seq data of ZS11 as above mentioned [40,41]. Sixty-seven genes were observed to be induced in response to at least one analyzed treatment, while nine genes were observed to be repressed in response to at least one analyzed treatment (Figure 4 and Table S8). There were 33 genes that could not be detected in response to any of the analyzed treatments. Moreover, there are some genes significantly induced by certain treatments, such as *BnCYP94B3a* and *BnCYP96A15l* by dehydration and cold or *BnCYP94B1d* by ABA. Consistent with a previous study [30], the expression of some *BnCYP704* genes, such as *BnCYP704B1b*, *BnCYP704A2e*, *BnCYP704A2b*, *BnCYP704A2d*, *BnCYP704A2a*, and *BnCYP704B1a*, are affected by abiotic stresses (Figure 4 and Table S8).

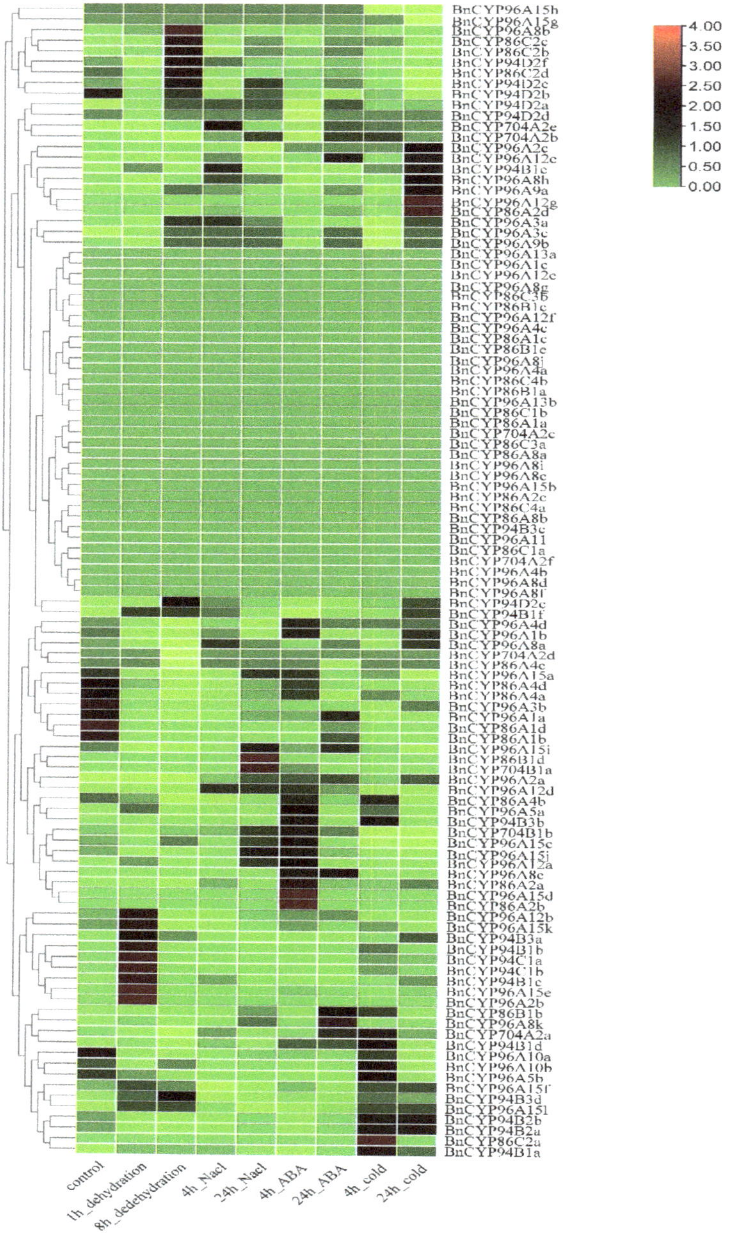

Figure 4. Expression profiles of the clan 86 *BnCYP450* genes under different stress conditions. The expression level is equal to the mean values and transforms $\log_2$ values for normalization. The color scale represents relative expression levels from low (green colored) to high (red colored).

2.7. The Brassica Napus Genome Encodes Two Putative ZmMS26-Like Genes

As displayed in Figure 5a and Table S1, there are two orthologous sequences for AtMS26 found in the rapeseed genome, *BnCYP704B1a* (ZS11A07G034230) and *BnCYP704B1b* (ZS11C06G043970), both of which contained six exons and five introns (Figure 5b). To remain consistent with previous MS26 nomenclature, we renamed *BnCYP704B1a* as *BnMS26a* and *BnCYP704B1b* as *BnMS26b*, respectively. Both *BnCYP704B1a* and *BnCYP704B1b* encode predicted proteins with 519 amino acids, estimated molecular masses of 59.8 kDa, and predicted pI of 8.58. Comparison of the predicted amino acid sequences and coding sequences of *BnCYP704B1a* and *BnCYP704B1b* revealed that they are highly similar, displaying about 98% amino acid identity (Figure S3) and 97% nucleotide identity (Figure S4). All of these indicate that *BnCYP704B1a* and *BnCYP704B1b* are segmental duplicating genes in rapeseed, and they originated from different progenitor species according to the U triangle [2]. In other words, *BnCYP704B1a* originated from *Brassica rapa* and *BnCYP704B1b* originated from *Brassica oleracea*. Phylogenetic analysis shows that ZmMS26 homologs in monocots (rice, maize, sorghum, Brachypodium, tomato and wheat) and dicot (the Brassicaceae species Arabidopsis, cabbage and rapeseed) are divided into two subfamilies (Figure 5a).

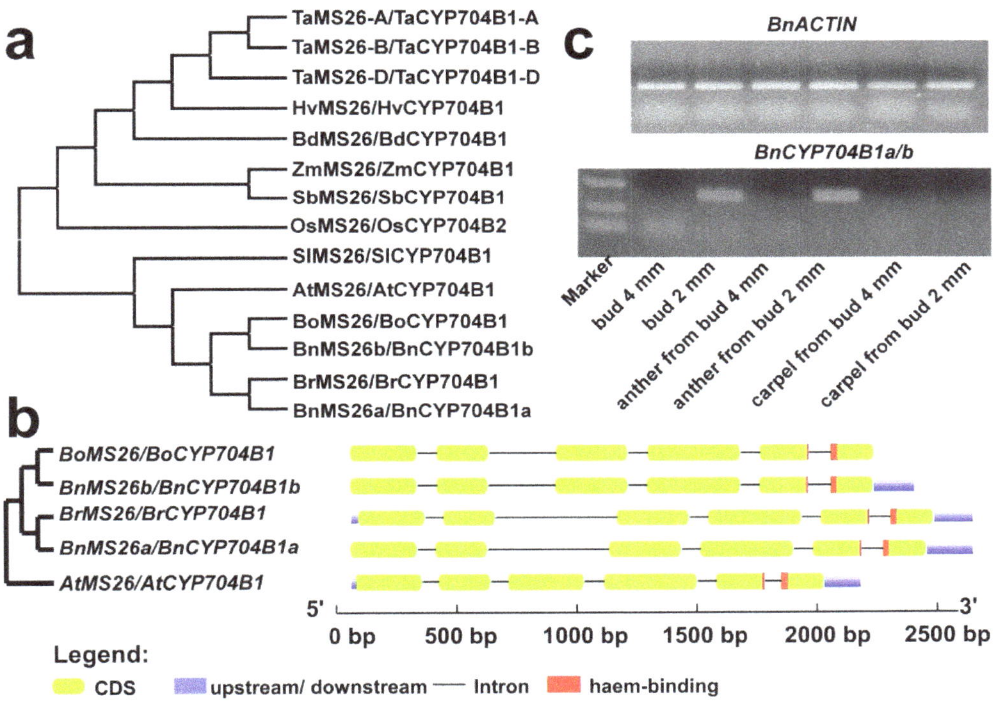

Figure 5. Phylogenetic analysis of MS26 orthologs and anther-specific expression of *BnCYP704B1a* and *BnCYP704B1b*. (**a**) Phylogenetic analysis of MS26 orthologs. The protein sequences of MS26 orthologs were aligned by the MUSCLE tool; the maximum likelihood tree was generated using MEGA 6.0. (**b**) Genetic architecture of *BoMS26*, *BnCYP704B1b*, *BrMS26*, *BnCYP704B1a* and *AtMS26*. (**c**) RT-PCR analysis of *BnCYP704B1a* and *BnCYP704B1b* in buds, anther and carpel released from buds using *BnACTIN* as control. The exon/intron distribution of selected *MS26* genes was determined using the online Gene Structure Display Server 2.0. The CDS (corresponding to exons) is represented by green boxes. The introns are indicated as black lines. The haem-binding motif region on the genomic sequence is denoted by red box.

2.8. *BnCYP704B1a and BnCYP704B1b Are Mainly Expressed in Young Anther*

The expression of the *ZmMS26* gene is limited to developing anthers [11,15,16], which is consistent with its role in pollen development. Moreover, the *ZmMS26* orthologs in rice, sorghum, and wheat also confer male sterility when targeted by mutagenesis, strongly supporting the conserved function in anthers [11,15,16]. To examine the expression patterns of the *BnCYP704B1a* and *BnCYP704B1b* genes, we first compared their expression levels in different tissues by analyzing available RNA-seq data from rapeseed [42]. The expression of *BnCYP704B1a* and *BnCYP704B1b* was only significantly detected in young buds (around 2 mm in diameter, Figure S4). To gain an insight into whether they also have a conserved anther-expressed pattern, we determined the expression levels of *Bn-CYP704B1a* and *BnCYP704B1b* in different flower tissues using RT-PCR. Concerning their high sequence identity, we used degenerate primers to detect their combined expression. Consistent with the transcriptome data, the expression of *BnCYP704B1a* and *BnCYP704B1b* could be only detected in young buds (Figure 5c). In flower parts isolated from young buds, the expression of *BnCYP704B1a* and *BnCYP704B1b* was only detected in anther, but not in carpel (Figure 5b and Figure S5), suggesting that they may also have a role at the very early stage of pollen development.

2.9. *BnCYP704B1a and BnCYP704B1b Simultaneous Knockout Mutants Are Male Sterile*

To determine whether the *BnCYP704B1a* and *BnCYP704B1b* could be edited by CRISPR/Cas9 system, we designed six sgRNA (Target-1~Target-6) to simultaneously target the first or the second exon (Figure 6a, Table S9). The six sgRNAs were divided into three groups and inserted into pHSE401 expression vectors as two gRNA expression cassettes [43]. The resulting three recombinant plasmids were transformed into *Agrobacterium tumefaciens* and co-cultured with the hypocotyls of rapeseed. As previously reported, CRISPR/Cas9-induced editing events could take place in the callus cells or in the regenerated shoots derivate from transformed hypocotyl before regeneration in rapeseed [44]. As expected, homozygous mutants with gRNA-directed mutation were acquired in the regenerated shoots (Figure S6). The editing efficiency of each sgRNA expression cassette was measured in the regenerated shoots. The sgRNA expression cassette harboring Target-1 and Target-2 was more likely to edit at both target sites (Figure S6). The sgRNA expression cassette harboring Target-5 and Target-6 was prone to edit at target site 5 (Figure S6). No editing event was observed at target sites 3 and 4 (Figure S6). Moreover, editing events with deletions were frequently observed in *BnCYP704B1a* and *BnCYP704B1b* in all of the regenerated shoots (Figure S6). We obtained two T_2 plants harboring mutations in both *BnCYP704B1a* and *BnCYP704B1b* from the edited shoots (Figure 6b). All of these lines exhibited a pollenless, sterile phenotype in mature anthers (Figure 6). Sequencing results confirmed that all of these lines harbored the editing events at the targeted sites observed in the regenerated shoots (Figure 6b). To investigate whether the deletions generated by the CRISPR/Cas9 system could be transmitted to the next generation, they were cross-pollinated with the recipient line K407. The fertility of the corresponding F_1 and F_2 progenies were determined. All F_1 progenies developed normal pollen grain in the mature anthers and produced seeds via self-fertilization. Around a quarter of the F_2 progeny plants displayed the pollenless, sterile phenotype, indicating that *BnCYP704B1a* and *BnCYP704B1b* are two duplicate, recessive genes. Sequencing results also confirmed that the observed editing events were able to be transmitted to the next generations. Furthermore, the sterile edited *BnCYP704B1a* and *BnCYP704B1b* plants were able to set seed via artificial pollination with pollen from the recipient line K407, indicating normal female organ development and fertilization potential.

To examine whether any off-target event occurs, we determined the potential off-target sites for *BnCYP704B1a* and *BnCYP704B1b*, respectively. After sequencing the amplified fragments of those potential off-target sites, no genome editing was observed in any examined samples (Table S10). These results indicated that the two sgRNA expression cassettes harboring Target-1 and Target-2 or Target-5 and Target-6 had the highest editing efficiency and an undetectable off-target effect.

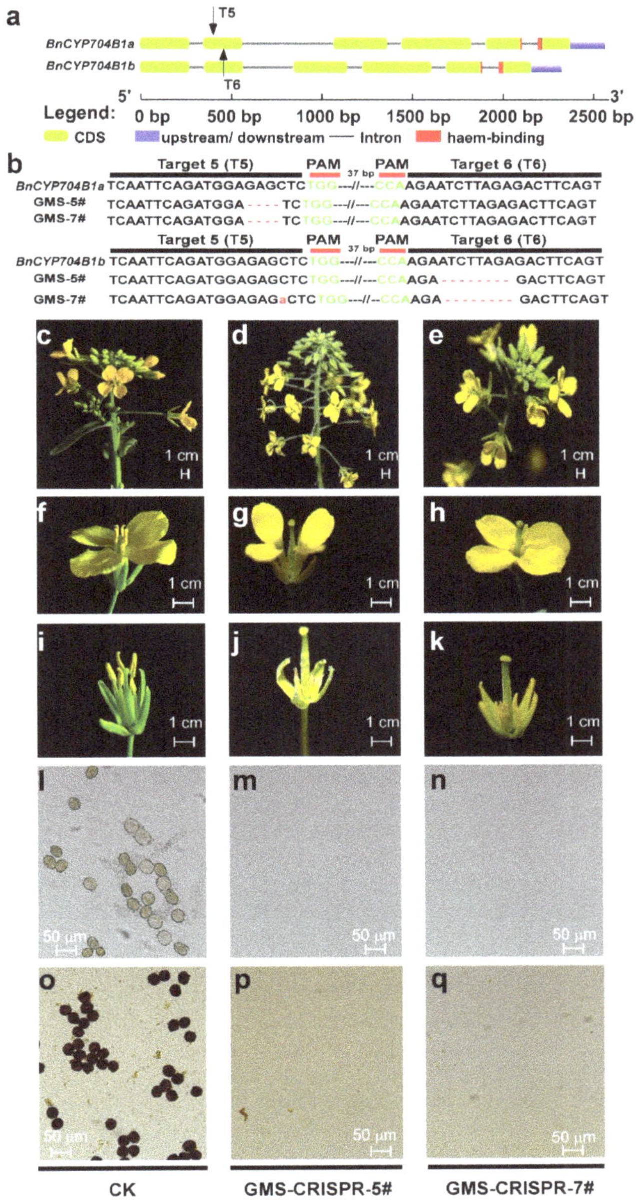

Figure 6. Male sterility phenotypes of CRISPR/Cas9-edited GMS lines. (**a**) Gene structures of *BnCYP704B1a* and *BnCYP704B1b* and the corresponding CRISPR/Cas9 guide RNA target sites, marked by black arrows. (**b**) Mutated genotypes of *BnCYP704B1a* and *BnCYP704B1b* after CRISPR/Cas9 editing. The target sites sequences are marked in *BnCYP704B1a* and *BnCYP704B1b* sequences, with DEFINE (PAM) highlighted in green. (**c**–**e**) inflorescence; (**f**–**h**) flower; (**i**–**k**) flower without petal; (**l**–**q**) pollen fertility in CK (represents the recipient line K407) and CRISPR/Cas9-edited plants; (**o**–**q**) the pollen was stained with I_2—KI solution. No detectable pollen was observed from CRISPR/Cas9-edited GMS plants. (**c**–**k**), scale bars, 1 cm; (**l**–**q**), scale bars, 50 μm.

3. Discussion

In the present study, a total of 687 *BnCYP450* genes were identified from *Brassica napus* cultivar ZS11, which is assigned into 22 subfamilies and clustered into nine clans (Table 1). The number of *BnCYP450s* was greater than those of Arabidopsis (251), *Brassica rapa* (354), *Brassica oleracea* (343), rice (355), apple (348), tomato (457), cotton (461), model legume (346), grape (236), and fewer than in wheat (1700) [22–25,27,28]. Some subfamilies and clans showed deep phylogenetic relationships. For example, the CYP720 subfamily was inside the CYP90 subfamily on the CYP85 clan branch. The CYP704 subfamily was separated into two clades by the CYP94 and CYP86 subfamily on the CYP86 clan branch. The CYP89 subfamily was clustered inside the CYP77 subfamily on the CYP71 clan branch. The CYP76 subfamily from the CYP71 clan was separated into two clades by the CYP72 clan on the phylogenetic tree (Figure S1). Based on the above analysis, we speculated that the CYP450 superfamily has a significant degree of evolutionary extension in *Brassica napus*. Moreover, the large range of Mw and pI of BnCYP450s may determine their functional diversity in distinct biosynthetic pathways.

It is conceivable that gene duplication represents a major force to gene expansion [21,40,45]. Compared with those of *AtCYP450s* (251), *BrCYP450s* (354), and *BoCYP450s* (343), the number of *BnCYP450s* (687) was significantly expanded in *Brassica napus*, which may be attributed to the genome polyploidy event in terms of the U′ triangle [1,2,46]. Indeed, there were 32 tandem duplication events and 608 segmental duplications observed across AA and CC subgenomes (Table S3). Moreover, there were 88 and 91 duplication events observed within AA or CC subgenomes, respectively (Tables S4 and S5). So, tetraploid, segmental, and tandem duplication are the main types of gene duplication events taking place in the *BnCYP450s* superfamily. Under evolutionary pressures, duplication events have extended the gene family members and mutations in the upstream regions can alter the expression patterns of new members [47,48].

As mentioned above, some members of the CYP86 clan (CYP704, CYP86 and CYP94 subfamilies) are recognized as fatty acid hydroxylases responsible for the biosynthesis of pollen exine and anther cutin [7,21,29–32]. It is tempting to speculate that gene expression patterns are clues to their biological significance [21,39,49]. The expression patterns of the *CYP86* clan *BnCYP450s* genes from 12 tissues and four types of abiotic stresses were determined. There were only a few of the *CYP86* clan *BnCYP450s* genes displayed tissue-specific expression (Figure 3 and Table S7), such as *BnCYP96A8c*, *BnCYP96A5b* and *BnCYP96A5a*. It is interesting that most of the *CYP86* clan *BnCYP450s* genes are induced by at least one analyzed treatment (Figure 4 and Table S8). Further RT-PCR analysis revealed that *BnCYP704B1a* and *BnCYP704B1b* are specifically expressed in young anther (Figure 5b), neither of which could be detected in any of the analyzed tissue or treatment (Figures 3 and 4 and Tables S7 and S8). Concerning the importance of *BnCYP450* superfamily members, the expression patterns of the *CYP86* clan *BnCYP450s* examined in the present study may provide an important way to uncover the biological functions of other *BnCYP450s* in future.

Gene editing by the CRISPR/Cas9 system provides a rapid and specific way to manipulate the genetic material used in crop breeding. Since its first application in plants in 2013, targeted mutagenesis of interested genes has been used to create new germplasms in many crops [11,12,14,17]. One of these strategies is to target the mutagenesis of male-sterile control genes to create new male sterile lines in crops [9,11–14,18,20]. One of such male-sterile control genes is *ZmMS26* [7,11,16,20,29,31,32]. Targeted mutagenesis of orthologues of MS26 have created male sterile lines in maize, rice, wheat, sorghum, and bread wheat [11,16,20]. In the current work, gene editing directed the knock-out of two orthologues of *MS26* in *Brassica napus* is capable of generating a pure male sterile line (Figure 6). Moreover, editing events with deletions were frequently observed upon *BnCYP704B1a* and *BnCYP704B1b* in the obtained male sterile plants. This may be mainly attributed to the error-prone character of the two sgRNA expression cassettes, in which sgRNA directed deletion is observed to occur at high frequencies [43,50]. Target sequences with high editing and low off-target efficiencies will improve the breeding efficiency and reduce the breeding cost associated with the development of commercially applied genic male sterility lines.

We have also established male sterile lines in tomato via CRISPR/Cas9 directed knock-out of tomato *CYP704B1* homolog [14]. As previously described, the conserved haem-binding loop in the C-terminal is essential for catalytic activity of haem-thiolate cytochrome P450s [11,16,20]. Targeted mutagenesis of the conserved haem domain of ZmMS26 (mainly located in exon 5 or exon 6) orthologs confers male sterility in monocots (such as maize, rice, sorghum, wheat) [11,16,20]. As mentioned above, our designed gRNAs were mainly located in exon 1 and exon 2 and conferred male sterility when simultaneously edited (Figure 6). Taken together, these results indicated that CRISPR/Cas9 directed simultaneously knock-out of *ZmMS26* orthologs could be used as a universal strategy to establish male sterile lines in allotetraploid plants (such as rapeseed and wheat).

It is well-known that propagating a pure male sterile line is essential for commercial hybrid seed production [10]. The created male sterile line in the present study can serve as the foundation for applications in the two-line hybrid breeding of rapeseed. Taken together, our findings suggest that practical pure male sterile lines can be developed by the simultaneous, targeted knock-out of *BnCYP704B1a* and *BnCYP704B1b* using the CRISPR/Cas9 system, which is an important step towards capturing heterosis in commercial hybrid rapeseed production. Isolation of the inherited mutations in Cas9-free (also known as transgene clean) plants can ensure the stable transmission of the identified mutations to next generations. Moreover, the screening process for the male sterile line is time-consuming, laborious, and inefficient (50% fertile plants need to be removed before fertilizing). A long period of plant regeneration could be avoided with the selection of pure male sterile plants in advance, which are almost indistinguishable from classical male sterile lines during the molecular breeding period. Recently, a fluorescence-based seed sorting strategy has been adopted in the streamlined identification of male sterile lines in rice [10]. A convenient and efficient DsRed-based visual screening method has also been established in *Brassica napus* [51]. To further simplify the screening procedure for pure male sterile plants, a similar strategy also needs to be developed in hybrid rapeseed production in future.

4. Materials and Methods

4.1. Identification of the Cytochrome P450 Gene Superfamily in Brassica napus

Protein sequences of the rapeseed cultivar ZS11 were obtained from National Genomics Data Center (NGDC, accession number PRJCA002883) [1]. The *cytochrome P450* genes of *Brassica napus* (*BnCYP450s*) were isolated from ZS11 through the HMM profile corresponding to the Pfam cytochrome P450 family PF00067 by running the HUMMER3.1 software of Linux version, with the threshold was e-value < e^{-10}. A total of 251 AtCYP450 protein sequences download from the cytochrome P450 homepage (https://drnelson.uthsc.edu/, accessed on 11 November 2022) [22] were used as queries to perform a BLASTP search in the local protein database of ZS11. Then, the putative BnCYP450s were obtained via taking the intersection of the HUMMER and BLASTP methods. Finally, these proteins were submitted to the NCBI-CDD server (http://www.ncbi.nlm.nih.gov/Structure/cdd/wrpsb.cgi, accessed on 11 November 2022) and the SMART (Simple Modular Architecture Research Tool, http://smart.embl-heidelberg.de/, accessed on 11 November 2022) database to perform the cytochrome P450 domain predictions as described [40]. The theoretical molecular weight and isoelectric points of BnCYP450s were calculated by DNAstar as previously described [49].

4.2. Gene Duplication and Genomic Synteny of BnCYP450s

Gene duplication and genomic synteny of *BnCYP450s* were determined as previously described [40]. All of the BnCYP450 protein sequences were aligned using BLASTP with the threshold was e-value of e^{-100}. Then, the duplication pattern of *BnCYP450* genes was determined through the MCScanX software with default parameters and divided into tandem and segmental duplication as previously described [40,52]. Similarly, all of the BnCYP450 and AtCYP450 protein sequences were aligned using BLASTP with the threshold e-value of e^{-100}, and all putative syntenic blocks were mapped with JCVI software as described [40,46].

4.3. Phylogenetic and Gene Structure Analysis

Protein sequences of the clan 86 BnCYP450 members were aligned with the FFT-NS-I method of the MAFFT software at first. The phylogenetic tree was constructed via FastTree software with the maximum likelihood method as previously described [40,53]. Then, the phylogenetic tree was visualized by Figtree software [39]. The overall intron/exon organization of the clan 86 *BnCYP450* genes were displayed based on GFF annotation files by TBtools software as previously described [40,54]. Several MS26 orthologs, namely AtMS26 (*Arabidopsis thaliana* MS26, At1g69500) [29], BoMS26 (*Brassica oleracea* MS26, Bo6g108940) [31], BrMS26 (*Brassica rapa* MS26, Bra004386), BnMS26a (*Brassica napus* MS26a), BnMS26b (*Brassica napus* MS26b), SlMS26 (*Solanum lycopersicum* MS26, XM_004228326.4) [14], OsMS26 (*Oryza sativa* MS26, XP_015629295.1), HvMS26 (*Hordeum vulgare* MS26, BAK08270.1), ZmMS26 (*Zea mays* MS26, NP_001130648.10) [11,15,20], SbMS26 (*Sorghum bicolor* MS26, XP_002465796.1), BdMS26 (*Brachypodium distachyon* MS26, XP_003558727.1), TaMS26-A (*Triticum aestivum* MS26-A, TraesCS4A03G0032700.1), TaMS26-B (*Triticum aestivum* MS26-B, TraesCS4B03G0752700.1), and TaMS26-D (*Triticum aestivum* MS26-D, TraesCS4D03G0673200.1) [16], were aligned using the MUSCLE tool, and the maximum likelihood trees were generated using MEGA 6.0. The online Gene Structure Display Server (GSDS2.0, http://gsds.cbi.pku.edu.cn/ accessed on 30 November 2022) was used to decipher the intron-exon distribution of *AtMS26*, *BoMS26*, *BrMS26*, *BnMS26a*, and *BnMS26b* genes as previously described [31].

4.4. Expression Profiles of the Clan 86 BnCYP450 Genes

RNA-seq data from different tissues and under abiotic stresses of ZS11 were downloaded from the NGDC (accession numbers PRJNA394926 and CA001775) [2]. All of these RNA-seq data were mapped to the reference genome of ZS11 with HISAT2 software [55]. Transcript abundance of the clan 86 *BnCYP450* genes was calculated by the TPM (Transcripts Per Million) values with FeatureCounts R package and a histogram was generated via TBtools software as previously described [54].

4.5. Determination of the Expression Profiles of BnCYP704B1a and BnCYP704B1b

Brassica napus RNA-seq data derived from different tissues (http://yanglab.hzau.edu.cn/BnTIR, accessed on 11 November 2022) [42] was used to profile the expression levels of *BnCYP704B1a* and *BnCYP704B1b*. To examine the anther-specific expression of *BnCYP704B1a* and *BnCYP704B1b*, semi-quantitative RT-PCR was used. Briefly, a total RNA was extracted from young buds (around 2 mm or 4 mm in diameter), anther (isolated from young buds), and carpel (isolated from young buds) with TRIzol reagent (TaKaRa). The complementary DNA (cDNA) was obtained after reverse transcription with M-MLV Reverse transcriptase (Promega). Semi-quantitative RT-PCR was performed to assess the expression levels of *BnCYP704B1a* and *BnCYP704B1b* and was normalized to ZS11C02G003910, which is an *actin*-like gene in *Brassica napus*.

4.6. Guide RNA Design and CRISPR/Cas9 Vector Construction

Guide RNAs targeting *BnCYP704B1a* and *BnCYP704B1b* were designed by CRISPR RGEN Tools (http://www.rgenome.net/cas-designer/, accessed on 11 November 2022), which provides both bulge-allowed RNA-guided Endonuclease (RGEN) targets via Cas-Designer [56] and potential off-targets within a 2-nt mismatch and optional 3-nt bulges via Cas-OFFinder [57]. It is advised to design at least two guide RNAs when intending to knock-out a single gene [43,56,57]. Since we were targeting two homologous genes with high sequence similarity, guide RNAs targeting either *BnCYP704B1a* or *BnCYP704B1b* were chosen to co-edit them. The CRISPR/Cas9 toolkit was set to design two guide RNA expression cassettes for genome editing in rapeseed as described previously [44]. Briefly, two guide RNAs were incorporated into two forward or two reverse primers, respectively, with the two forward or two reverse primers partially overlapping. The expression cassettes were amplified from pCBC-DT1T2 with these guide RNA-incorporated primers. Purified expression cassettes were digested with *Bsa*I and inserted into pHSE401 via T4 DNA ligase (TaKaRa) reactions. The oligos used to construct the CRISPR/Cas9 vector for *BnCYP704B1a* and *BnCYP704B1b* (pHSE401-*BnCYP704B1*) are listed in Table S9.

4.7. Plant Material and Plant Transformation

Seeds of the winter rapeseed selfing line K407 were obtained from Hybrid Rapeseed Research Centre of Shaanxi Province, China. K407 shows high transformation efficiency (from our preliminary data, not shown). Rapeseed hypocotyl transformation followed the protocol as previously described [44,51], with modifications to *Agrobacterium tumefaciens* depletion and hygromycin selection. Briefly, seeds were surface sterilized and germinated in the dark. After 5–6 days, the etiolated hypocotyls were cut into 0.8–1-cm segments and co-cultivated with *Agrobacterium tumefaciens* harboring the pHSE401-*BnCYP704B1* construct for 2 days in a solid M1 medium. Following co-cultivation, the hypocotyl explants were placed on M2 medium with 300 μg/mL Timentin and 10 μg/mL hygromycin to remove *Agrobacterium tumefaciens* and select transgenic plants respectively. Further sub-culturing of the explants was conducted on M3 shooting medium to induce callus and shoot regeneration at intervals of 2–3 weeks. After 6–8 weeks of selection and when normal shoots appeared, explants were transferred to M4 rooting medium and kept at 4 °C for 2–4 weeks for vernalization. The resulting plants were placed into soil after the formation of roots.

4.8. Determination of CRISPR/Cas9-Mediated Editing Events

Genomic DNA was extracted from regenerated shoots or plants with DNA easy extraction solution [40]. Fragments covering the sgRNA target region were amplified using genotyping primers (Table S9). The PCR products from regenerated shoots or plants were sent out for sequencing (Invitrogen). Moreover, PCR products from T_2 plants were cloned into a pMD19 T-easy cloning vector (TaKaRa). Bacterial colony PCR was conducted, and positive single colonies were chosen for sequencing. To identify editing events for the target site, the generated sequences were aligned with *BnCYP704B1a* and *BnCYP704B1b* gene sequences. The editing events were verified via checking the corresponding peaks incorporated into the sequencing reports.

4.9. Phenotypic Analysis

The anthers of the edited plants were collected just before anthesis. Pollen grains were released by breaking anthers and suspending them in a 1% iodine/potassium iodide solution (KI/I_2) to assess pollen viability. The released and stained pollen was examined and photographed under a microscope (BX53, Olympus Corporation, Tokyo, Japan) as previously described [12,14].

5. Conclusions

The cytochrome P450 (CYP450) monooxygenase superfamily plays essential roles in plant growth and development via producing many primary and secondary metabolites. This study isolated a total of 687 *BnCYP450* genes in *Brassica napus* cultivar ZS11. The *Bn-CYP450* genes were categorized into 47 subfamilies, clustered into nine clans, and grouped into non-A-type and A-type. Gene duplication and syntenic analysis provided a clear collinear relationship between *BnCYP450s* and *AtCYP450s*. Phylogeny, gene architecture, expression profiles of the *CYP86* clan genes were determined in detail. Two *ZmMS26* orthologous genes, *BnCYP704B1a* and *BnCYP704B1b*, were observed to be specifically expressed in young anthers. Simultaneously targeted knocked-out of these two genes through CRISPR/Cas9 brought about a male-sterile line *Brassica napus*. Taken together, a global investigation of the *BnCYP450s* and targeted editing of two anther-expressed *ZmMS26* homologous genes were conducted in the current study, which could provide valuable solutions to investigate the biological significance of other *BnCYP450s* in future.

Supplementary Materials: The following supporting information can be downloaded at: https://www.mdpi.com/article/10.3390/plants12020365/s1. Table S1. Details of the identified *BnCYP450* genes in *Brassica napus*; Table S2. List of tandem duplicated gene pairs of *BnCYP450s*; Table S3. List of segmental duplication of the *BnCYP450* genes across AA/CC subgenome in *Brassica napus*; Table S4. List of segmental duplication of the *BnCYP450* genes in AA subgenome; Table S5. List of segmental duplication of the *BnCYP450* genes in CC subgenome; Table S6. One-to-one orthologous relationships of the *BnCYP450* genes between *Brassica napus* and *Arabidopsis thaliana*; Table S7. Expression (TPM) of *BnCYP450* genes in different tissues; Table S8. Expression (TPM) of *BnCYP450* genes under different treatments; Table S9. Primers used in this study; Table S10. Potential off-target sites of Target-5 and Target-6 detected in GMS plants; Figure S1. Phylogenetic analysis of BnCYP450s; Figure S2. Synteny analysis of *AtCYP450* and *BnCYP450* genes; Figure S3. Alignment the protein sequences of *BnCYP704B1a* and *BnCYP704B1b*; Figure S4. Alignment the CDS sequences of *BnCYP704B1a* and *BnCYP704B1b*; Figure S5. The expression profiles of *BnCYP704B1a* and *BnCYP704B1b*; Figure S6. The sgRNA target sites and induced mutations upon *BnCYP704B1a* and *BnCYP704B1b* in regenerated shoots.

Author Contributions: C.X., J.M. and Y.Z. conceived the project and designed the experiment plans; Z.W., Y.Z., M.S., X.T., S.H., B.L., P.J., W.G., F.L., L.X. and R.A. conducted the experiments and performed bioinformatics analyses; C.X. and J.M. analyzed the data, prepared Figures 1–6 and wrote the article; X.Z., W.H. and Y.Z. reviewed and edited the manuscript. All authors have read and agreed to the published version of the manuscript.

Funding: This work was financially supported by the Innovation Capability Support Program of Shaanxi Province (2020TD-051); by the State Basic Research Project from the Department of Science and Technology of Shaanxi Province (2021JM-086); by the Key Research and Development Program of Yangling Seed Industry Innovation Center (Ylzy-yc-2021-02); by the Scientific Research and Sharing Platform Construction project of Shaanxi Province (2021PT-036); by the State Key Research and Development Project from the Department of Science and Technology of Shaanxi Province (2018NY-097); by the National Key Research and Development Project from the Ministry of Science and Technology of China (2016YFD0101900).

Institutional Review Board Statement: Not applicable.

Informed Consent Statement: Not applicable.

Data Availability Statement: Not applicable.

Acknowledgments: We are grateful to Chen Qijun (China Agricultural University) for providing the CRISPR/Cas9 system and Wang Qinhu (Northwest A&F University) for help with bioinformatics analyses.

Conflicts of Interest: The authors declare no conflict of interest.

References

1. Chen, X.; Tong, C.; Zhang, X.; Song, A.; Hu, M.; Dong, W.; Chen, F.; Wang, Y.; Tu, J.; Liu, S.; et al. A high-quality *Brassica napus* genome reveals expansion of transposable elements, subgenome evolution and disease resistance. *Plant Biotechnol. J.* **2021**, *19*, 615–630. [CrossRef] [PubMed]
2. Sun, F.; Fan, G.; Hu, Q.; Zhou, Y.; Guan, M.; Tong, C.; Li, J.; Du, D.; Qi, C.; Jiang, L.; et al. The high-quality genome of *Brassica napus* cultivar 'ZS11' reveals the introgression history in semi-winter morphotype. *Plant J. Cell Mol. Biol.* **2017**, *92*, 452–468. [CrossRef] [PubMed]
3. Li, S.; Shen, L.; Hu, P.; Liu, Q.; Zhu, X.; Qian, Q.; Wang, K.; Wang, Y. Developing disease-resistant thermosensitive male sterile rice by multiplex gene editing. *J. Integr. Plant Biol.* **2019**, *61*, 1201–1205. [CrossRef] [PubMed]
4. Deng, Z.; Li, X.; Wang, Z.; Jiang, Y.; Wan, L.; Dong, F.; Chen, F.; Hong, D.; Yang, G. Map-based cloning reveals the complex organization of the *BnRf* locus and leads to the identification of *BnRf(b)*, a male sterility gene, in *Brassica napus*. *TAG Theor. Appl. Genet. Theor. Angew. Genet.* **2016**, *129*, 53–64. [CrossRef]
5. Dun, X.; Zhou, Z.; Xia, S.; Wen, J.; Yi, B.; Shen, J.; Ma, C.; Tu, J.; Fu, T. BnaC.Tic40, a plastid inner membrane translocon originating from *Brassica oleracea*, is essential for tapetal function and microspore development in *Brassica napus*. *Plant J. Cell Mol. Biol.* **2011**, *68*, 532–545. [CrossRef]

6. Li, J.; Hong, D.; He, J.; Ma, L.; Wan, L.; Liu, P.; Yang, G. Map-based cloning of a recessive genic male sterility locus in *Brassica napus* L. and development of its functional marker. *TAG Theor. Appl. Genet. Theor. Angew. Genet.* **2012**, *125*, 223–234. [CrossRef]
7. Yi, B.; Zeng, F.; Lei, S.; Chen, Y.; Yao, X.; Zhu, Y.; Wen, J.; Shen, J.; Ma, C.; Tu, J.; et al. Two duplicate CYP704B1-homologous genes *BnMs1* and *BnMs2* are required for pollen exine formation and tapetal development in *Brassica napus*. *Plant J. Cell Mol. Biol.* **2010**, *63*, 925–938. [CrossRef]
8. Zhou, Z.; Dun, X.; Xia, S.; Shi, D.; Qin, M.; Yi, B.; Wen, J.; Shen, J.; Ma, C.; Tu, J.; et al. *BnMs3* is required for tapetal differentiation and degradation, microspore separation, and pollen-wall biosynthesis in *Brassica napus*. *J. Exp. Bot.* **2012**, *63*, 2041–2058. [CrossRef]
9. Barman, H.N.; Sheng, Z.; Fiaz, S.; Zhong, M.; Wu, Y.; Cai, Y.; Wang, W.; Jiao, G.; Tang, S.; Wei, X.; et al. Generation of a new thermo-sensitive genic male sterile rice line by targeted mutagenesis of *TMS5* gene through CRISPR/Cas9 system. *BMC Plant Biol.* **2019**, *19*, 109. [CrossRef]
10. Chang, Z.; Chen, Z.; Wang, N.; Xie, G.; Lu, J.; Yan, W.; Zhou, J.; Tang, X.; Deng, X.W. Construction of a male sterility system for hybrid rice breeding and seed production using a nuclear male sterility gene. *Proc. Natl. Acad. Sci. USA* **2016**, *113*, 14145–14150. [CrossRef]
11. Djukanovic, V.; Smith, J.; Lowe, K.; Yang, M.; Gao, H.; Jones, S.; Nicholson, M.G.; West, A.; Lape, J.; Bidney, D.; et al. Male-sterile maize plants produced by targeted mutagenesis of the cytochrome P450-like gene (MS26) using a re-designed I-CreI homing endonuclease. *Plant J. Cell Mol. Biol.* **2013**, *76*, 888–899.
12. Jiang, B.; Chen, L.; Yang, C.; Wu, T.; Yuan, S.; Wu, C.; Zhang, M.; Gai, J.; Han, T.; Hou, W.; et al. The cloning and CRISPR/Cas9-mediated mutagenesis of a male sterility gene *MS1* of soybean. *Plant Biotechnol. J.* **2021**, *19*, 1098–1100. [CrossRef]
13. Li, J.; Zhang, H.; Si, X.; Tian, Y.; Chen, K.; Liu, J.; Chen, H.; Gao, C. Generation of thermosensitive male-sterile maize by targeted knockout of the *ZmTMS5* gene. *J. Genet. Genom. Yi Chuan Xue Bao* **2017**, *44*, 465–468. [CrossRef]
14. Liu, J.; Wang, S.; Wang, H.; Luo, B.; Cai, Y.; Li, X.; Zhang, Y.; Wang, X. Rapid generation of tomato male-sterile lines with a marker use for hybrid seed production by CRISPR/Cas9 system. *Mol. Breed.* **2021**, *41*, 25. [CrossRef]
15. Qi, X.; Zhang, C.; Zhu, J.; Liu, C.; Huang, C.; Li, X.; Xie, C. Genome editing enables next-generation hybrid seed production technology. *Mol. Plant* **2020**, *13*, 1262–1269. [CrossRef] [PubMed]
16. Singh, M.; Kumar, M.; Thilges, K.; Cho, M.J.; Cigan, A.M. MS26/CYP704B is required for anther and pollen wall development in bread wheat (*Triticum aestivum* L.) and combining mutations in all three homeologs causes male sterility. *PLoS ONE* **2017**, *12*, e0177632. [CrossRef]
17. Zhan, X.; Lu, Y.; Zhu, J.K.; Botella, J.R. Genome editing for plant research and crop improvement. *J. Integr. Plant Biol.* **2021**, *63*, 3–33. [CrossRef] [PubMed]
18. Zhou, H.; He, M.; Li, J.; Chen, L.; Huang, Z.; Zheng, S.; Zhu, L.; Ni, E.; Jiang, D.; Zhao, B.; et al. Development of commercial thermo-sensitive genic male sterile rice accelerates hybrid rice breeding using the CRISPR/Cas9-mediated *TMS5* editing system. *Sci. Rep.* **2016**, *6*, 37395. [CrossRef] [PubMed]
19. Chaubal, R.; Zanella, C.; Trimnell, M.R.; Fox, T.W.; Albertsen, M.C.; Bedinger, P. Two male-sterile mutants of *Zea mays* (Poaceae) with an extra cell division in the anther wall. *Am. J. Bot.* **2000**, *87*, 1193–1201. [CrossRef]
20. Cigan, A.M.; Singh, M.; Benn, G.; Feigenbutz, L.; Kumar, M.; Cho, M.J.; Svitashev, S.; Young, J. Targeted mutagenesis of a conserved anther-expressed P450 gene confers male sterility in monocots. *Plant Biotechnol. J.* **2017**, *15*, 379–389. [CrossRef]
21. Mizutani, M.; Ohta, D. Diversification of P450 genes during land plant evolution. *Annu. Rev. Plant Biol.* **2010**, *61*, 291–315. [CrossRef]
22. Nelson, D.R. The cytochrome p450 homepage. *Hum. Genom.* **2009**, *4*, 59–65. [CrossRef] [PubMed]
23. Yu, J.; Tehrim, S.; Wang, L.; Dossa, K.; Zhang, X.; Ke, T.; Liao, B. Evolutionary history and functional divergence of the cytochrome P450 gene superfamily between *Arabidopsis thaliana* and Brassica species uncover effects of whole genome and tandem duplications. *BMC Genom.* **2017**, *18*, 733. [CrossRef] [PubMed]
24. Xiong, R.; He, T.; Wang, Y.; Liu, S.; Gao, Y.; Yan, H.; Xiang, Y. Genome and transcriptome analysis to understand the role diversification of cytochrome P450 gene under excess nitrogen treatment. *BMC Plant Biol.* **2021**, *21*, 447. [CrossRef]
25. Jiu, S.; Xu, Y.; Wang, J.; Wang, L.; Liu, X.; Sun, W.; Sabir, I.A.; Ma, C.; Xu, W.; Wang, S.; et al. The cytochrome P450 monooxygenase inventory of grapevine (*Vitis vinifera* L.): Genome-wide identification, evolutionary characterization and expression analysis. *Front. Genet.* **2020**, *11*, 44. [CrossRef] [PubMed]
26. Ren, J.; Yang, L.; Li, Q.; Zhang, Q.; Sun, C.; Liu, X.; Yang, N. Global investigation of cytochrome P450 genes in the chicken genome. *Genes* **2019**, *10*, 617. [CrossRef] [PubMed]
27. Sun, K.; Fang, H.; Chen, Y.; Zhuang, Z.; Chen, Q.; Shan, T.; Khan, M.K.R.; Zhang, J.; Wang, B. Genome-wide analysis of the cytochrome *P450* gene family involved in salt tolerance in *Gossypium hirsutum*. *Front. Plant Sci.* **2021**, *12*, 685054. [CrossRef] [PubMed]
28. Xia, Y.; Yang, J.; Ma, L.; Yan, S.; Pang, Y. Genome-wide identification and analyses of drought/salt-responsive cytochrome *P450* genes in *Medicago truncatula*. *Int. J. Mol. Sci.* **2021**, *22*, 9957. [CrossRef]
29. Dobritsa, A.A.; Shrestha, J.; Morant, M.; Pinot, F.; Matsuno, M.; Swanson, R.; Moller, B.L.; Preuss, D. CYP704B1 is a long-chain fatty acid omega-hydroxylase essential for sporopollenin synthesis in pollen of Arabidopsis. *Plant Physiol.* **2009**, *151*, 574–589. [CrossRef]
30. Shi, J.X.; Cui, M.H.; Yang, L.; Kim, Y.J.; Zhang, D.B. Genetic and biochemical mechanisms of pollen wall development. *Trends Plant Sci.* **2015**, *20*, 741–753. [CrossRef]

31. Ji, J.L.; Yang, L.M.; Fang, Z.Y.; Zhuang, M.; Zhang, Y.Y.; Lv, H.H.; Liu, Y.M.; Li, Z.S. Recessive male sterility in cabbage (*Brassica oleracea* var. capitata) caused by loss of function of BoCYP704B1 due to the insertion of a LTR-retrotransposon. *TAG Theor. Appl. Genet. Theor. Angew. Genet.* **2017**, *130*, 1441–1451. [CrossRef]
32. Li, H.; Pinot, F.; Sauveplane, V.; Werck-Reichhart, D.; Diehl, P.; Schreiber, L.; Franke, R.; Zhang, P.; Chen, L.; Gao, Y.; et al. Cytochrome P450 family member CYP704B2 catalyzes the {omega}-hydroxylation of fatty acids and is required for anther cutin biosynthesis and pollen exine formation in rice. *Plant Cell* **2010**, *22*, 173–190. [CrossRef] [PubMed]
33. Morrone, D.; Chen, X.; Coates, R.M.; Peters, R.J. Characterization of the kaurene oxidase CYP701A3, a multifunctional cytochrome P450 from gibberellin biosynthesis. *Biochem. J.* **2010**, *431*, 337–344. [CrossRef] [PubMed]
34. Regnault, T.; Daviere, J.M.; Heintz, D.; Lange, T.; Achard, P. The gibberellin biosynthetic genes AtKAO1 and AtKAO2 have overlapping roles throughout Arabidopsis development. *Plant J. Cell Mol. Biol.* **2014**, *80*, 462–474. [CrossRef] [PubMed]
35. Kushiro, T.; Okamoto, M.; Nakabayashi, K.; Yamagishi, K.; Kitamura, S.; Asami, T.; Hirai, N.; Koshiba, T.; Kamiya, Y.; Nambara, E. The Arabidopsis cytochrome P450 CYP707A encodes ABA 8′-hydroxylases: Key enzymes in ABA catabolism. *EMBO J.* **2004**, *23*, 1647–1656. [CrossRef]
36. Chen, K.; Li, G.J.; Bressan, R.A.; Song, C.P.; Zhu, J.K.; Zhao, Y. Abscisic acid dynamics, signaling, and functions in plants. *J. Integr. Plant Biol.* **2020**, *62*, 25–54. [CrossRef]
37. Wasternack, C.; Feussner, I. The oxylipin pathways: Biochemistry and function. *Annu. Rev. Plant Biol.* **2018**, *69*, 363–386. [CrossRef]
38. Zhao, B.; Li, J. Regulation of brassinosteroid biosynthesis and inactivation. *J. Integr. Plant Biol.* **2012**, *54*, 746–759. [CrossRef]
39. Bai, Y.; Zhu, W.; Hu, X.; Sun, C.; Li, Y.; Wang, D.; Wang, Q.; Pei, G.; Zhang, Y.; Guo, A.; et al. Genome-wide analysis of the bZIP gene family identifies two ABI5-Like bZIP transcription factors, BrABI5a and BrABI5b, as positive modulators of ABA signalling in Chinese cabbage. *PLoS ONE* **2016**, *11*, e0158966. [CrossRef]
40. Song, M.; Linghu, B.; Huang, S.; Li, F.; An, R.; Xie, C.; Zhu, Y.; Hu, S.; Mu, J.; Zhang, Y. Genome-wide survey of leucine-rich repeat receptor-like protein kinase genes and CRISPR/Cas9-targeted mutagenesis *BnBRI1* in *Brassica napus*. *Front. Plant Sci.* **2022**, *13*, 865132. [CrossRef]
41. Zhu, W.; Guo, Y.; Chen, Y.; Wu, D.; Jiang, L. Genome-wide identification, phylogenetic and expression pattern analysis of *GATA* family genes in *Brassica napus*. *BMC Plant Biol.* **2020**, *20*, 543. [CrossRef] [PubMed]
42. Liu, D.; Yu, L.; Wei, L.; Yu, P.; Wang, J.; Zhao, H.; Zhang, Y.; Zhang, S.; Yang, Z.; Chen, G.; et al. BnTIR: An online transcriptome platform for exploring RNA-seq libraries for oil crop *Brassica napus*. *Plant Biotechnol. J.* **2021**, *19*, 1895–1897. [CrossRef] [PubMed]
43. Xing, H.L.; Dong, L.; Wang, Z.P.; Zhang, H.Y.; Han, C.Y.; Liu, B.; Wang, X.C.; Chen, Q.J. A CRISPR/Cas9 toolkit for multiplex genome editing in plants. *BMC Plant Biol.* **2014**, *14*, 327. [CrossRef] [PubMed]
44. Braatz, J.; Harloff, H.J.; Mascher, M.; Stein, N.; Himmelbach, A.; Jung, C. CRISPR-Cas9 targeted mutagenesis leads to simultaneous modification of different homoeologous gene copies in polyploid oilseed rape (*Brassica napus* L.). *Plant Physiol.* **2017**, *174*, 935–942. [CrossRef]
45. Freeling, M. Bias in plant gene content following different sorts of duplication: Tandem, whole-genome, segmental, or by transposition. *Annu. Rev. Plant Biol.* **2009**, *60*, 433–453. [CrossRef]
46. Tang, H.; Bowers, J.E.; Wang, X.; Ming, R.; Alam, M.; Paterson, A.H. Synteny and collinearity in plant genomes. *Science* **2008**, *320*, 486–488. [CrossRef]
47. Heidari, P.; Abdullah, S.F.; Poczai, P. Magnesium transporter gene family: Genome-wide identification and characterization in *Theobroma cacao*, *Corchorus capsularis*, and *Gossypium hirsutum* of Family Malvaceae. *Agronomy* **2021**, *11*, 1651. [CrossRef]
48. Heidari, P.; Puresmaeli, F.; Mora-Poblete, F. Genome-wide identification and molecular evolution of the magnesium transporter (MGT) gene family in *Citrullus lanatus* and *Cucumis sativus*. *Agronomy* **2022**, *12*, 2253. [CrossRef]
49. Li, G.; Hu, X.; Hou, L.; Cao, L.; Wang, Q.; Wang, D.; Mu, X.; Zhang, Y.; Zhou, X.; Zhao, Y.; et al. Molecular identification of BrHAB2a, one of the two AtHAB2-like proteins in *Brassica rapa*, is an important component of ABA signaling. *Biochem. Biophys. Res. Commun.* **2018**, *503*, 495–500. [CrossRef]
50. Lu, Y.; Ye, X.; Guo, R.; Huang, J.; Wang, W.; Tang, J.; Tan, L.; Zhu, J.K.; Chu, C.; Qian, Y. Genome-wide targeted mutagenesis in rice using the CRISPR/Cas9 system. *Mol. Plant* **2017**, *10*, 1242–1245. [CrossRef]
51. Zhang, K.; He, J.; Liu, L.; Xie, R.; Qiu, L.; Li, X.; Yuan, W.; Chen, K.; Yin, Y.; Kyaw, M.M.M.; et al. A convenient, rapid and efficient method for establishing transgenic lines of *Brassica napus*. *Plant Methods* **2020**, *16*, 43. [CrossRef] [PubMed]
52. Wang, Y.; Tang, H.; Debarry, J.D.; Tan, X.; Li, J.; Wang, X.; Lee, T.H.; Jin, H.; Marler, B.; Guo, H.; et al. MCScanX: A toolkit for detection and evolutionary analysis of gene synteny and collinearity. *Nucleic Acids Res.* **2012**, *40*, e49. [CrossRef] [PubMed]
53. Price, M.N.; Dehal, P.S.; Arkin, A.P. FastTree: Computing large minimum evolution trees with profiles instead of a distance matrix. *Mol. Biol. Evol.* **2009**, *26*, 1641–1650. [CrossRef] [PubMed]
54. Chen, C.; Chen, H.; Zhang, Y.; Thomas, H.R.; Frank, M.H.; He, Y.; Xia, R. TBtools: An integrative toolkit developed for interactive analyses of big biological data. *Mol. Plant* **2020**, *13*, 1194–1202. [CrossRef]
55. Kim, D.; Langmead, B.; Salzberg, S.L. HISAT: A fast spliced aligner with low memory requirements. *Nat. Methods* **2015**, *12*, 357–360. [CrossRef] [PubMed]

56. Park, J.; Bae, S.; Kim, J.S. Cas-Designer: A web-based tool for choice of CRISPR-Cas9 target sites. *Bioinformatics* **2015**, *31*, 4014–4016. [CrossRef] [PubMed]
57. Bae, S.; Park, J.; Kim, J.S. Cas-OFFinder: A fast and versatile algorithm that searches for potential off-target sites of Cas9 RNA-guided endonucleases. *Bioinformatics* **2014**, *30*, 1473–1475. [CrossRef]

plants

MDPI

Article

Identification and Phylogenetic Analysis of the *R2R3-MYB* Subfamily in *Brassica napus*

Dingfan Luo [1,2], Desheng Mei [2,3], Wenliang Wei [1,*] and Jia Liu [2,3,*]

[1] College of Agriculture, Yangtze University, Jingzhou 434023, China
[2] Oil Crops Research Institute, Chinese Academy of Agricultural Sciences, No. 2 Xudong 2nd Rd., Wuhan 430062, China
[3] Key Laboratory of Biology and Genetic Improvement of Oil Crops, Ministry of Agriculture and Rural Affairs, Wuhan 430062, China
* Correspondence: whwenliang@163.com (W.W.); liujia02@caas.cn (J.L.)

Abstract: The R2R3-MYB sub-family proteins are composed of most members of MYB (v-Myb avian myeloblastosis viral oncogene homolog) protein, a plant-specific transcription factor (TF) that is classified into four classes depending on the number of MYB repeats. R2R3-MYB TFs are involved in physiological and biochemical processes. However, the functions of the *Brassica napus R2R3-MYB* genes are still mainly unknown. In this study, 35 *Brassica napus MYB* (*BnaMYB*) genes were screened in the genome of *Brassica napus*, and details about their physical and chemical characteristics, evolutionary relationships, chromosome locations, gene structures, three-dimensional protein structures, cis-acting promoter elements, and gene duplications were uncovered. The *BnaMYB* genes have undergone segmental duplications and positive selection pressure, according to evolutionary studies. The same subfamilies have similar intron–exon patterns and motifs, according to the genes' structure and conserved motifs. Additionally, through cis-element analysis, many drought-responsive and other stress-responsive cis-elements have been found in the promoter regions of the *BnaMYB* genes. The expression of the *BnaMYB* gene displays a variety of tissue-specific patterns. Ten lignin-related genes were chosen for drought treatment. Our research screened four genes that showed significant upregulation under drought stress, and thus may be important drought-responsive genes. The findings lay a new foundation for understanding the complex mechanisms of *BnaMYB* in multiple developmental stages and pathways related to drought stress in rapeseed.

Keywords: *Brassica napus*; drought stress; *R2R3-MYB* genes

Citation: Luo, D.; Mei, D.; Wei, W.; Liu, J. Identification and Phylogenetic Analysis of the *R2R3-MYB* Subfamily in *Brassica napus*. *Plants* **2023**, *12*, 886. https://doi.org/10.3390/plants12040886

Academic Editor: Ryo Fujimoto

Received: 10 January 2023
Revised: 9 February 2023
Accepted: 10 February 2023
Published: 16 February 2023

1. Introduction

In land plants, drought is often a major stress that can reduce crop productivity severely [1]. Rapeseed plants suffer terrible consequences when there is a water shortage, resulting in changed physiological processes (photosynthesis rates, osmotic protection, and oil contents) and reduced plant growth [2]. Rapeseed is quite vulnerable to drought stress during the flowering stage, and it has been projected that global climatic oscillations, which are the cause of severe and prolonged drought in some areas of the biosphere, will reduce rapeseed output [3]. Therefore, finding the methods that can mitigate drought stress is critical for *B. napus* (*Brassica napus* L.). A method of accumulating secondary metabolites are formed through accumulating lignin [4–9]. In poplar (*Populus*), overexpression of the transcription factor PdNF-YB21 promotes root growth, along with drought tolerance [9]. Furthermore, overexpression of *VlbZIP30* in grapevine (*Vitis vinifera*) [8], *poCCoAOMT* in tobacco (*Nicotiana benthamiana*) [10], *IbLEA14* in sweet potato (*Ipomoea batatas*) [11], *osTF1L* in rice [12], and *PeLAC10*, *CmCAD2* and *CmCAD2* in Arabidopsis (*Arabidopsis thaliana*) has been found [7,12,13]. All these genes enhance the synthesis of lignin and drought resistance. Inbred lines of maize (*Zea mays*) that are more lignified than drought-sensitive lines imply that lignification is a key mechanism for drought stress adaptation [14].

There are several genes within the *MYB* gene family that act as TFs for plants. There is a conserved DBD (DNA-binding domain) in the MYB transcription factor, which can be composed of up to four imperfect repeats, known as R repeats [15,16]. Each R repeat's second and third helices join with specific gene-specific promoter sequences to create a helix-turn-helix (HTH) structure, which is 50–55 amino acids long [15,17]. In addition, the third α-helix is usually responsible for recognizing short DNA sequences [18]. Sequence similarity determines whether the MYB domain repeats belong to the R1, R2, or R3 category [16,19]. Members of the plant MYB superfamily generally consist of one to four imperfect MYB repeats, and there are four types of repetitions: 1R-MYB (one or two separated repeats), 2R-MYB (R2R3-MYB, two adjacent repeats), 3R-MYB (three adjacent repeats), and 4R-MYB (consisting of four adjacent repeats) [16]. The R2R3-MYB family has considerably grown in size and now dominates in the plant lineage [20].

In a recent study [21], 2120 *R2R3-MYB* genes were identified in six brassica plants, including 130 in *Arabidopsis*, 236 in *B. rapa*, 247 in *B. oleracea*, 248 in *B. nigra*, 425 in *B. napus*, 422 in *B. juncea* and 412 in *B. carinata*. However, the number of *R2R3-MYBs* in *Arabidopsis*, *B. rapa* and *B. napus* were different from those found in previous studies, namely, 126 in *Arabidopsis* [18], 256 in *B. rapa* [22], and 249 in *B. napus* [23]. The main reason may be that with the development of third-generation sequencing technology, the information has become more complete.

R2R3-MYB TFs have been shown to synthesize hormones, participate in signal transduction, and be involved in physiological and biochemical processes, particularly in reaction to a variety of biotic and abiotic stimuli [16,24–26]. In addition, R2R3-MYB also contributes significantly to lignin synthesis [16]. The R2R3-MYB transcription factor PbrMYB169 protein of pear (*Pyrus bretschneideri*) significantly activated the promoters of lignin genes *C3H1, CCR1, CCOMT2, CAD, 4CL1, 4CL2, HCT2* and *LAC18* by binding with the AC element in the promoter, thus positively regulating the lignification in the cell matrix of pear fruit [27]. The constitutive expression of three members of MYB transcription factor family (PtrMYB92, PtrMYB3 and PtrMYB20) in poplar activated the expression of lignin biosynthesis gene and induced the ectopic deposition of lignin. At the same time, they can also activate the promoter of poplar wood biosynthesis gene [28,29]. The overexpression of TaMYB4 leads to the reduction of the transcription of *CAD* and *CCR* genes involved in the lignin biosynthesis pathway, which has a negative regulatory effect on the lignin biosynthesis in wheat (*Triticum aestivum* L.) [30]. ZmMYB31 and ZmMYB42 can downregulate the *COMT* genes of maize (*Zea mays* L.) and Arabidopsis, and overexpression results in a reduction of lignin content in transgenic plants [31]. The overexpression of MusaMYB31 can downregulate many common phenylpropane genes and lignin biosynthesis pathways, and lignin biosynthesis is negatively regulated in banana (*Musa acuminata*) [32]. According to these results, we found that the R2R3-MYB transcription factor can bind to the promoter and significantly activate the lignin gene. However, the expression of some R2R3-MYB transcription factors can downregulate the genes in the lignin synthesis pathway, thus negatively regulating the lignin biosynthesis in plants. We have learned that lignin deposition has a great impact on drought responses, so *R2R3-MYB* is responsive to drought stress while regulating lignin synthesis.

In *Arabidopsis*, there are about 126 members of the *R2R3-MYB* subfamily, which is roughly divided into 25 subgroups and distributed across 11 branches [18]. The functions of most members have been identified, and they have been found to be involved in the control of plant-specific processes, including (i) primary and secondary metabolism, (ii) cell fate and identity, (iii) development processes, and (iv) the response to biological and abiotic stresses. We found in the literature that *AtMYB26* can control the deposition of secondary walls in the anthers [33]. *AtMYB46* is a positive regulator of lignin biosynthesis in fibers and vessels [34]. These two genes may be able to cope with drought stress by regulating the production of lignin. These two genes are not included in the 25 subgroups but are in the same branch as Subgroups 13 and 16. Therefore, we preliminarily selected the Brassica napus homologs to this branch for further analysis.

In this work, 35 *BnMYB* genes of rapeseed were discovered by comparing the genomic sequences of the S13 and S16 subfamilies of the known *R2R3-MYB* from *Arabidopsis thaliana*. Additionally, we used the method of bioinformatic analysis to effectively process, analyze, and visualize a large amount of data from the field of life science. The physical and chemical characteristics of the *BnMYB* genes' cis-acting promoter and their gene duplication were clarified, as well as their evolutionary relationships, genomic positions, gene structures, and three-dimensional membrane proteins. We also looked at how the *BnaMYB* gene family was expressed in various tissues and how they responded to drought treatment. Further research into the functional characterization of *BnMYBs* may benefit from our findings.

2. Results

2.1. Identification of the R2R3-MYB Transcription Factor Family in Rapeseed

To identify *R2R3-MYB* family genes in rapeseed, 12 AtTCP protein sequences were used as queries for a BlastP search against the rapeseed genome. As a result, 35 *BnMYB* genes were consistent with the R2R3-MYB domain (Table 1). Hereafter, these genes are named *BnMYB1* to *BnMYB35*. The physical and chemical characteristics of each member were then examined and predicted. The proteins that the *BnaMYB* genes encoded ranged in size from 239 amino acids (*BnaMYB35*) to 370 amino acids (*BnaMYB17, BnaMYB18*); the MW ranged from 77,433.37 kDa (*BnaMYB8*) to 195,053.44 kDa (*BnaMYB23*); and the PI ranged from 4.89 (*BnaMYB23*) to 5.11 (*BnaMYB8*). The predicted subcellular location of each BnaMYB protein was in the nucleus. Table 1 also contains the length of the coding sequences.

Table 1. The position and molecular information of *MYB* gene family in *B. napus*.

Gene Name	Gene ID	Chromosomes Position	CDS (bp)	Length (bp)	Protein MW (kDa)	pI
BnaMYB1	BnaA01G0357100ZS	32256891-32258168(+)	1074	358	108816.87	5.01
BnaMYB2	BnaA01G0365200ZS	32786604-32787938(+)	885	295	112433.86	5.06
BnaMYB3	BnaA02G0045300ZS	2546354-2548355(-)	840	280	167892.18	4.92
BnaMYB4	BnaA02G0378200ZS	33155658-33156863(-)	1011	337	102768.22	5.06
BnaMYB5	BnaA03G0050300ZS	2408564-2410538(-)	843	281	166460.81	4.9
BnaMYB6	BnaA03G0135900ZS	6915762-6916844 (-)	804	268	91357.94	5
BnaMYB7	BnaA03G0312000ZS	16483091-16484321 (-)	1023	341	104361.56	4.96
BnaMYB8	BnaA03G0487400ZS	27023236-27024166 (+)	753	251	77433.37	5.11
BnaMYB9	BnaA05G0412600ZS	39809095-39810464 (+)	1047	349	116330.89	5.01
BnaMYB10	BnaA05G0423500ZS	40421096-40422152 (+)	885	295	88892.42	5.08
BnaMYB11	BnaA05G0452900ZS	41865529-41866766 (+)	1008	336	102341.73	5.07
BnaMYB12	BnaA06G0057600ZS	3572905-3574580 (+)	996	332	142233.5	4.97
BnaMYB13	BnaA06G0163800ZS	10091574-10092862 (-)	777	259	110035.56	4.97
BnaMYB14	BnaA06G0344100ZS	42258187-42259525 (-)	1071	357	112803.96	5.05
BnaMYB15	BnaA09G0021000ZS	1330928-1332132 (-)	1011	337	100979.61	4.97
BnaMYB16	BnaA09G0060200ZS	3683531-3684746 (+)	1008	336	101112.82	5.06
BnaMYB17	BnaA09G0135900ZS	8220775-8222850 (-)	1110	370	175135.25	4.93
BnaMYB18	BnaA09G0147200ZS	8944567-8946278 (+)	1110	370	143838.31	5
BnaMYB19	BnaA10G0226500ZS	23076072-23078201 (+)	843	281	179215	4.96
BnaMYB20	BnaC01G0191600ZS	14174638-14175619 (+)	807	269	82001.37	5.1
BnaMYB21	BnaC01G0443900ZS	50997955-50999227 (+)	1074	358	108353.21	5.02
BnaMYB22	BnaC01G0455800ZS	52367618-52368963 (+)	885	295	113600.18	5.05
BnaMYB23	BnaC02G0052100ZS	3261244-3263562 (-)	837	279	195053.44	4.89
BnaMYB24	BnaC02G0506000ZS	61058311-61059530 (-)	1023	341	103681.14	5.06
BnaMYB25	BnaC03G0058300ZS	3046411-3048356 (-)	840	280	163856.16	4.91
BnaMYB26	BnaC03G0373800ZS	25163139-25164386 (-)	1023	341	105884.4	4.96
BnaMYB27	BnaC05G0072600ZS	4097294-4098491 (+)	1005	335	100976.31	5.02
BnaMYB28	BnaC05G0464600ZS	51456439-51457659 (+)	987	329	103719.42	5.04
BnaMYB29	BnaC05G0475500ZS	52295048-52296128 (+)	885	295	90920.47	5.08
BnaMYB30	BnaC05G0511100ZS	54454866-54456100 (+)	1008	336	102098.47	5.07
BnaMYB31	BnaC07G0351100ZS	48234276-48235576 (+)	1074	358	108612.43	5.04
BnaMYB32	BnaC07G0465900ZS	56085074-56086043 (+)	786	262	80884.41	5.09
BnaMYB33	BnaC09G0004300ZS	266734-267940 (-)	1011	337	101103.74	4.97
BnaMYB34	BnaC09G0048600ZS	3138803-3140038 (+)	1017	339	103080.97	5.06
BnaMYB35	BnaC09G0531900ZS	62845844-62847979 (+)	837	239	179779.67	4.96

2.2. Location of the Chromosomes and Phylogenetic Analysis

Additionally, the 35 *BnaMYB* genes were found on the chromosomes (Figure 1). The rapeseed genome had 35 *R2R3-MYB* genes that were widely distributed throughout 19 chromosomes. On chromosome A10, there was just one gene; on chromosomes A01, A02, C02,

C03, C07 and C09, there were two genes; on chromosomes A05, A06, and C01, there were three; and on chromosomes A03, A09 and C05, there were four. Table S1 shows the position of each gene on the chromosomes.

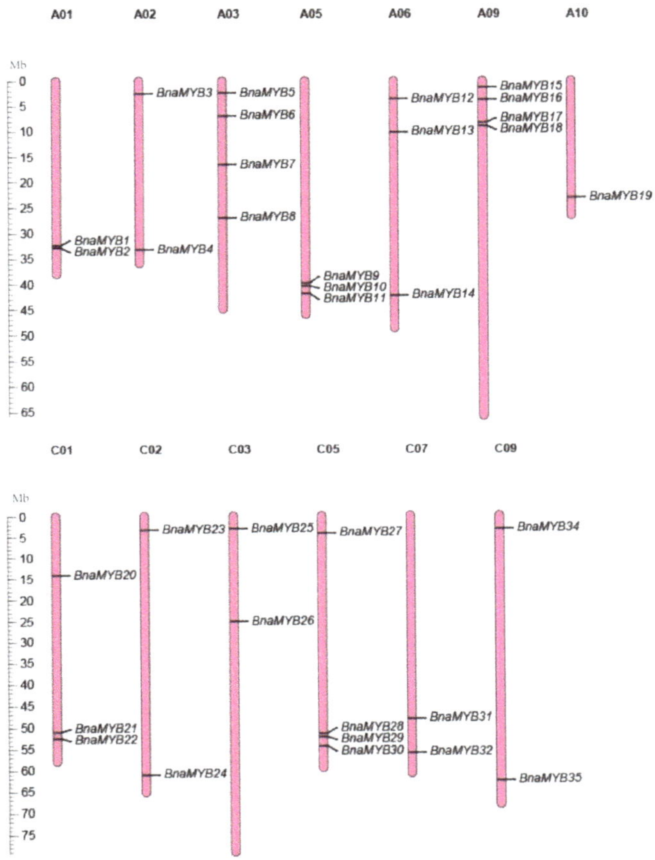

Figure 1. Chromosomal localization of *BnaMYB* genes in *B. napus*. The reference genome used was ZS11. The chromosomes with different sizes are represented by the pink vertical bars with different lengths. The locations of genes are shown from top to bottom.

We built a phylogenetic tree using 133 MYB proteins from seven different species, including *Arabidopsis*, *Brassica napus*, *Brassica oleracea*, *Brassica rapa*, *Capsella rubella*, *Oryza sativa*, *Raphanus sativus* and *Zea mays*, to better understand the evolutionary relationships among the *MYB* genes of the different species. Table S2 shows the protein sequences. These *MYB* genes were divided into three groups, as seen in Figure 2A. The majority of *MYB* members were found in the third group, which consisted of seven species, while the second group had the fewest *MYB* genes. Groups A, B, and C contained 10, 5 and 20 *BnaMYB* gene members, respectively (marked with asterisks in Figure 2A). Members of *BnaMYB* that clustered together may have a special connection and carry out similar functions.

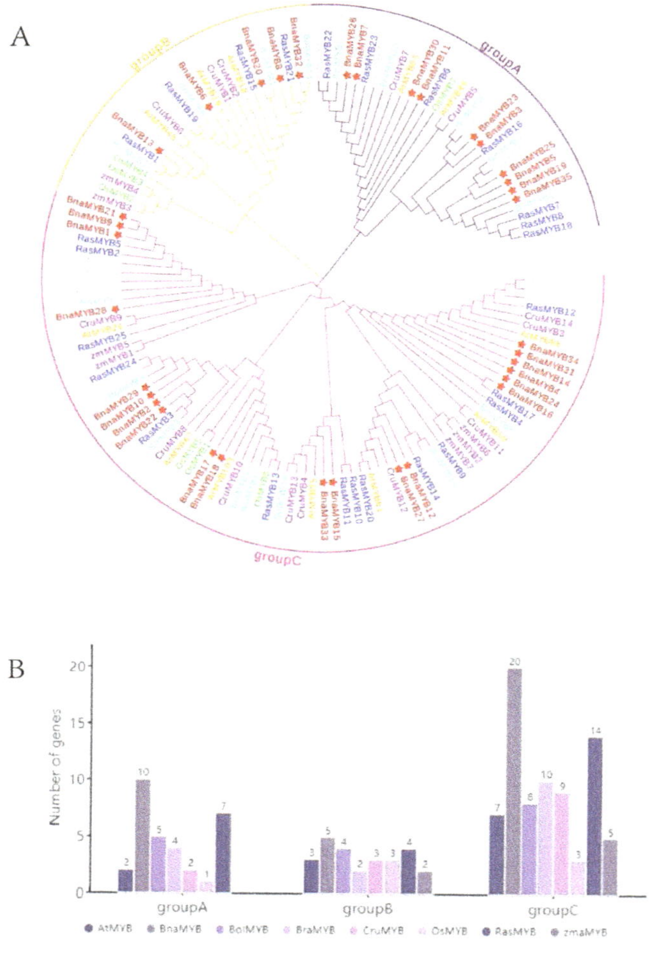

Figure 2. Phylogenetic analysis and number of *R2R3-MYB* homologs in different species. (**A**) Phylogenetic tree using 133 R2R3-MYB proteins from 7 species, including rice, maize, *Arabidopsis*, *B. napus*, *B. oleracea*, *B. rapa* and radish. The clades of Groups A, B and C are marked in purple, yellow, and pink, respectively. Among them, *BnaMYBs* are represented by red five-pointed stars. The abbreviations represent the species as follows: Os, *Oryza sativa*; Zma, *Zea mays*; At, *Arabidopsis thaliana*; Bna, *Brassica napus*; Bol, *Brassica oleracea*; Bra, *Brassica rapa*; Cru, *Capsella rubella*; Rsa, *Raphanus sativus*. (**B**) The number of *R2R3-MYB* genes in each group. The ordinate is the number of genes, and the abscissa shows the *R2R3-MYB* genes of each species.

The number of *R2R3-MYB* genes in each subspecies within each group was then tallied (Figure 2B). Although the first group had no examples of *ZmMYB*, the seven species were distributed throughout the three groups. Each group of species had two to four copies of the *R2R3-MYB* genes, depending on the quantity of these genes in *A. thaliana*. We found that the *BnaMYB* genes in Group A were all homologs of *AtMYB46*, and the *BnaMYB* genes in Group C were all homologs of *AtMYB86*.

2.3. Conserved Motifs and Gene Structure

Because Arabidopsis is a model plant, the research on its genome and gene function is the most complete. Rape and Arabidopsis belong to the same family of Cruciferae, with the closest genetic relationships and the clearest comparisons of genome homology. Therefore, we created an evolutionary tree with 35 BnaMYB protein sequences from *B. napus* and 12 AtMYB protein sequences from *A. thaliana* to examine the structural variety of *BnaMYB* genes. All of the *R2R3-MYB* genes were divided into Groups I, II and III (Figure 3A). Their conserved protein domains and gene structure were then investigated further.

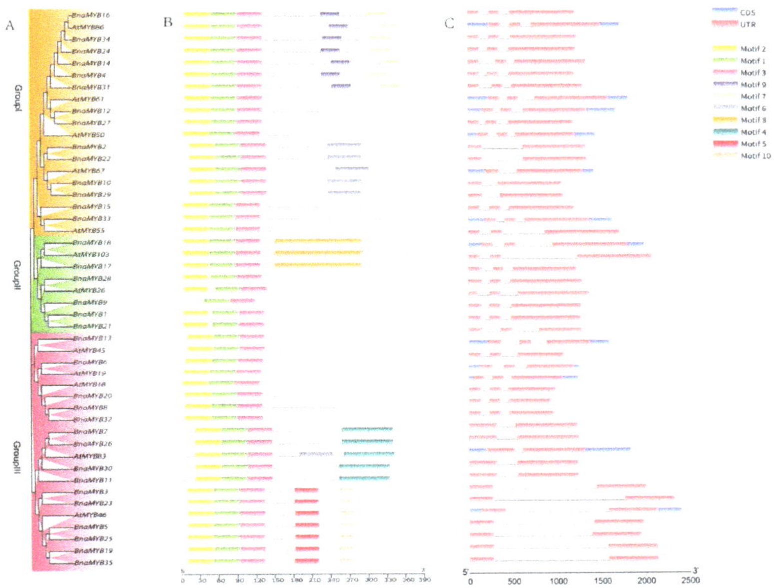

Figure 3. Comparative analysis of conserved domains and gene structure between *AtMYB* genes and *BnaMYB* genes. (**A**) The phylogenetic analysis of 12 *AtMYB* genes and 35 *BnaMYB* genes divided them into three groups: I, II and III. (**B**) The conserved domains of AtMYB and BnaMYB proteins. Ten conserved motifs of the BnaMYB protein were identified through the MEME website (https://meme-suite.org/meme/, accessed on 30 March 2022). Different colored boxes represent different motifs. (**C**) The gene structure of *AtMYB* and *BnaMYB* genes. The green box represents exons, the yellow box represents the UTR, and the black line represents introns. All results were visualized with TBtools software (v1.098769).

The full-length protein sequences were examined to identify their conserved motifs (Figure 3B). The *R2R3-MYB* genes' conserved motifs ranged from two to five. Ten conserved motifs in all were found; their widths ranged from 26 to 141 amino acids (Table S3). In addition, individuals of the same group had common conserved motifs. For instance, Motifs 7 and 9 were particular to Class I, Motif 8 to Class II and Motifs 4, 5 and 10 to Class III. The quantity and organization of the *MYB* motifs in the various groups, however, varied slightly. In Group II, *BnaMYB9* had fewer conserved motifs than other members, whilst *BnaMYB17*, *BnaMYB18* and *AtMYB103* all had conserved motifs that were different from the other members.

To further understand the development of the *R2R3-MYB* gene family in rape, according to the findings, there were between one and either two or three exons and introns, respectively (Figure 3C). Sixteen genes had two exons and one intron, whereas 31 genes had three exons and two introns. Exon gains and losses were discovered during the evolution of the *MYB* family of genes. Our findings suggested that over the evolution of the rapeseed genome, the exon–intron architecture of *MYB* genes remained relatively stable. Additionally, the *BnaMYB* gene members within a class had gene architectures that were quite similar and consistent with their evolutionary groups. In summary, according to our analysis of the conserved motif composition, gene structures, and phylogenetic interactions, the consistency of the class organization was convincingly sustained, showing that the MYB proteins have extraordinarily well-maintained amino acid residues and that the members within a class may play parallel roles.

2.4. Multi-Sequence Alignment and Predicted 3d Protein Structures

We used 3D structure prediction analysis and multiple sequence alignment to clarify the structural properties of the BnaMYB proteins (Figures S1 and S2). These findings led us to the conclusion that the MYB DNA-binding domains in these proteins are highly conserved. The BnaMYB proteins were split into three groups on the basis of their genetic compatibility.

2.5. Collinearity within B. napus and among Different Species

Analysis of genome-wide replication is crucial for understanding the genesis, evolution, and genome-wide expansion of organisms. To further understand the reasons behind gene replication events in *BnaMYB*, we therefore examined the replication events of the *R2R3-MYB* gene family in *B. napus*. A04, A07, A08, C04, C06 and C08 were the only chromosomes without fragment repeats, along with 48 pairs of genes with large fragment repeats (Figure 4). These findings suggested that the amplification and evolution of the *R2R3-MYB* genes in the *B. napus* genome may have been significantly influenced by the replication of large fragments.

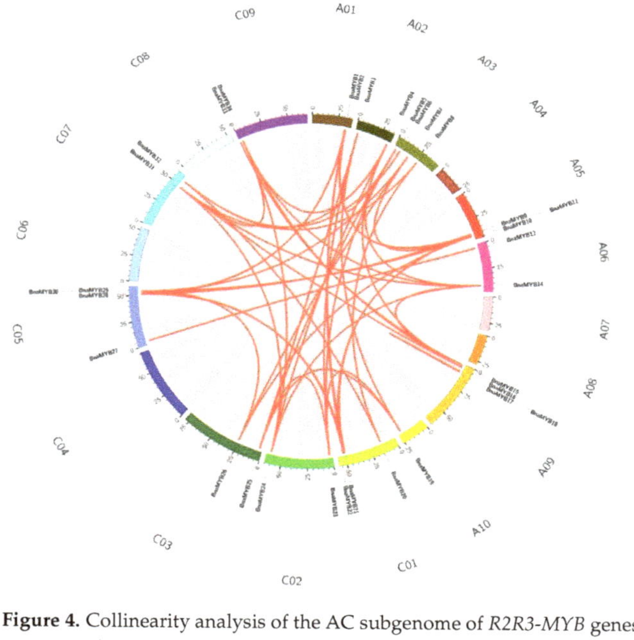

Figure 4. Collinearity analysis of the AC subgenome of *R2R3-MYB* genes in *B. napus*. The gray lines represent the replication events of all genes in *B. napus* and the red lines represent tandem repeat events within the *BnaMYB* genes.

The evolution of the *R2R3-MYB* gene family in *Brassica* has been tracked in previous research [21]. In order to specifically understand the evolutionary relationship of target genes, we analyzed the homologous relationship among *Arabidopsis*, *B. napus* (A and C subgenomes), *B. rapa* (A genome), and *B. oleracea* (C genome) (Figure 5). The collinearity analysis of *MYB* showed that there were a large number of orthologous *MYB* genes in *Arabidopsis*, *B. rapa*, *B. oleracea*, and *B. napus*. There were 16 pairs of genes in *Arabidopsis* and

B. rapa that showed collinearity, and 15 *B. rapa R2R3-MYB* genes had homologous genes in *Arabidopsis*, 10 of which were multi-copy genes and 5 were single-copy genes. *B. rapa* lacked *AtMYB50* homologs, indicating that gene loss occurred in *B. rapa* during evolution. Furthermore, 16 *B. oleracea R2R3-MYB* genes contained similar genes to *Arabidopsis*. The A and C subgenomes of *B. napus* mostly overlapped those of the related diploids *B. rapa* and *B. oleracea*. Fourteen similar gene pairs were discovered in the A genome of *B. napus* and *B. rapa*, and 21 in the C genome of *B. napus* and *B. oleracea*. Even if gene loss has occurred, the great majority of *R2R3-MYB* genes in *B. napus* are still present and have contributed to the evolution of the *MYB* gene family.

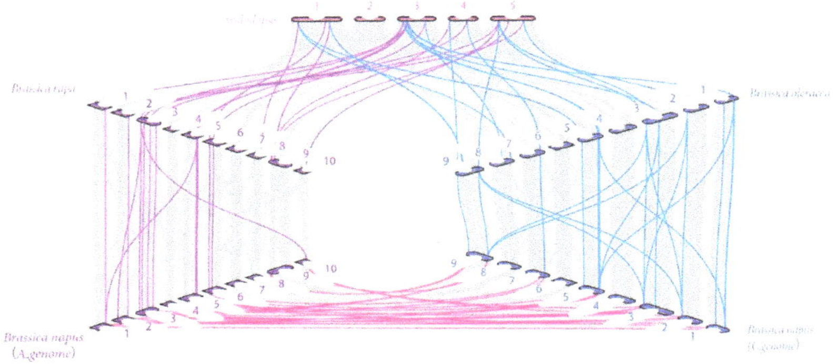

Figure 5. Syntenic relationships of R2R3-MYB genes in *B. napus* and three ancestral plant species. The figure shows the collinearity between *Arabidopsis* (*A. thaliana*), *Brassica rapa* (*B. rapa*), *Brassica oleracea* (*B. oleracea*) and *Brassica napus* (*B. napus*).

2.6. Synteny of BnMYB Genes

We compared the CDS sequences (Figure S3), which showed the similarity and identity of these sequences more clearly. We estimated the synonymous (Ks) and nonsynonymous (Ka) replacement rates (Ka/Ks) of 44 segmented repeated pairs (Table S4). The average Ka/Ks ratio of fragmented repeated gene pairs was 0.61. The calculation results showed that the Ka/Ks ratio of all duplicate BnMYB gene pairs was less than 1, which means that they are undergoing a purification selection process.

2.7. Cis-Acting Elements and Functional Annotation

Transcriptional regulators bind to the cis-acting elements, which control gene transcription. We examined the cis-acting elements of the 1500 bp upstream sequence of *BnaMYB* promoters in order to determine the potential function of the *BnaMYB* genes, excluding components with uncertain functions and general transcriptional regulatory elements (Table 2, Figure 6). All details of cis elements of each gene are listed in Supplementary Table S5. ABRE, the AuxRR core, a TGA element, a TGA box, a TGACG motif, a CGTCA motif, a P box, a GARE motif, TCA and a TCA element were distinguished as being responsive elements related to four phytohormones (abscisic acid (ABA), auxin, methyl jasmonate (MeJA), gibberellin (GA), and salicylic acid (SA)). Overall, we focused on and identified eight types. We also found four stress-related (drought, low-temperature, anaerobic, defense, and stress) responsive elements, including MBS, MYB, MYC, LTR, ARE and TC-rich repeats, suggesting their involvement in stress responses. The drought stress-responsive elements MBS, MYC and MYB were primarily found in 13 genes, 31 genes and 35 genes, respectively. These findings revealed that the majority of *BnaMYB* genes have drought-responsive elements, indicating that *BnaMYB* genes may be extremely important in responding to the stress caused by drought.

Table 2. The *cis*-elements identified in more than three *BnaMYB* genes.

Site Name	Sequence	Function of the Cis-Elements
TGA element	AACGAC	Auxin-responsive element
TC-rich repeats	GTTTTCTTAC	Cis-acting element involved in defense and stress responses
LTR	CCGAAA	Cis-acting element involved in low-temperature responses
TCA element	CCATCTTTTT	Cis-acting element involved in salicylic acid responses
ABRE	ACGTG	Cis-acting element involved in abscisic acid responses
AuxRR core	GGTCCAT	Cis-acting regulatory element involved in auxin responses
CGTCA-motif	CGTCA	Cis-acting regulatory element involved in the MeJA responses
TGACG motif	TGACG	Cis-acting regulatory element involved in MeJA responses
GARE motif	TCTGTTG	Gibberellin-responsive element
P box	CCTTTTG	Gibberellin-responsive element
MBS	CAACTG	MYB binding site involved in drought response
TGA box	TGACGTAA	Part of an auxin-responsive element
ARE	AAACCA	Cis-acting regulatory element essential for anaerobic induction
MYB	TAACCA	Drought response element
MYC	CAATTG	Drought and MYBd reaction element
TCA	TCATCTTCAT	Cis-acting element involved in salicylic acid responses

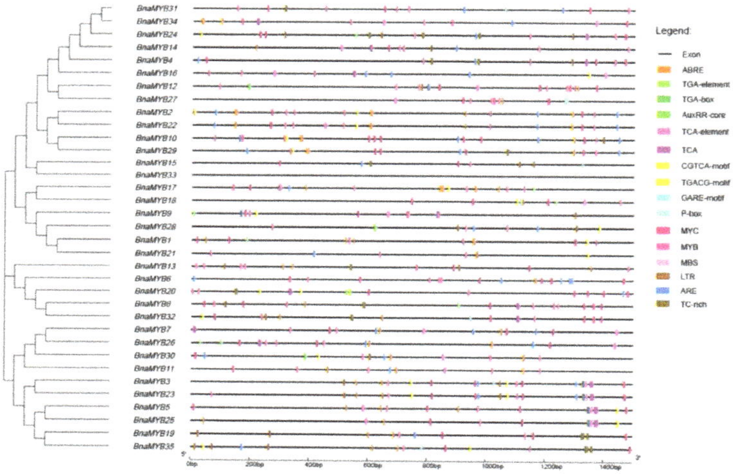

Figure 6. Cis-acting element analysis of *MYB* gene subfamily in *B. napus.* The boxes with different colors on the black line represent different cis-elements.

2.8. Tissue-Specific Expression Patterns of BnaMYB Genes

Through the use of RNA-seq data from *Brassica napus* (ZS11 variant) (BioProject ID: PRJCA001495), the histo-specific expression profiles of *BnMYB* genes were detected in nine different tissues and organs (i.e., stems, roots, siliques, cotyledons, sepals, seeds, buds, leaves, and petals). The expression patterns of all *BnaMYB* genes varied in the different tissues, as shown in Figure 7 and Table S6. For instance, the majority of genes displayed greater expression levels in the stems, including *BnaMYB3*, *BnaMYB5*, *BnaMYB7*, *BnaMYB11*, *BnaMYB14*, *BnaMYB15*, *BnaMYB16*, *BnaMYB17*, *BnaMYB18*, *BnaMYB19*, *BnaMYB23*, *BnaMYB25*, *BnaMYB26*, *BnaMYB27*, *BnaMYB28*, *BnaMYB30*, *BnaMYB31*, *BnaMYB33*, *BnaMYB34* and *BnaMYB35*. In the cotyledons, a few genes showed higher expression levels, including *BnaMYB2*, *BnaMYB4*, *BnaMYB6*, *BnaMYB7*, *BnaMYB8*, *BnaMYB10*, *BnaMYB14*, *BnaMYB20*, *BnaMYB22*, *BnaMYB24*, *BnaMYB26*, *BnaMYB29*, *BnaMYB31* and *BnaMYB32*. In the roots, *BnaMYB4*, *BnaMYB11*, *BnaMYB14*, *BnaMYB15*, *BnaMYB16*, *BnaMYB17*, *BnaMYB18*, *BnaMYB27*, *BnaMYB31* and *BnaMYB33* displayed higher expression patterns. In the sepals, *BnaMYB1*, *BnaMYB9*, *BnaMYB21* and *BnaMYB28* exhibited higher expression patterns. In the siliques,

BnaMYB5, *BnaMYB15* and *BnaMYB33* displayed higher expression patterns. Likewise, in the seeds, many genes, including *BnaMYB12* and *BnaMYB27*, showed higher expression patterns. These findings led us to the conclusion that the *BnaMYB* gene family was essential for rapeseed's development at all stages, and that other members with similar expression traits might serve related purposes.

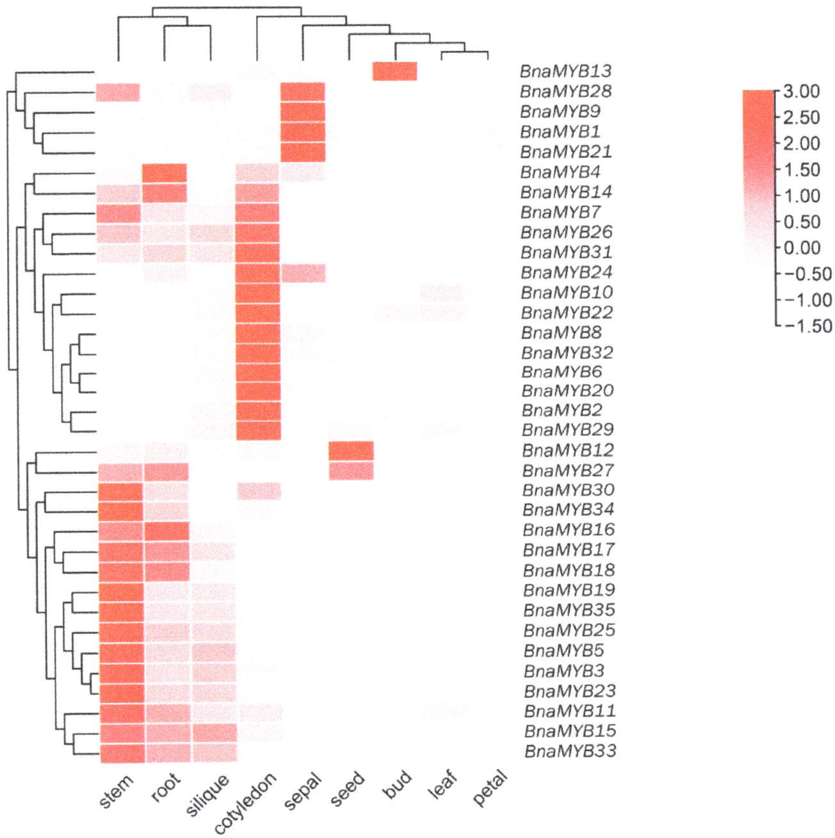

Figure 7. Heat map showing the expression profiles of *BnaMYB* genes in different tissues. The sampling time for the seeds, leaves, and siliques was the 10th day. The red, pink, and white colors display high to low expression levels. The heat map of expression was created by taking the log10 of transcripts per million (TPM). The transcriptomic data were sourced online from the BnTIR: *Brassica napus* information resource (http://yanglab.hzau.edu.cn, accessed on 2 May 2022).

2.9. Expression Profiles of BnMYB Genes under Drought Stress

The majority of the *BnaMYB* genes had drought-responsive elements, which suggests that they may be involved in the response to drought, according to an analysis of the cis-acting elements in the upstream sequence of the *BnaMYB* promoters we conducted before. We chose 10 lignin-related genes (*BnaMYB3*, *BnaMYB5*, *BnaMYB7*, *BnaMYB11*, *BnaMYB19*, *BnaMYB23*, *BnaMYB25*, *BnaMYB26*, *BnaMYB30* and *BnaMYB35*) and submitted the 10 genes to drought treatment for 7 days to further discover their potential function (Figure 8). The primers used for gene expression analysis are shown in Table S7.

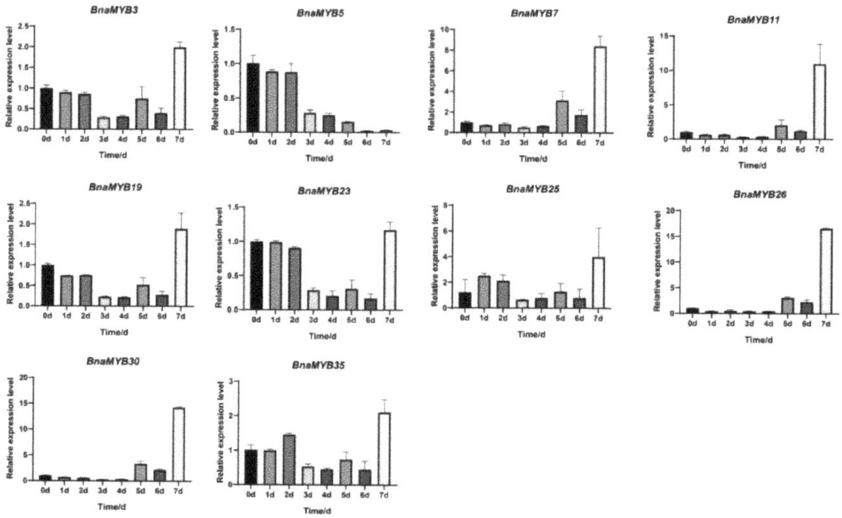

Figure 8. The expression patterns of *BnaMYB* genes under drought conditions. The 0 h (CK), 2, 4, 6, and 8 d labels indicate the time points (days) when the samples were harvested for studying gene expression after the drought treatment.

The expression patterns of *BnaMYB* genes under drought stress showed three patterns. First, the expression of *BnaMYB3*, *BnaMYB19*, *BnaMYB23*, *BnaMYB25* and *BnaMYB35* decreased gradually under the drought treatment and then decreased to the lowest level at 3 d or 4 d, then reached the peak on Day 7, showing a U-shaped curve. However, the expression of *BnaMYB5* decreased gradually after 0 h and reached the lowest point at 6 d and 7 d. Lastly, the expression levels of *BnaMYB7*, *BnaMYB11*, *BnaMYB26*, and *BnaMYB30* were very low in the first 4 days, significantly increased on Days 5 and 6, and rapidly increased on Day 7 to reach the peak, indicating that drought treatment could activate the expression of these four genes.

The majority of *BnaMYB* genes generally had various drought expression patterns, indicating that these *MYB* genes were engaged in the pathway that causes drought stress. Four genes (*BnaMYB7*, *BnaMYB11*, *BnaMYB26* and *BnaMYB30*) showed significantly upregulated expression under drought stress.

3. Discussion

The *R2R3-MYB* gene family of *B. napus* and related species was biologically analyzed in this work, including their geographic position, phylogenetic analysis, gene structure, homologous protein domains, and predicted 3D protein structure. The findings demonstrated that the 19 rapeseed chromosomes contained an unequal distribution of 35 *BnaMYB* family members. These individuals were divided into three groups through phylogenetic analysis, and the majority of the genes within each subgroup shared similar gene structures, conserved protein domains, and three-dimensional protein structures, which is indicative of the *BnaMYB* gene family's evolutionary stability.

3.1. Characterization and Evolution of the R2R3-MYB Gene Family in Rapeseed

Allotetraploid rapeseed is a crop that has been engaged in extensive genome replication and combination activities [35]. One of the biggest obstacles to agricultural production in the modern world is dry soils. Crop plants have been bred conventionally and genetically to increase tolerance or resistance to drought and salinity in an effort to boost agricultural productivity in impacted areas. Future breeding and genetic engineering plans are likely to be greatly aided by a knowledge of the genomic mechanisms underlying plants' responses to drought stress [36].

In plants, the *MYB* family has a significant function. The function of *MYBs* has been systematically researched in *Arabidopsis* [16,37], *Setaria italica* (Muthamilarasan et al., 2014), *Vitis vinifera* [38], *Zea mays* [25], *Populus trichocarpa* [39], *Gossypium raimondii* [40], and other plants, especially *Arabidopsis* and *Oryza sativa* [24]. The largest *MYB* gene subfamily in plants is the *R2R3-MYB* subfamily. Genetic methods have been used to characterize a large number of R2R3-MYB proteins, and it has been discovered that these proteins are involved in the regulation of a variety of plant-specific processes, such as (i) primary and secondary metabolism, (ii) cell fate and identity, (iii) developmental processes, and (iv) responses to biotic and abiotic stresses [16]. To the best of our knowledge, the rapeseed genome does not yet have a complete description of the *R2R3-MYB* gene family. The availability of rapeseed's genome sequences makes it possible to identify rapeseed genes across the entire genome [35]. Here, we identified 35 *BnaMYB* genes in the genome of rapeseed.

The *BnaMYB* gene family may have developed in the rapeseed genome through segmental events and many tandem duplications. A key strategy for gene expansion is tandem duplication. Genes that have been enlarged by tandem duplication are always dispersed throughout chromosomes as a group [41,42]. Exon-intron connections and numbers, in addition to duplication events, can help to explain how the gene family has evolved [43–45]. When the gene structures of *BnaMYBs* and members of the same group were studied, it was shown that *MYB* genes contributed equally to exon/intron allocation in terms of exon and intron numbers, while *BnaMYBs* showed similar motif patterns. These findings suggest that these *R2R3-MYB* members could play similar roles in a range of abiotic stressors.

3.2. Diversity in the Functional Expression of BnaMYB Genes in Brassica napus

We looked at the expression patterns of 35 *BnaMYB* genes in diverse organs using publicly accessible data. The results showed that they varied in their patterns of expression in different tissues and that they were expressed in different tissues. In stems and cotyledons, 85.7% of *BnaMYB* genes were primarily expressed, while the primary expression of 10 genes was detected in the roots. *BnaMYB12* was only strongly expressed in the seeds and less so in other tissues. *BnaMYB1*, *BnaMYB9*, *BnaMYB21* and *BnaMYB28* displayed higher transcription levels in the sepals. These findings suggest that the *BnaMYB* gene family might well be important for rapeseed's development and expansion at all stages. Twenty of the *BnaMYB* genes were mostly expressed in the stems; therefore, they may play a crucial role in how the stems react to environmental influences. Little research has been conducted on *Brassica napus*, despite the fact that several *R2R3-MYB* genes have been linked to drought adaptations and endurance in some species, particularly in rice and *Arabidopsis* [46–49]. Hence, our study of *BnaMYBs* is helpful to analyze the response of *R2R3-MYB* to drought stress in *Brassica napus*.

Numerous members of the plant *R2R3-MYB* family have been identified as transcription factors that are triggered by abiotic stress to date [47,50,51]. Cotton's resilience to drought can be improved by transgenic *Arabidopsis* overexpressing the *R2R3-MYB* gene *GaMYB85* [52]. *Arabidopsis* overexpressing *LpMYB1* is more tolerant to salt and drought because of this R2R3-MYB factor [53]. Under prolonged drought stress, *TaMYBsdu1* increases noticeably in the roots and leaves of wheat [47]. *TaMYB33* improves tolerance to salt and drought, partly by restoring the osmotic equilibrium and eliminating superoxide radicals [54,55]. *TaMYB30-B* and *TaMYB19-B* can increase the ability of transgenic *Arabidopsis* plants to withstand drought stress [56,57]. These studies indicate that *R2R3-MYB* has the function of dealing with drought stress in wheat, cotton, and *Arabidopsis* and their function during drought stress is affected by the ABA synthesis and signal transduction pathway. For instance, the accumulation of 35S:*GaMYB85* resulted in the transcription of *RD22*, *RD29A*, *ADH1*, *AB15*, and *P5CS*. The *AB15* gene, which codes for a bZIP TF, is known to be an ABA-responsive gene that is important for seed germination under drought stress [58,59]. In light of the previous analysis, we speculated that drought stress in *Brassica napus* is affected by the lignin content. MYB46 controls not only the signaling molecules but also the genes that produce all three of the primary components of the secondary walls: grain, hemicellulose, and lignin [60]. Additionally, *MYB85* controls the production of lignin by regulating the lignin-synthesizing genes and, when overexpressed, results in ectopic lignin deposition [61]. This indicates that the *BnaMYB* gene family may affect lignin synthesis, and lignin synthesis responds to drought stress. We will pay attention to this in future research to deeply understand the relationship between the *BnaMYB* gene family and lignin.

The tissue-specific expression patterns reflected which tissues of the plant the gene mainly acts on. We found that 20 genes were highly expressed mainly in the stem, indicating that these genes were probably involved in lignin synthesis. Ten of these genes were selected for drought treatment, and their expression profiles were detected under drought stress. Our RT-qPCR data showed that the *BnaMYB* gene had significant changes under drought conditions, which proved that 10 genes were responsive to drought stress. The expression of *BnaMYB5* decreased significantly under drought stress, indicating that this gene could not cope with drought; *BnaMYB3, BnaMYB19, BnaMYB23, BnaMYB25* and *BnaMYB35* showed U-shaped curves, indicating that these five genes did not show drought resistance under mild drought stress but showed drought resistance under severe drought. Four genes (*BnaMYB7, BnaMYB11, BnaMYB26* and *BnaMYB30*) were significantly upregulated under drought stress, indicating that they are probably drought-resistant genes. According to our understanding, this is the first evidence indicating that *R2R3-MYB* genes in *B. napus* are implicated in the response to drought stress. This work is helpful for understanding the characteristics and functions of *R2R3-MYB* genes in different species.

4. Methods

4.1. Identification of Rapeseed's R2R3-MYB Family Members

On the BnPIR website (http://cbi.hzau.edu.cn/bnapus/, accessed on 14 February 2022) [35], the genome sequences, protein sequences, and gene annotation data of rapeseed were downloaded. The Markov model of the conserved domains of *R2R3-MYB* (Myb_DNA-binding, PF00249) was downloaded from the Pfam database (http://pfam.xfam.org/, accessed on 14 February 2022). The cut-off E-value was chosen to 1e-10 when using the Markov model on HMMER software to preliminary screen the protein sequences of rapeseed. In order to exclude possible R2R3-MYB proteins without two conserved MYB DNA-binding domains, all candidate proteins were then uploaded to three online sources, SMART (http://www.omicsclass.com/article/681, accessed on 17 February 2022), NCBICDD (http://www.omicsclass.com/article/310, accessed on 17 February 2022), and PFAM (http://pfam.xfam.org/, accessed on 17 February 2022). For the MW and pI prediction analyses, ExPASy (http://web.expasy.org/protparam, accessed on 18 February 2022) received the identified *R2R3-MYB* candidate genes. Additionally, the subcellular localization was predicted by WoLF PSORT (https://www.genscript.com/wolf-psort.html, accessed on 18 February 2022).

4.2. Chromosome Location and Phylogenetic Analysis

The BnPIR website (http://cbi.hzau.edu.cn/bnapus, accessed on 5 March 2022) provided information on the chromosome location of *BnaMYBs*. After that, the distribution status of the *BnaMYBs* discovered on the chromosomes was shown using MapChart software. With the aid of Adobe Illustrator, the outcomes were improved.

We downloaded the protein sequences of the MYB family members of *Brassica oleracea, Brassica rapa, Capsella rubella, Oryza sativa, Raphanus sativus* and *Zea mays*. The sequences of the protein on the *Brassica oleracea* and *Brassica rapa* were collected from the Phytozome website (https://phytozome.jgi.doe.gov/pz/portal.html, accessed on 10 March 2022). The sequences of the protein in *Oryza sativa* and *Zea mays* were collected from the Ensembl website (http://plants.ensembl.org/index.html, accessed on 10 March 2022). The protein sequences of *Capsella rubella* and *Raphanus sativus* were collected from the NCBI website (https://www.ncbi.nlm.nih.gov/, accessed on 10 March 2022). OrthoMCL software (v2.0.3) [62] was used to search for orthologous, co-orthologous, and paralogous genes in *Arabidopsis thaliana, Brassica oleracea, Brassica rapa, Capsella rubella, Oryza sativa, Raphanus sativus* and *Zea mays* using entire R2R3-MYB protein sequences. The E-value cut-off of an all-against-all BLASTP alignment process was set at 1×10^{-10}, and the alignments with a match cut-off value lower than 50 were eliminated.

These plants' MYB protein sequences were analyzed using Clustal W after the protein sequences of *MYB* family members from *B. napus, Arabidopsis, B. oleracea, B. rapa, Capsella rubella, Oryza sativa, Raphanus sativus* and *Zea mays* were retrieved. Additionally, MEGA 11.0 software was used to create the evolutionary tree using the neighbor joining method (NJ) once the results of sequence alignment were obtained [63,64]. Finally, we used Evolview (http://www.omicsclass.com/article/671, accessed on 13 March 2022) to visualize the evolutionary tree.

4.3. *Distinct Gene Structure and Conserved Protein Domain of R2R3-MYBs*

The gene structure of the *R2R3-MYB* genes contains intron and exon information [65]. MEME Suite (https://meme-suite.org/meme/, accessed on 30 March 2022) was used to examine the homologous domains of the genes using the BnaMYB protein sequences, increasing the maximum motif count to 10, the maximum motif amino acid count to 20, and the minimum motif width to 6, and leaving all other variables at their normal values. Finally, the similar motifs of BnaMYB proteins were viewed using TBtools software.

4.4. *Predicted 3D Structure of R2R3-MYB Proteins*

Thirty-five BnaMYB protein sequences were entered into the DNAMAN8.0 program for triple sequence comparison. We next predicted the 3D structure of the proteins using the internet tool Phyre2 (http://www.sbg.bio.ic.ac.uk/phyre2/html/page.cgi?id=index, accessed on 25 April 2022).

4.5. *Analysis of Collinearity within B. napus and Related Species*

McScanx analysis was used to analyze the *BnaMYB* genes' intra-species collinearity in *B. napus* and Circos software (v0.69-8) was used to visualize the associations. Additionally, the McScanX software (Python version) was used to visualize the collinearity.

4.6. *Analysis of the Synteny of BnMYB Genes*

The CDS sequences of 35 *BnaMYB* genes were input into DNAMAN8.0 for sequence alignment. We selected the CDS sequences of homologous gene pairs, uploaded them to the ALTER website (http://www.sing-group.org/ALTER/, accessed on 28 January 2023), and outputted the results in ALN format. We converted ALN format to AXT format with AXT Converter and used KaKs Calculator 2.0 software to calculate the value of Ka/KS.

4.7. *Cis-Acting Elements and Functional Annotation*

The publicly available online whole-genome information on *B. napus* (http://cbi.hzau.edu.cn/bnapus, accessed on 28 April 2022) provided 1500 bp of the upstream sequences of the *BnaMYB* genes. The cis-acting components were extracted using the online tool Plant CARE (http://bioinformatics.psb.ugent.be/webtools/plantcare/html, accessed on 29 April 2022) and then visualized using the online tool DSGS.

4.8. *Spatial and Temporal Expression Patterns of R2R3-MYB Genes*

The BnTIR: *Brassica napus* genomic resource is a website (http://yanglab.hzau.edu.cn, accessed on 2 May 2022) [66]. The stems, roots, siliques, cotyledons, sepals, seeds, buds, leaves, and petals were among the tissues for which we obtained RNA-seq data. To create the heat map of expression, the data were entered into the Heatmapper program (http://www.heatmapper.ca, accessed on 2 May 2022).

4.9. Plant Materials and Treatment Methods

In this study, the rapeseed genotype "ZS11", a typical cultivated variety, was used for the stress treatments. The seeds of the ZS11 genotype were furnished by OCRI, CAAS, China. Before the stress treatments, some seeds were randomly selected from the same batch of seeds to determine the germination rate. The seeds with a 100% germination rate were considered to be vigorous seeds. The vigorous seeds were carefully chosen and sterilized with a 10% hypochlorous acid solution for 5 min. The seeds were grown on water-saturated filter paper in a chamber (25 °C day and night, and a 16 h/8 h light/dark cycle) until the plants grew to the five-leaf stage. When the plants had grown to the five-leaf stage, we stopped watering to impose drought stress. Samples from the stem were collected 0, 1, 2, 3, 4, 5, 6 and 7 days after the initiation of the drought treatment [67,68]. Moreover, three biological replicates of each sample were collected. The gathered stems were immediately put into liquid nitrogen and kept in a freezer for later use at −80 °C.

4.10. RNA Extraction and RT-PCR Analysis

Using a carbohydrate phenolic total RNA extraction kit, the total RNA was isolated from leaves that had undergone various drought treatments (Tiangen Biochemical Technology Co., Ltd.). An ultramicroscopic spectrophotometer was used to measure the quantity and quality of the RNA (Thermo Fisher, Waltham, MA, USA, Nanodrop One). As samples for subsequent RT-qPCR experiments, we synthesized cDNA using a reverse transcription kit and diluted it 100 times with ddH_2O. Specific primers were created using the online qPCR Primer Database (https://biodb.swu.edu.cn/qprimerdb/, accessed on 27 May 2022) based on the coding sequences of the *BnaMYB* genes. The specific primers were amplified, and the primers with a ct value of no more than 30, a normal amplification curve, and no obvious impurity peak in the dissolution curve were selected for subsequent experiments. The quantitative experiment was conducted in real-time using SYBR Premix Ex TaqTM (TaKaRa). Three separate replicates were collected for this experiment, and the samples that had not been subjected to drought were used as standards. The $2^{-\Delta\Delta CT}$ technique is often used in investigations of the relative expression of genes [69]. The graphs were developed using GraphPad Prism 9.0.0 software.

5. Conclusions

In total, 35 *BnaMYB* genes were found in *B. napus*, dispersed unevenly across 19 chromosomes. We analyzed how the MYB proteins are related, their gene structure, their conserved motifs, and their three-dimensional structures. These genes were divided into three subfamilies, each of which had reasonably traditional gene architectures and patterns. The *BnaMYBs* promoter regions also contained cis-acting hormones and abiotic stress response components, along with drought-responsive genes. There were 48 pairs of large-segment repetitive genes discovered in *B. napus* through the analysis of collinearity. The *R2R3-MYB* genes of *B. napus* had undergone polyploidization and various degrees of loss and expansion, according to comparative genomic research. We also looked at the gene expression patterns of *BnaMYB* in various rapeseed tissues, and the results showed that the members of the *BnaMYB* gene family were crucial at different phases of the development of *B. napus*. In addition, we examined the expression patterns of 10 *BnaMYB* genes under drought. Overall, by carefully examining the conservation and divergence of the *BnaMYB* gene family's activities, our work lays the biological groundwork for the subsequent functional discovery of *R2R3-MYB* genes in cruciferous plants.

Supplementary Materials: The following supporting information can be downloaded at: https://www.mdpi.com/article/10.3390/plants12040886/s1, Figure S1: Multiple sequence alignment of BnaMYB proteins; Figure S2: Three-dimensional structure prediction of Myb_DNA-binding domain of BnaMYB proteins. The images are from dark blue to dark red, indicating from the N end to the C end; Figure S3: CDS sequence alignment of 35 BnaMYB genes; Table S1: The position of genes in chromosomes; Table S2: Protein sequence of MYB genes in each species; Table S3: The information of identified 10 motifs in BnMYB proteins; Table S4: Information on Ka, Ks and Ka/Ks ratios of Brassica napus. See the abbreviations at the end of the table [70]; Table S5:Cis-element of genes; Table S6: Absolute gene expression values in thirty-five BnaMYBs; Table S7:Primers used for gene expression analysis.

Author Contributions: D.L., W.W. and J.L. designed the experiments; D.L. performed the experiments; D.L. and J.L. analyzed the data; D.M. provided technical support; D.L. wrote the paper; W.W. and J.L. revised the manuscript. All authors have read and agreed to the published version of the manuscript.

Funding: This work was supported by the Central Public-Interest Scientific Institution Basal Research Fund (No. 161017202203 and 1610172020001) and an open project (KF2020007) of the Key Laboratory of Biology and Genetic Improvement of Oil Crops, Ministry of Agriculture and Rural Affairs, China. We thank the Science and Technology Innovation Project of the Chinese Academy of Agricultural Sciences for computational and experimental support.

Informed Consent Statement: This study's collection of material and experimental research complied with the relevant institutional national and international guidelines and legislation.

Data Availability Statement: All data generated or analyzed during this study are included in this published article and its Supplementary Materials. Zhongshuang11 is an elite conventional variety of rapeseed bred and preserved by OICR-CAAS (Oil Crops Research Institute).

Conflicts of Interest: The authors declare no conflict of interest.

References

1. Pandey, V.; Shukla, A. Acclimation and Tolerance Strategies of Rice under Drought Stress. *Rice Sci.* **2015**, *22*, 147–161. [CrossRef]
2. Rizwan, M.; Atta, B.; Bilal, M.; Noushahi, H.A.; Ali, M.Y.; Shabbir, M.A.; Salim, M.A.; Hussain, M.; Liaqat, N.; Ahmar, S. Effect of Abiotic Stresses on Brassica Species and Role of Transgenic Breeding for Adaptation. *Asian J. Res. Crop Sci.* **2019**, *3*, 1–10.
3. Sabagh, A.E.; Hossain, A.; Barutçular, C.; Islam, M.S.; Ratnasekera, D.; Kumar, N.; Meena, R.S.; Gharib, H.S.; Saneoka, H.; da Silva, J.A. Drought and salinity stress management for higher and sustainable canola (*Brassica napus* L.) production: A critical review. *Aust. J. Crop Sci.* **2019**, *13*, 88–97. [CrossRef]
4. Yoshimura, K.; Masuda, A.; Kuwano, M.; Yokota, A.; Akashi, K. Programmed proteome response for drought avoidance/tolerance in the root of a C(3) xerophyte (wild watermelon) under water deficits. *Plant Cell Physiol.* **2008**, *49*, 226–241. [CrossRef]
5. Moura, J.C.; Bonine, C.A.; de Oliveira Fernandes Viana, J.; Dornelas, M.C.; Mazzafera, P. Abiotic and biotic stresses and changes in the lignin content and composition in plants. *J. Integr. Plant Biol.* **2010**, *52*, 360–376. [CrossRef]
6. Liu, Q.; Zheng, L.; He, F.; Zhao, F.-J.; Shen, Z.; Zheng, L. Transcriptional and physiological analyses identify a regulatory role for hydrogen peroxide in the lignin biosynthesis of copper-stressed rice roots. *Plant Soil* **2014**, *387*, 323–336. [CrossRef]
7. Liu, W.; Jiang, Y.; Wang, C.; Zhao, L.; Jin, Y.; Xing, Q.; Li, M.; Lv, T.; Qi, H. Lignin synthesized by CmCAD2 and CmCAD3 in oriental melon (*Cucumis melo* L.) seedlings contributes to drought tolerance. *Plant Mol. Biol.* **2020**, *103*, 689–704. [CrossRef]
8. Tu, M.; Wang, X.; Yin, W.; Wang, Y.; Li, Y.; Zhang, G.; Li, Z.; Song, J.; Wang, X. Grapevine VlbZIP30 improves drought resistance by directly activating VvNAC17 and promoting lignin biosynthesis through the regulation of three peroxidase genes. *Hortic. Res.* **2020**, *7*, 150. [CrossRef]
9. Zhou, Y.; Zhang, Y.; Wang, X.; Han, X.; An, Y.; Lin, S.; Shen, C.; Wen, J.; Liu, C.; Yin, W.; et al. Root-specific NF-Y family transcription factor, PdNF-YB21, positively regulates root growth and drought resistance by abscisic acid-mediated indoylacetic acid transport in Populus. *New Phytol.* **2020**, *227*, 407–426. [CrossRef]
10. Zhao, D.; Luan, Y.; Shi, W.; Zhang, X.; Meng, J.; Tao, J. A Paeonia ostii caffeoyl-CoA O-methyltransferase confers drought stress tolerance by promoting lignin synthesis and ROS scavenging. *Plant Sci.* **2021**, *303*, 110765. [CrossRef]
11. Park, S.C.; Kim, Y.H.; Jeong, J.C.; Kim, C.Y.; Lee, H.S.; Bang, J.W.; Kwak, S.S. Sweetpotato late embryogenesis abundant 14 (IbLEA14) gene influences lignification and increases osmotic- and salt stress-tolerance of transgenic calli. *Planta* **2011**, *233*, 621–634. [CrossRef] [PubMed]
12. Bang, S.W.; Lee, D.K.; Jung, H.; Chung, P.J.; Kim, Y.S.; Choi, Y.D.; Suh, J.W.; Kim, J.K. Overexpression of OsTF1L, a rice HD-Zip transcription factor, promotes lignin biosynthesis and stomatal closure that improves drought tolerance. *Plant Biotechnol. J.* **2019**, *17*, 118–131. [CrossRef]
13. Li, L.; Yang, K.; Wang, S.; Lou, Y.; Zhu, C.; Gao, Z. Genome-wide analysis of laccase genes in moso bamboo highlights PeLAC10 involved in lignin biosynthesis and in response to abiotic stresses. *Plant Cell Rep.* **2020**, *39*, 751–763. [CrossRef] [PubMed]
14. Hu, Y.L.W.; Xu, Y.Q.; Li, G.J.; Liao, Y.; Fu, F.L. Differential expression of candidate genes for lignin biosynthesis under drought stress in maize leaves. *J. Appl. Genet.* **2009**, *50*, 213–223. [CrossRef] [PubMed]
15. Lipsick, J.S. One billion years of Myb. *Oncogene* **1996**, *13*, 223–235. [PubMed]
16. Dubos, C.; Stracke, R.; Grotewold, E.; Weisshaar, B.; Martin, C.; Lepiniec, L. MYB transcription factors in Arabidopsis. *Trends Plant Sci.* **2010**, *15*, 573–581. [CrossRef]

17. Ogata, K.; Hojo, H.; Aimoto, S.; Nakai, T.; Nakamura, H.; Sarai, A. Solution structure of a DNA-binding unit of Myb: A helix-turn-helix-related motif with conserved tryptophans forming a hydrophobic core. *Proc. Natl. Acad. Sci. USA* **1992**, *89*, 6428–6432. [CrossRef]
18. Stracke, R.; Werber, M.; Weisshaar, B. The R2R3-MYB gene family in Arabidopsis thaliana. *Curr. Opin. Plant Biol.* **2001**, *4*, 447–456. [CrossRef]
19. Jin, H.; Martin, C. Multifunctionality and diversity within the plant MYB-gene family. *Plant Mol. Biol.* **1999**, *41*, 577–585. [CrossRef]
20. Chen-Kun, J. Insights into the Diversification and Evolution of R2R3-MYB Transcription Factors in Plants. *Plant Physiol.* **2020**, *183*, 637–655.
21. Chen, D.; Chen, H.; Dai, G.; Zhang, H.; Liu, Y.; Shen, W.; Zhu, B.; Cui, C.; Tan, C. Genome-wide identification of R2R3-MYB gene family and association with anthocyanin biosynthesis in Brassica species. *BMC Genom.* **2022**, *23*, 441. [CrossRef] [PubMed]
22. Wang, Z.; Tang, J.; Hu, R.; Wu, P.; Hou, X.L.; Song, X.M.; Xiong, A.S. Genome-wide analysis of the R2R3-MYB transcription factor genes in Chinese cabbage (*Brassica rapa* ssp. pekinensis) reveals their stress and hormone responsive patterns. *BMC Genom.* **2015**, *16*, 17. [CrossRef] [PubMed]
23. Hajiebrahimi, A.; Owji, H.; Hemmati, S. Genome-wide identification, functional prediction, and evolutionary analysis of the R2R3-MYB superfamily in Brassica napus. *Genome Biol.* **2017**, *60*, 797–814. [CrossRef] [PubMed]
24. Katiyar, A.; Smita, S.; Lenka, S.K.; Rajwanshi, R.; Chinnusamy, V.; Bansal, K.C. Genome-wide classification and expression analysis of MYB transcription factor families in rice and Arabidopsis. *BMC Genom.* **2012**, *13*, 544. [CrossRef]
25. Du, H.; Feng, B.R.; Yang, S.S.; Huang, Y.B.; Tang, Y.X. The R2R3-MYB transcription factor gene family in maize. *PLoS ONE* **2012**, *7*, e37463. [CrossRef]
26. Baldoni, E.; Genga, A.; Cominelli, E. Plant MYB Transcription Factors: Their Role in Drought Response Mechanisms. *Int. J. Mol. Sci.* **2015**, *16*, 15811–15851. [CrossRef]
27. Xue, C.; Yao, J.L.; Xue, Y.S.; Su, G.Q.; Wang, L.; Lin, L.K.; Allan, A.C.; Zhang, S.L.; Wu, J. PbrMYB169 positively regulates lignification in fruit stone cells of pear (*Pyrus bretschneideri*). *J. Exp. Bot.* **2019**, *70*, 1801–1814. [CrossRef]
28. Li, C.; Wang, X.; Ran, L.; Tian, Q.; Fan, D.; Luo, K. PtoMYB92 is a transcriptional activator of the lignin biosynthetic pathway during secondary cell wall formation in Populus tomentosa. *Plant Cell Physiol.* **2015**, *56*, 2436–2446. [CrossRef]
29. McCarthy, R.L.; Zhong, R.; Fowler, S.; Lyskowski, D.; Piyasena, H.; Carleton, K.; Spicer, C.; Ye, Z.H. The Poplar MYB Transcription Factors, PtrMYB3 and PtrMYB20, are Involved in the Regulation of Secondary Wall Biosynthesis. *Plant Cell Physiol.* **2010**, *63*, 1757–1760. [CrossRef]
30. Ma, Q.-H.; Wang, C.; Zhu, H.-H. TaMYB4 cloned from wheat regulates lignin biosynthesis through negatively controlling the transcripts of both cinnamyl alcohol dehydrogenase and cinnamoyl-CoA reductase genes. *Biochimie* **2011**, *93*, 1179–1186. [CrossRef]
31. Fornalé, S.; Sonbol, F.M.; Maes, T.; Capellades, M.; Puigdomènech, P.; Rigau, J.; Caparros-Ruiz, D. Down-regulation of the maize and Arabidopsis thaliana caffeic acid O-methyl-transferase genes by two new maize R2R3-MYB transcription factors. *Plant Mol. Biol.* **2006**, *62*, 809–823. [CrossRef] [PubMed]
32. Tak, H.; Negi, S.; Ganapathi, T.R. Overexpression of MusaMYB31, a R2R3 type MYB transcription factor gene indicate its role as a negative regulator of lignin biosynthesis in banana. *PLoS ONE* **2017**, *12*, e0172695. [CrossRef] [PubMed]
33. Yang, C.; Xu, Z.; Song, J.; Conner, K.; Vizcay Barrena, G.; Wilson, Z.A. Arabidopsis MYB26/MALE STERILE35 regulates secondary thickening in the endothecium and is essential for anther dehiscence. *Plant Cell* **2007**, *19*, 534–548. [CrossRef] [PubMed]
34. Zhong, R.; Richardson, E.A.; Ye, Z.H. The MYB46 transcription factor is a direct target of SND1 and regulates secondary wall biosynthesis in Arabidopsis. *Plant Cell* **2007**, *19*, 2776–2792. [CrossRef]
35. Song, J.M.; Guan, Z.; Hu, J.; Guo, C.; Yang, Z.; Wang, S.; Liu, D.; Wang, B.; Lu, S.; Zhou, R.; et al. Eight high-quality genomes reveal pan-genome architecture and ecotype differentiation of Brassica napus. *Nat. Plants* **2020**, *6*, 34–45. [CrossRef]
36. Oh, J.E.; Kwon, Y.; Kim, J.H.; Noh, H.; Hong, S.W.; Lee, H. A dual role for MYB60 in stomatal regulation and root growth of Arabidopsis thaliana under drought stress. *Plant Mol. Biol.* **2011**, *77*, 91–103. [CrossRef]
37. Mondal, S.K.; Roy, S. Genome-wide sequential, evolutionary, organizational and expression analyses of phenylpropanoid biosynthesis associated MYB domain transcription factors in Arabidopsis. *J. Biomol. Struct. Dyn.* **2018**, *36*, 1577–1601. [CrossRef]
38. Wong, D.C.J.; Schlechter, R.; Vannozzi, A.; Holl, J.; Hmmam, I.; Bogs, J.; Tornielli, G.B.; Castellarin, S.D.; Matus, J.T. A systems-oriented analysis of the grapevine R2R3-MYB transcription factor family uncovers new insights into the regulation of stilbene accumulation. *DNA Res.* **2016**, *23*, 451–466. [CrossRef]
39. Wilkins, O.; Nahal, H.; Foong, J.; Provart, N.J.; Campbell, M.M. Expansion and diversification of the Populus R2R3-MYB family of transcription factors. *Plant Physiol.* **2009**, *149*, 981–993. [CrossRef]
40. He, Q.; Jones, D.C.; Li, W.; Xie, F.; Ma, J.; Sun, R.; Wang, Q.; Zhu, S.; Zhang, B. Genome-Wide Identification of R2R3-MYB Genes and Expression Analyses During Abiotic Stress in Gossypium raimondii. *Sci. Rep.* **2016**, *6*, 22980. [CrossRef]
41. Sémon, M.; Wolfe, K.H. Consequences of genome duplication. *Curr. Opin. Genet. Dev.* **2007**, *17*, 505–512. [CrossRef]
42. Fang, L.; Cheng, F.; Wu, J.; Wang, X. The Impact of Genome Triplication on Tandem Gene Evolution in Brassica rapa. *Front. Plant Sci.* **2012**, *3*, 261. [CrossRef] [PubMed]
43. Xu, G.; Guo, C.; Shan, H.; Kong, H. Divergence of duplicate genes in exon-intron structure. *Proc. Natl. Acad. Sci. USA* **2012**, *109*, 1187–1192. [CrossRef] [PubMed]

44. Raza, A.; Su, W.; Gao, A.; Mehmood, S.S.; Hussain, M.A.; Nie, W.; Lv, Y.; Zou, X.; Zhang, X. Catalase (CAT) Gene Family in Rapeseed (*Brassica napus* L.): Genome-Wide Analysis, Identification, and Expression Pattern in Response to Multiple Hormones and Abiotic Stress Conditions. *Int. J. Mol. Sci.* **2021**, *22*, 4281. [CrossRef]
45. Su, W.; Raza, A.; Gao, A.; Jia, Z.; Zhang, Y.; Hussain, M.A.; Mehmood, S.S.; Cheng, Y.; Lv, Y.; Zou, X. Genome-Wide Analysis and Expression Profile of Superoxide Dismutase (SOD) Gene Family in Rapeseed (*Brassica napus* L.) under Different Hormones and Abiotic Stress Conditions. *Antioxidants* **2021**, *10*, 1182. [CrossRef] [PubMed]
46. Lee, T.G.; Jang, C.S.; Kim, J.Y.; Kim, D.S.; Park, J.H.; Kim, D.Y.; Seo, Y.W. A Myb transcription factor (TaMyb1) from wheat roots is expressed during hypoxia: Roles in response to the oxygen concentration in root environment and abiotic stresses. *Physiol. Plant.* **2006**, *129*, 375–385. [CrossRef]
47. Rahaie, M.; Xue, G.P.; Naghavi, M.R.; Alizadeh, H.; Schenk, P.M. A MYB gene from wheat (*Triticum aestivum* L.) is up-regulated during salt and drought stresses and differentially regulated between salt-tolerant and sensitive genotypes. *Plant Cell Rep.* **2010**, *29*, 835–844. [CrossRef] [PubMed]
48. Liu, H.; Zhou, X.; Dong, N.; Liu, X.; Zhang, H.; Zhang, Z. Expression of a wheat MYB gene in transgenic tobacco enhances resistance to Ralstonia solanacearum, and to drought and salt stresses. *Funct. Integr. Genom.* **2011**, *11*, 431–443. [CrossRef] [PubMed]
49. Mao, X.; Jia, D.; Li, A.; Zhang, H.; Tian, S.; Zhang, X.; Jia, J.; Jing, R. Transgenic expression of TaMYB2A confers enhanced tolerance to multiple abiotic stresses in Arabidopsis. *Funct. Integr. Genom.* **2011**, *11*, 445–465. [CrossRef]
50. Prabu, G.; Kawar, P.G.; Pagariya, M.C.; Prasad, D.T. Identification of Water Deficit Stress Upregulated Genes in Sugarcane. *Plant Mol. Biol. Report.* **2010**, *29*, 291–304. [CrossRef]
51. Yang, A.; Dai, X.; Zhang, W.H. A R2R3-type MYB gene, OsMYB2, is involved in salt, cold, and dehydration tolerance in rice. *J. Exp. Bot* **2012**, *63*, 2541–2556. [CrossRef] [PubMed]
52. Butt, H.I.; Yang, Z.; Gong, Q.; Chen, E.; Wang, X.; Zhao, G.; Ge, X.; Zhang, X.; Li, F. GaMYB85, an R2R3 MYB gene, in transgenic Arabidopsis plays an important role in drought tolerance. *BMC Plant Biol.* **2017**, *17*, 142. [CrossRef] [PubMed]
53. Yao, L.; Jiang, Y.; Lu, X.; Wang, B.; Zhou, P.; Wu, T. A R2R3-MYB transcription factor from Lablab purpureus induced by drought increases tolerance to abiotic stress in Arabidopsis. *Mol. Biol. Rep.* **2016**, *43*, 1089–1100. [CrossRef] [PubMed]
54. Qin, Y.; Wang, M.; Tian, Y.; He, W.; Han, L.; Xia, G. Over-expression of TaMYB33 encoding a novel wheat MYB transcription factor increases salt and drought tolerance in Arabidopsis. *Mol. Biol. Rep.* **2012**, *39*, 7183–7192. [CrossRef]
55. Zhong, R.; Ye, Z.H. MYB46 and MYB83 bind to the SMRE sites and directly activate a suite of transcription factors and secondary wall biosynthetic genes. *Plant Cell Physiol.* **2012**, *53*, 368–380. [CrossRef]
56. Zhang, Z.; Liu, X.; Wang, X.; Zhou, M.; Zhou, X.; Ye, X.; Wei, X. An R2R3 MYB transcription factor in wheat, TaPIMP1, mediates host resistance to Bipolaris sorokiniana and drought stresses through regulation of defense- and stress-related genes. *New Phytol.* **2012**, *196*, 1155–1170. [CrossRef]
57. Zhang, L.; Liu, G.; Zhao, G.; Xia, C.; Jia, J.; Liu, X.; Kong, X. Characterization of a wheat R2R3-MYB transcription factor gene, TaMYB19, involved in enhanced abiotic stresses in Arabidopsis. *Plant Cell Physiol.* **2014**, *55*, 1802–1812. [CrossRef]
58. Lopez-Molina, L.; Mongrand, S.; Chua, N.-H. A postgermination developmental arrest checkpoint is mediated by abscisic acid and requires the ABI5 transcription factor in Arabidopsis. *Proc. Natl. Acad. Sci. USA* **2001**, *98*, 4782–4787. [CrossRef]
59. Mittal, A.; Gampala, S.S.; Ritchie, G.L.; Payton, P.; Burke, J.J.; Rock, C.D. Related to ABA-Insensitive3(ABI3)/Viviparous1 and AtABI5 transcription factor coexpression in cotton enhances drought stress adaptation. *Plant Biotechnol. J.* **2014**, *12*, 578–589. [CrossRef]
60. Ko, J.H.; Jeon, H.W.; Kim, W.C.; Kim, J.Y.; Han, K.H. The MYB46/MYB83-mediated transcriptional regulatory programme is a gatekeeper of secondary wall biosynthesis. *Ann. Bot.* **2014**, *114*, 1099–1107. [CrossRef]
61. Zhong, R.; Lee, C.; Zhou, J.; McCarthy, R.L.; Ye, Z.H. A battery of transcription factors involved in the regulation of secondary cell wall biosynthesis in Arabidopsis. *Plant Cell* **2008**, *20*, 2763–2782. [CrossRef] [PubMed]
62. Li, L.; Stoeckert, C.J., Jr.; Roos, D.S. OrthoMCL: Identification of ortholog groups for eukaryotic genomes. *Genome Res.* **2003**, *13*, 2178–2189. [CrossRef] [PubMed]
63. Mailund, T.; Brodal, G.S.; Fagerberg, R.; Pedersen, C.N.; Phillips, D. Recrafting the neighbor-joining method. *BMC Bioinform.* **2006**, *7*, 29. [CrossRef] [PubMed]
64. Kumar, S.; Stecher, G.; Tamura, K. MEGA7: Molecular Evolutionary Genetics Analysis Version 7.0 for Bigger Datasets. *Mol. Biol. Evol.* **2016**, *33*, 1870–1874. [CrossRef]
65. Hu, B.; Jin, J.; Guo, A.Y.; Zhang, H.; Luo, J.; Gao, G. GSDS 2.0: An upgraded gene feature visualization server. *Bioinformatics* **2015**, *31*, 1296–1297. [CrossRef]
66. Liu, D.; Yu, L.; Wei, L.; Yu, P.; Wang, J.; Zhao, H.; Zhang, Y.; Zhang, S.; Yang, Z.; Chen, G.; et al. BnTIR: An online transcriptome platform for exploring RNA-seq libraries for oil crop Brassica napus. *Plant Biotechnol. J.* **2021**, *19*, 1895–1897. [CrossRef]
67. Phung, T.H.; Jung, H.I.; Park, J.H.; Kim, J.G.; Back, K.; Jung, S. Porphyrin biosynthesis control under water stress: Sustained porphyrin status correlates with drought tolerance in transgenic rice. *Plant Physiol.* **2011**, *157*, 1746–1764. [CrossRef]
68. Morabito, C.; Secchi, F.; Schubert, A. Grapevine TPS (trehalose-6-phosphate synthase) family genes are differentially regulated during development, upon sugar treatment and drought stress. *Plant Physiol. Biochem.* **2021**, *164*, 54–62. [CrossRef]

69. Wang, Z.; Chen, Y.; Fang, H.; Shi, H.; Chen, K.; Zhang, Z.; Tan, X. Selection of reference genes for quantitative reverse-transcription polymerase chain reaction normalization in Brassica napus under various stress conditions. *Mol. Genet. Genom.* **2014**, *289*, 1023–1035. [CrossRef]
70. Nei, M.; Gojobori, T. Simple methods for estimating the numbers of synonymous and nonsynonymous nucleotide substitutions. *Mol. Biol. Evol.* **1986**, 3, 418–426. [CrossRef]

plants MDPI

Article

Genome-Wide Analysis of MYB Transcription Factor Gene Superfamily Reveals BjPHL2a Involved in Modulating the Expression of *BjCHI1* in *Brassica juncea*

Chang Gen Xie [1,†], Ping Jin [1,†], Jiamin Xu [1], Shangze Li [1], Tiantian Shi [2], Rui Wang [2], Shuangwei Jia [2,‡], Zixuan Zhang [1], Weike Guo [1], Wenfang Hao [1], Xiaona Zhou [1], Jun Liu [2,*] and Ying Gao [2,*]

1 State Key Laboratory of Crop Stress Biology for Arid Areas, College of Life Sciences, Northwest A&F University, Xianyang 712100, China
2 National Key Facility for Crop Gene Resources and Genetic Improvement (NFCRI), Institute of Crop Sciences, Chinese Academy of Agricultural Sciences (CAAS), Beijing 100081, China
* Correspondence: gaoying@caas.cn (Y.G.); liujun@caas.cn (J.L.); Tel.: +86-010-82105869 (Y.G.)
† These authors contributed equally to this work.
‡ Current address: Beijing Bohui Innovation Biotechnology Co., Ltd., No.9, Shengmingyuan, 15 Road, Changping District, Bejing 102206, China.

Abstract: *Brassica juncea* is an economically important vegetable and oilseed crop. The MYB transcription factor superfamily is one of the largest transcription factor families in plants, and plays crucial roles in regulating the expression of key genes involved in a variety of physiological processes. However, a systematic analysis of the MYB transcription factor genes in *Brassica juncea (BjMYB)* has not been performed. In this study, a total of 502 BjMYB superfamily transcription factor genes were identified, including 23 1R-MYBs, 388 R2R3-MYBs, 16 3R-MYBs, 4 4R-MYBs, 7 atypical MYBs, and 64 MYB-CCs, which is approximately 2.4-fold larger than that of AtMYBs. Phylogenetic relationship analysis revealed that the MYB-CC subfamily consists of 64 *BjMYB-CC* genes. The expression pattern of members of PHL2 subclade homologous genes in *Brassica juncea (BjPHL2)* after *Botrytis cinerea* infection were determined, and *BjPHL2a* was isolated from a yeast one-hybrid screen with the promoter of *BjCHI1* as bait. *BjPHL2a* was found to localize mainly in the nucleus of plant cells. An EMSA assay confirmed that *BjPHL2a* binds to the Wbl-4 element of *BjCHI1*. Transiently expressed *BjPHL2a* activates expression of the GUS reporter system driven by a *BjCHI1* mini-promoter in tobacco (*Nicotiana benthamiana*) leaves. Taken together, our data provide a comprehensive evaluation of *BjMYBs* and show that *BjPHL2a*, one of the members of BjMYB-CCs, functions as a transcription activator by interacting with the Wbl-4 element in the promoter of *BjCHI1* for targeted gene-inducible expression.

Keywords: MYB transcription factors; *BjMYB*; *BjPHL2a*; *BjCHI1*; *Brassica juncea*; *Botrytis cinerea*

Citation: Xie, C.G.; Jin, P.; Xu, J.; Li, S.; Shi, T.; Wang, R.; Jia, S.; Zhang, Z.; Guo, W.; Hao, W.; et al. Genome-Wide Analysis of MYB Transcription Factor Gene Superfamily Reveals BjPHL2a Involved in Modulating the Expression of *BjCHI1* in *Brassica juncea*. *Plants* **2023**, *12*, 1011. https://doi.org/10.3390/plants12051011

Academic Editors: Zanmin Hu, Han Xiao, Yi Ren and Chengming Fan

Received: 15 January 2023
Revised: 10 February 2023
Accepted: 17 February 2023
Published: 23 February 2023

1. Introduction

Brassica juncea is an economically important vegetable and oilseed crop widely cultivated in Asia and Europe [1]. In terms of the well-known "triangle of U" model, *Brassica juncea* is an allotetraploid (AABB, 2n = 4x = 36) improved through interspecific hybridization between the diploid progenitors *Brassica rapa* (AA, 2n = 20) and *Brassica nigra* (BB, 2n = 16), followed by spontaneous chromosome doubling [1,2]. The fungal pathogen *Botrytis cinerea* is a persistent threat to plants throughout the world, including *Brassica juncea*, causing massive yield loss in agriculture [3]. Chitin has been recognized as a general elicitor of plant immune responses, and is also a major structural component of the carbohydrate skeleton of the cell wall of *Botrytis cinerea* [4,5]. Plants have developed a sophisticated system to fight against *Botrytis cinerea* via catalyzing chitin cleavage in the cell wall [6]. Plant chitinases, a class of pathogenesis-related enzymes (PRs), play an essential role in this process via randomly cleaving internal β-1, 4 glycosidic linkages in chitin [6,7].

BjCHI1 is the first isolated chitinase gene whose expression is associated with defense against *Botrytis cinerea* attack in *Brassica juncea* [8–10]. BjCHI1 contains two chitin binding domains (CBDs) [11,12], and extensive investigation has shown that inducible expression of *BjCHI1* is required for defense against *Botrytis cinerea* [11,13]. Several *cis*-acting elements have been predicted in the promoter of *BjCHI1* (BiC-P), and a W box-like 4 element (Wbl-4 box) was recently verified as a fungus-responsive *cis*-acting element [14]. A yeast one-hybrid (Y1H) screen with the core fungus-inducible element as bait was used to identify a number of transcription factors from a cDNA library in *Brassica juncea,* which were isolated and verified [11,15]. BjMYB1, an MYB-CC transcription factor, displayed binding activity to this element and activated the expression of *BjCHI1* [15]. Moreover, Bj26, a C_2H_2-type zinc finger protein also displayed binding and activating activity toward the Wbl-4 box [11]. However, the protein factors involved in the hierarchical and dynamic induction of *BjCHI1* remain unclear.

The fungus-responsive Wbl-4 box in the promoter of *BjCHI1* is recognized by BjMYB1, providing novel insight into the transcriptional regulation of *BjCHI1* [11,15]. MYB-like transcription factors are one of the largest transcription factor families in plants, they play important roles in regulating plant growth and protecting plants from various sources of stress, and they are distinguished by the presence of conserved DNA-binding domains named R repeats. MYB transcription factors are thus classified into four subfamilies based on the number of R repeats: R2R3-MYB, 1R-MYB, 3R-MYB, and 4R-MYB [16]. The MYB–CC type transcription factors are a subtype within the MYB transcription factor superfamily that harbor a combination of MYB and coiled-coil (CC) motifs [17]. PHR1 and other MYB-CC genes modulate the phosphate starvation response in plants via binding to the PHR1-binding DNA sequence (P1BS) *cis*-regulatory element within either the 5′ untranslated regions (UTRs) or the promoter regions of phosphate starvation-induced (PSI) genes [17–22]. MYB and MYB-CC transcription factor genes have been well characterized in a number of plant species, such as *Arabidopsis thaliana* [16], *Zea mays* [17,23], rice (*Oryza sativa*) [24], *Brassica rapa* [25–27], *Brassica napus* [26–28], *Hedychium coronarium* [29], *Cymbidium ensifolium* [30], *Apocynum venetum* [31], carrot (*Daucus carota*) [32], mango (*Mangifera indica*) [33], chili pepper [34], and sweet cherry (*Prunus avium* L.) [35]. Nevertheless, relatively few members of the MYB transcription factors in *Brassica juncea* (BjMYBs) have been well explored [15], and there is still no comprehensive study of BjMYBs to date.

Our previous investigation revealed that BjMYB1 plays an important role in mediating the inducible expression of *BjCHI1* by *Botrytis cinerea* and its derived elicitor chitin [15]. In the current study, we conducted a genome-wide exploration of the MYB transcription factor superfamily in *Brassica juncea,* including 23 1R-MYBs, 388 R2R3-MYBs, 16 3R-MYBs, 4 4R-MYBs, 7 atypical MYBs, and 64 MYB-CCs; phylogenetic analysis and the chromosomal location of BjMYBs were determined. In addition, the gene structure and expression profiles of *BjMYB-CCs* were analyzed. Finally, *BjPHL2a* was isolated and verified to be required for inducible expression of *BjCHI1* via interaction with the fungus-responsive Wbl-4 box in the mini-promoter region of *BjCHI1*.

2. Results

2.1. Genome-Wide Identification of BjMYB Genes

A total of 502 putative BjMYBs with conserved Myb-like DNA binding domains or MYB-CC domains were isolated from the genome of *Brassica juncea*. This is approximatly 2.4-fold more than that of AtMYBs. We found 241 and 230 *BjMYB* genes in the A and B subgenome, respectively (Table S1). The BjMYBs were renamed in terms of their sequence similarity to AtMYBs (Table S1). The protein lengths of BjMYBs were between 96 and 1202 amino acids (aa), with predicted molecular weights (Mw) in the range of 11.58 to 133.39 kDa. The theoretical isoelectric points (pI) of BjMYBs ranged from 4.71 to 10.00 (Table S1). The pI of 309 BjMYBs (60.35%) showed values greater than seven. Other characteristics of BjMYBs, such as gene position and accession number, are presented in Table S1.

2.2. Classification and Phylogenetic Analyses of BjMYB Proteins

We divided BjMYB proteins into six subfamilies according to the AtMYB nomenclature (Table S1); these divisions include 23 1R-MYBs, 388 R2R3-MYBs, 16 3R-MYBs, 4 4R-MYBs, 7 atypical MYBs, and 64 MYB-CCs. To explore the evolutionary relationships among BjMYBs, an unrooted NJ tree with 502 BjMYBs was constructed. As shown in Figure 1, BjMYB proteins were then classified into 15 groups. Among the groups, groups C9 (56), C13 (61), C1 (64), C14 (66), and C15 (104) were the top largest groups, which represents around 70 % of BjMYB members. Groups C2 and C5 all contained only four members.

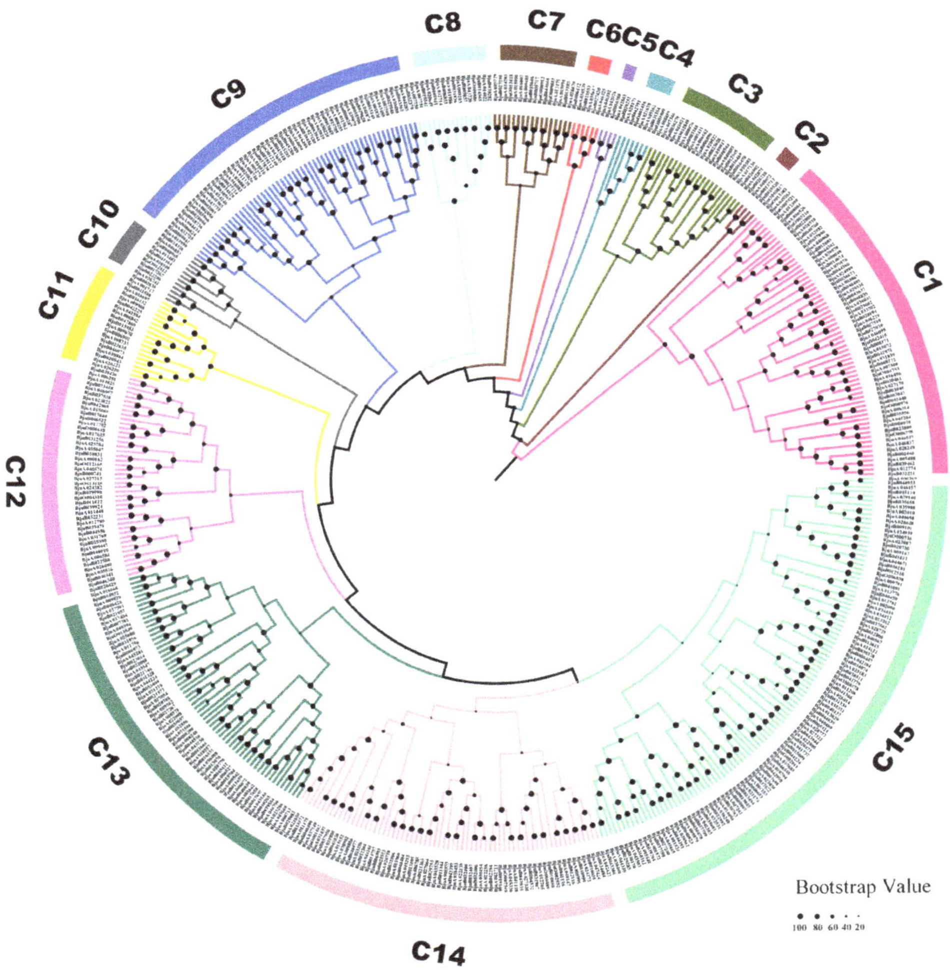

Figure 1. Phylogenetic analysis of MYB transcription factors from *Brassica juncea*. The protein sequences of 502 BjMYB transcription factors were aligned using the MUSCLE tool, and the neighbor-joining (NJ) method (bootstrap replications, n = 1000) was used to generate the phylogenetic tree with MEGA 11. The phylogenetic tree was highlighted with FigTree (version 1.4.4). The proteins are clustered into 15 distinct groups designated by a group number C1~C15 and labeled with different colored branches.

2.3. Chromosomal Location of *BjMYB* Genes in *Brassica juncea*

The chromosomal distribution of *BjMYB* genes was determined with MapChart as previously described [36] and is illustrated in Figure 2. The 471 identified *BjMYB* genes were mapped on 18 chromosomes of *Brassica juncea* (93.82%). Among them, 241 *BjMYB* genes were mapped to the A01 to A10 chromosomes and 230 *BjMYB* genes were mapped to the B01 to B08 chromosomes of *Brassica juncea*. In contrast, there remain 31 *BjMYB* genes on as yet unmapped scaffolds or contigs. *BjMYB* genes were found to be scattered on each chromosome in *Brassica juncea*, but their distribution density differed. Chromosomes A03 (45 members), B02 (38 members), and B03 (38 members) contained a higher density of *BjMYB* genes, while chromosomes A04 (12 members), A05 (16 members), and A10 (13 members) had fewer *BjMYB* genes anchored. We further analyzed the chromosomal distributions of *BjMYB-CC* subfamily genes. As a result, 28 *BjMYB-CC* subfamily genes were found on the A subgenome chromosomes and 30 on the B subgenome chromosomes. Surprisingly, no putative *BjMYB-CC* genes were located on chromosome A04. Moreover, there were six *BjMYB-CC* genes on as yet unmapped scaffolds or contigs.

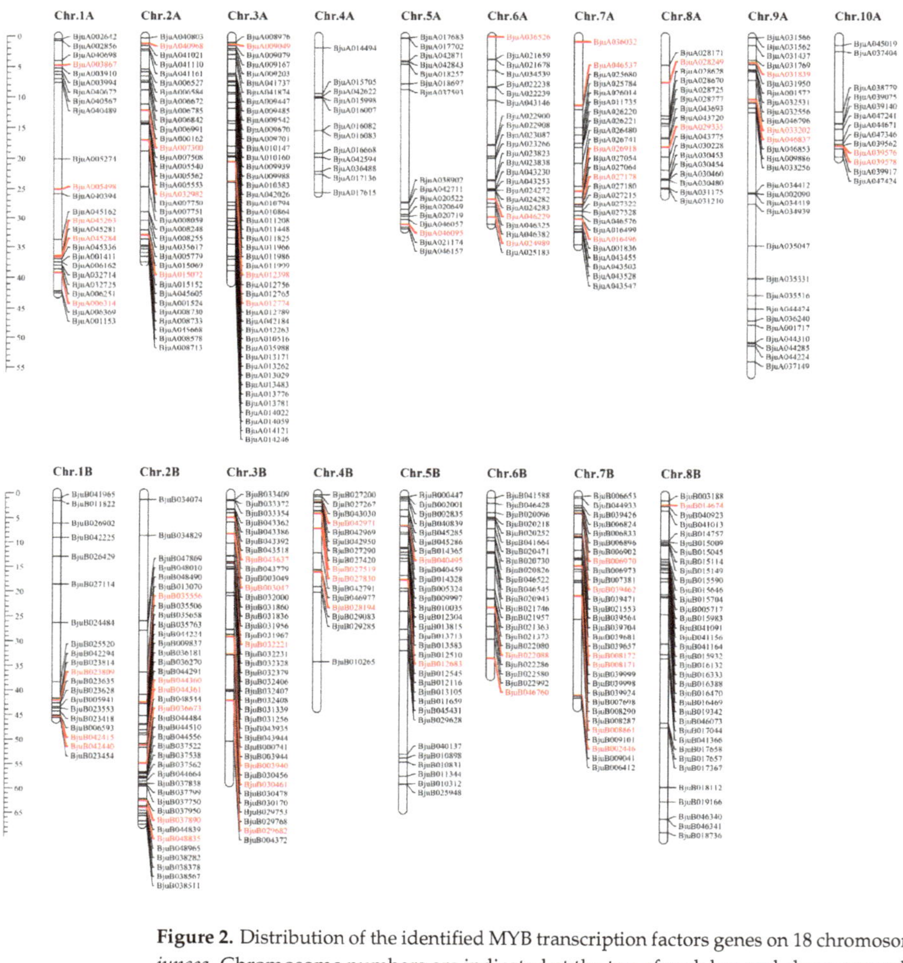

Figure 2. Distribution of the identified MYB transcription factors genes on 18 chromosomes of *Brassica juncea*. Chromosome numbers are indicated at the top of each bar and chromosomal distances are given in Mbp at the left of each bar. The *MYB* genes from the C1 group are labeled in red, while the other members of *MYB* genes are labeled in black.

2.4. Gene and Protein Structure Analyses of the BjMYB-CC Genes

We previously identified BjMYB1 that interact with the fungus-responsive *cis*-acting element Wbl-4 box in the promoter of *BjCHI1*, which is an MYB-CC transcription factor [15]. To further explore the phylogenetic relationships within the MYB-CC subfamily, we determined the overall exon/intron profile of *BjMYC-CC* and *AtMYB-CC* genes. Consistent with *AtMYB-CC* genes, all *BjMYB-CC* genes contained at least one intron. The exon and intron organization of *BjMYB-CC* genes was complicated (Figure 3). BjMYB48 (BjuB014674) contained the most introns (20 introns), while BjMYB23 (BjuA032982) and BjMYB20 (BjuB037890) had the fewest introns (one intron each). All of the four members of *BjPHL2* clade (*BjPHL2a*, *BjPHL2a1*, *BjPHL2b*, and *BjPHL2b1*) showed similar exon/intron arrangements to AtPHL2 (Figure 3), suggesting they may have resulted from gene duplication.

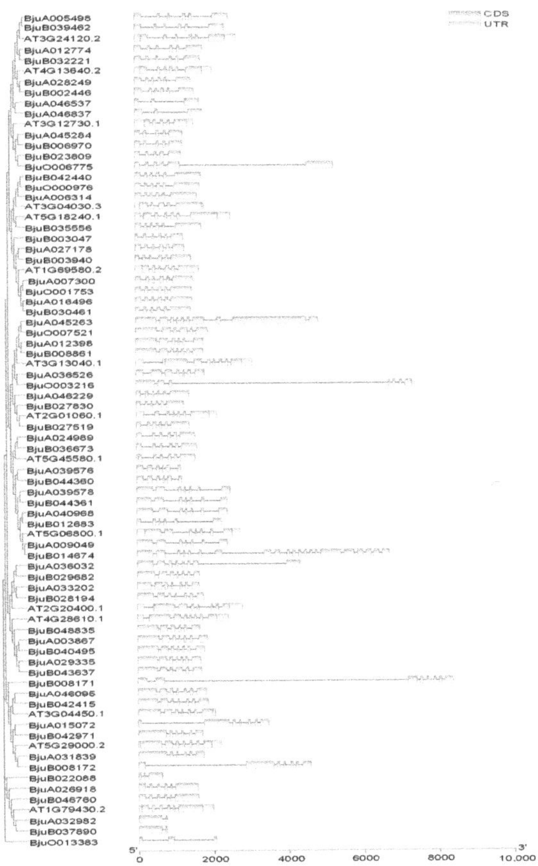

Figure 3. Schematic exon/intron structures of the *MYB-CC* genes from *Brassica juncea* and *Arabidopsis thalina*. Exon/intron organization of MYB-CC genes was visualized using the Gene Structure View (Advanced) from TB tools. The exons and introns are represented by green boxes and black lines, respectively. The UTR regions are indicated with yellow boxes. The length of CDS can be estimated by the scale at the bottom.

We next constructed phylogenetic trees of AtMYB-CCs and BjMYB-CCs (Figure 4), and found that all 64 putative BjMYB-CC transcription factors were named according to their sequence similarity to the corresponding AtMYB-CC orthologues (Figure 1 and

Table S1). There were four proteins (BjuB039462, BjuA005498, BjuA012774/BjPHL2a, and BjuB032221/BjuPHL2b) categorized into the same branch as AtPHL2 in our phylogenetic tree; BjuB039462 and BjuA005498 were designated as BjPHL2b1 and BjPHL2a1, respectively. We also observed that BjMYB1 was categorized into the same clade as BjPHL2b1 (Figure 4 and Figure S1). Moreover, the coding sequence (CDS) of BjMYB1 displayed high sequence identity (~98%) to BjPHL2b1 (Figure S2). Compared with BjPHL2b1, BjMYB1 was found to be missing 213-nt in the 5′ region, a 9-nt insertion in the central region, and a G to A replacement at the 3′ end, which resulted in a 71-amino acid-long deletion at the N-terminal and a YGQ insertion in the central region (Figures S1 and S3). To further investigate the evolutionary relationships, we found a total of 10 motifs (with E-value cutoff <e-1.0) [36], including the conserved MYB domain (motif seven), in BjMYBs (Figure S4 and Table S2). The distribution of the conserved MYB-CC motif in each member of MYB-CCs (15 AtMYB-CCs and 64 BjMYB-CCs) was also depicted individually (Figure 4 and Table S3). Consistent with the gene architecture, nearly all of the homologous proteins in the phylogenetic tree of the BjMYB-CC proteins displayed similar motif arrangement to their AtMYB-CC counterparts, such as the four members of *BjPHL2* clade (Figure 4 and Figure S4).

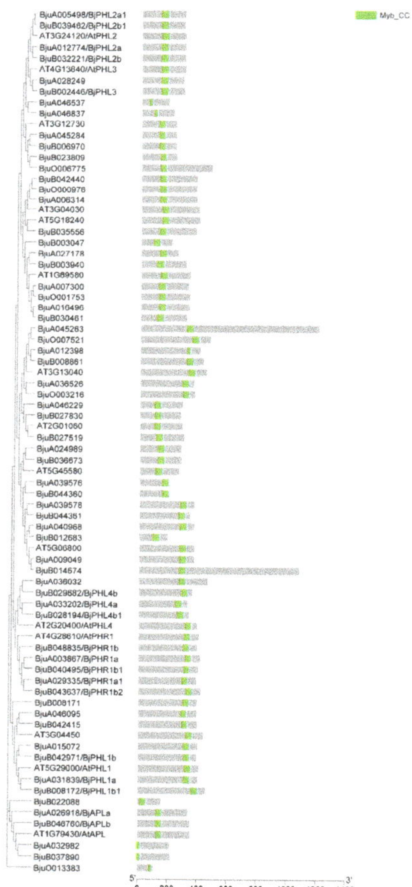

Figure 4. Protein architecture of the MYB-CC proteins in *Brassica juncea* and *Arabidopsis thalina*. The distribution of the conserved MYB-CC motif is represented by green-colored box. See Table S3 for detailed motif information.

2.5. Expression Profiles of BjMYB-CC Genes in Response to Abiotic Stresses

We next investigated the transcript abundance of 64 BjMYB-CC genes in response to abiotic stresses. The expression profiles under drought, high temperature, and salt conditions were determined using publicly available *Brassica juncea* RNA-seq data [37]. Surprisingly, only a few *BjMYB-CC* genes were observed to be induced in response to specific treatment (Figure 5 and Table S4). However, many genes were observed to be repressed in response to drought or high temperature treatment. The expression of *BjMYB-CC* genes under salt treatment displayed results similar to those of the control treatment (Figure 5 and Table S4).

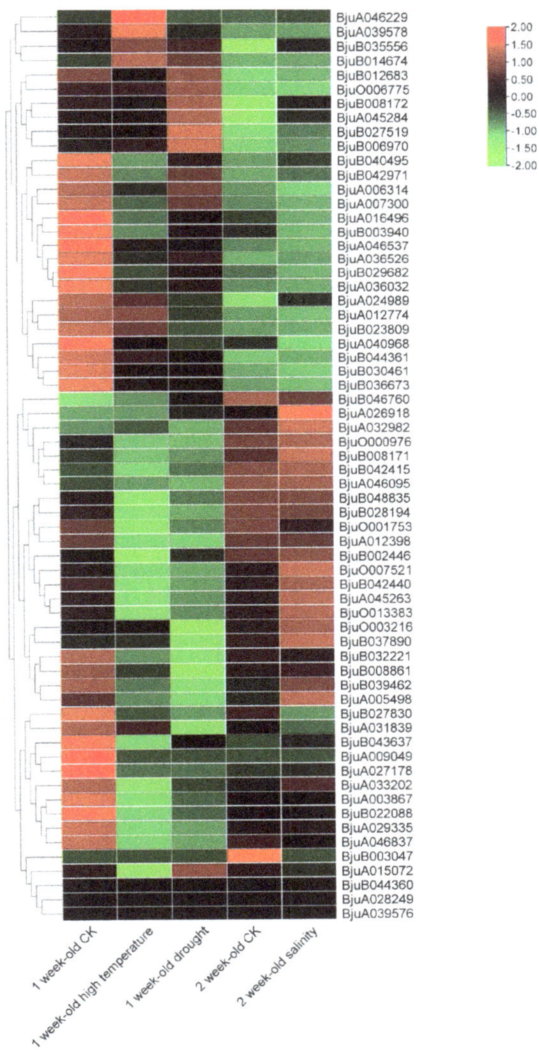

Figure 5. Expression profiles of *BjMYB-CC* genes under different stress conditions. Transcript abundance was normalized according to TPM (transcripts per million mapped reads) values, wherein the TPM values were log2 transformed. The color scale represents relative expression levels from low (green-colored) to high (red-colored).

2.6. Expression of *BjPHL2* Subclade Genes after *Botrytis cinerea* Infection

As mentioned above, there were four BjMYB proteins categorized into the same branch as the previously identified BjMYB1 in our phylogenetic tree (Figure 4). BjMYB1 was reported to be involved in host defense against fungal attack through activating the expression of *BjCHI1* by binding to the Wbl-4 element in the promoter [15]. It is tempting to speculate that gene expression patterns can provide important clues for paralogous genes function [33,36]. In order to give insight into the role of *BjPHL2* subclade members in plant resistance against fungal infection, we examined the expression of *BjPHL2a* and its paralogs after infection by *Botrytis cinerea*. Quantitative real-time PCR (qRT-PCR) analysis showed that only the transcription of *BjPHL2b1* was slightly induced by *Botrytis cinerea*, reaching a peak at 24 hours after inoculation (Figure 6). The expression pattern of *BjPHL2a1* and *BjPHL2b* after infection by *Botrytis cinerea* were similar to that of *BjPHL2a* (Figure 6), all of which were not significantly induced by *Botrytis cinerea* (no more than 1.5-fold).

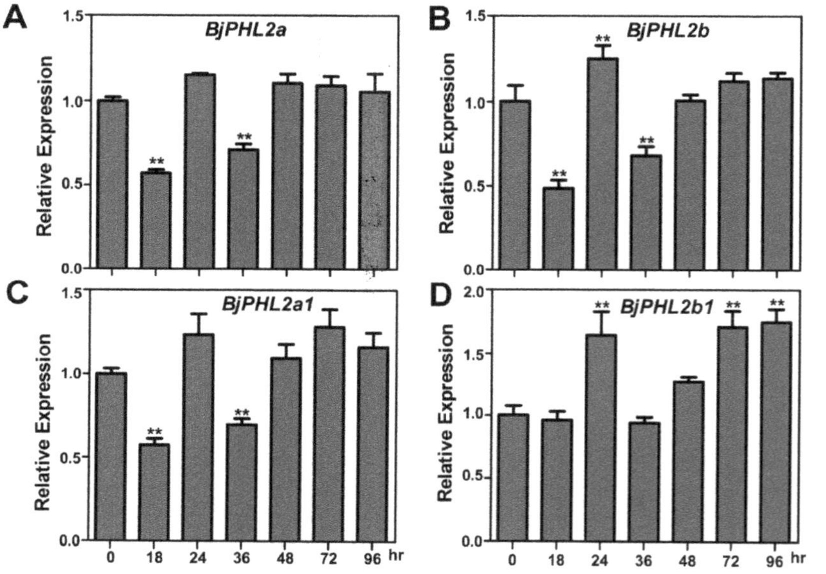

Figure 6. Expression of *BjPHL2* subclade genes in response to *Botrytis cinerea* infection. (**A**–**D**) qRT-PCR analysis of the expression pattern of *BjPHL2a, BjPHL2b, BjPHL2a1*, and *BjPHL2b1* in response to *Botrytis cinerea* infection. Seedlings of *Brassica juncea* were infected with *Botrytis cinerea* followed by sampling at 0, 18, 24, 36, 48, 72, and 96 hours after infection. The relative expression levels of *BjPHL2a, BjPHL2b, BjPHL2a1*, and *BjPHL2b1* were normalized to the expression of BjuB025449 (an AtACT2-like gene in *Brassica juncea*).

2.7. Isolation of *BjPHL2a*

We previously performed systemic Y1H screening of a cDNA library from *Brassica juncea* to identify putative transcriptional factors that interact with the fungus-responsive *cis*-acting element Wbl-4 box in the promoter of *BjCHI1* [11,14,15]. Sequencing analysis revealed that two of the isolated positive clones harbored BjcDNA (BjuA012774), which encodes a predicted protein with 297 amino acids (Figure 7A; 32.5 kDa, pI of 6.55). The CDS of BjuA012774 was used as a query for a local BLAST (basic local alignment search tool) search in *Brassica juncea* CDSs downloaded from the BGD (Brassica Genomics Database) [38]. BjuA012774 displayed high sequence identity (92%) to BjuB032221 [38]. The protein sequences of BjuA012774 and BjuB032221 were used as a query for a BLAST search in TAIR (The Arabidopsis Information Resource). BjuA012774 and BjuB032221 showed high sequence similarity to AtPHL2 (PHR1-LIKE2, At3g24120), which is an MYB-CC type transcription factor involved in the phosphate starvation response [18,39]. The online tool CD-Search was used to identify conserved domains in *BjPHL2a*. As shown in Figure 7A, *BjPHL2a* contains an Myb-CC motif followed by the conserved Myb DNA-binding motif (myb_SHAQKYF). To validate whether *BjPHL2a* actually possesses DNA binding and transactivation activity with the Wbl-4 box mini-promoter, we used a Y1H assay. The full-length coding region of *BjPHL2a* was introduced into the vector pGADT7, which contained a GAL4 activation domain. The resulting pGADT7-BjPHL2a plasmid was then transformed into the bait-reporter yeast strain Y1Hgold [pBait-AbAi] and mutated bait-reporter yeast strain Y1Hgold [pBait-m-AbAi], and growth of the transformants was observed. All of the transformants grew well on SD/-Leu (no AbA) medium (Figure 7B, upper panel). However, only yeast cells containing pGADT7-BjPHL2a grew well on the selective medium—SD/-Leu/+AbA^{550} (SD/-Leu media containing 550 ng/mL AbA) agar plates (Figure 7B, lower panel). Almost no DNA binding and transactivation activity were observed in transformants where the bait-reporter yeast strains [pBait-m-AbAi] were transformed with *BjPHL2a* (Figure 7B, lower panel). Taken together, these results suggest that *BjPHL2a* binds to the Wbl-4 box.

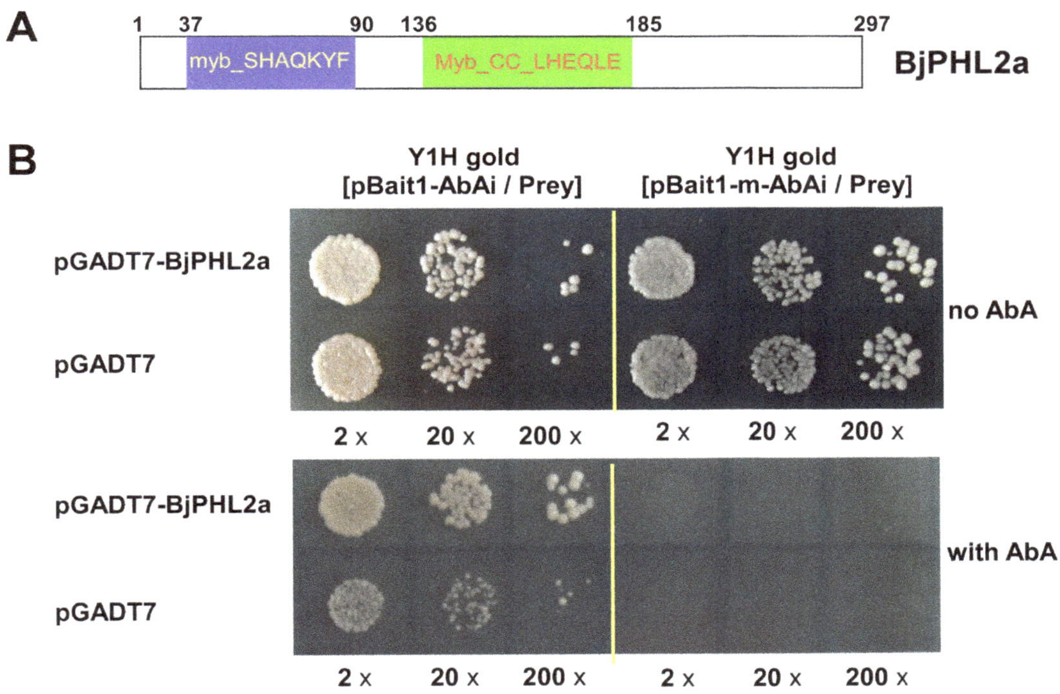

Figure 7. *BjPHL2a* interacts with the Wbl-4 element from *BjCHI1* promoter. (**A**) Schematic diagrams of *BjPHL2a*. The deduced 297 amino acids and the putative two MYB domains (in bold) of *BjPHL2a*. The myb_DNA binding motif (myb_SHAQKYF) is shown in the blue box; the MYB-CC domain (Myb_CC_LHEQLE) is shown in the green box. The numbers above the box indicate the amino acid positions in *BjPHL2a*. (**B**) Yeast one-hybrid analysis of *BjPHL2a*. The bait-reporter yeast strain Y1Hgold [pBait1-AbAi] (Bait) and bait mutated strains Y1Hgold [pBait1-m-AbAi] (Bait-m) expressing the indicated plasmids were grown on SD/-Leu (no AbA) and SD/-Leu/+AbA^{550} (with AbA) agar plates, respectively. Yeast cells were incubated until the optical density at 600 nm reached 0.8 and then were diluted 2-fold (2×), 20-fold (20×), or 200-fold (200×) and used for assays.

2.8. BjPHL2a Encodes an MYB-CC Protein Located in the Nucleus

Sequence analysis revealed that *BjPHL2a* has the same motif arrangement as AtPHL2, including an MYB DNA binding motif at the N-terminus and an MYB-CC motif in the central region (Figure 7A). However, no predicted NLS was observed in *BjPHL2a*, which is a typical characteristic of many transcription factors. To understand whether *BjPHL2a* is a functional transcription factor, we set out to determine the subcellular localization of *BjPHL2a*. The chimeric expression vector pCAMBIA1205-YFP-BjPHL2a and the control vector pCAMBIA1205-YFP were transiently expressed in *Nicotiana benthamiana* leaves via agroinfiltration. We observed fluorescent signals of YFP-BjPHL2a in epidermal cells by confocal fluorescence microscopy. As shown in Figure 8, the YFP-BjPHL2a fusion protein was found to localize predominantly in the nucleus of the cell, while YFP alone was present throughout the cell. These data indicate that *BjPHL2a* is a nuclear-localized protein, consistent with most transcription factors being present in the nucleus.

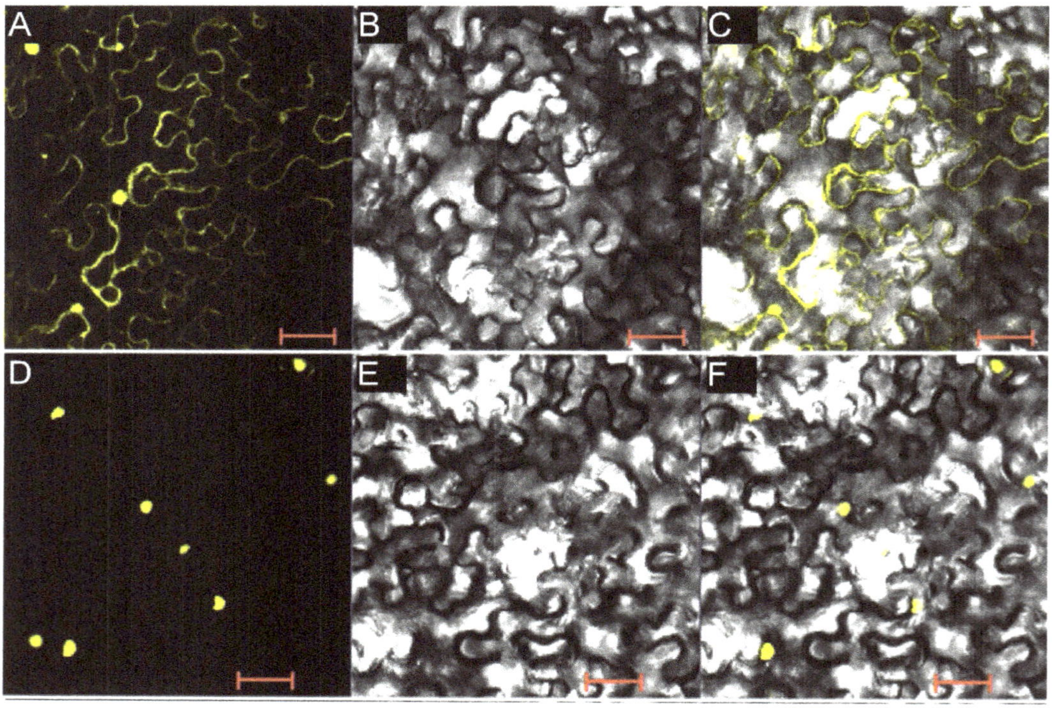

Figure 8. Subcellular localization of *BjPHL2a*. (**A–C**) YFP fluorescence from pCAMBIA1205-YFP empty vector. (**D–F**) YFP fluorescence from pCAMBIA1205-YFP-*BjPHL2a*. (**A**,**D**) YFP fluorescence in a dark field. (**B**,**E**) Cell morphology in a bright field. (**C**,**F**) Overlay of bright-field and YFP signals. Scale bars = 50 μm.

2.9. BjPHL2a Binds to the Wbl-4 Element In Vitro

Previous work has demonstrated that the MYB-core element [C/T]NGTT[G/A], AC-elements (ACC[A/T]A[A/C][T/C] and ACC[A/T][A/C/T][A/C/T]), and *cis*-regulatory elements are recognized by R2R3-type MYB transcription factors [15,16]. The MYB-CC protein BjMYB1 binds to W-box or W-box-like elements, but not to AC elements. To determine whether nuclear-localized *BjPHL2a* actually possesses DNA binding activity, we used an electrophoretic mobility shift assay (EMSA). His-tagged *BjPHL2a* recombinant protein was purified and used for EMSA. A shifted band was observed when His-*BjPHL2a* was incubated with a labeled Wbl-4 box (hot probe, Figure 9A, W4). This shift could be competed out with a huge amount (50× or 200×) of non-labeled Wbl-4 element (cold probe, Figure 9A, W4), indicating that *BjPHL2a* is capable of binding to the Wbl-4 element. Furthermore, when the Wbl-5 box was used, almost no shifted band was observed (Figure 9B, W5). In contrast, when the TGAC core sequence of Wbl-4 element was removed, the shifted band was not observed, indicating that deletion of the core sequence TGAC abolished the binding of *BjPHL2a* to the Wbl-4 element (Figure 9A, W4D). These results indicate that recombinant *BjPHL2a* proteins could interact with the Wbl-4 element.

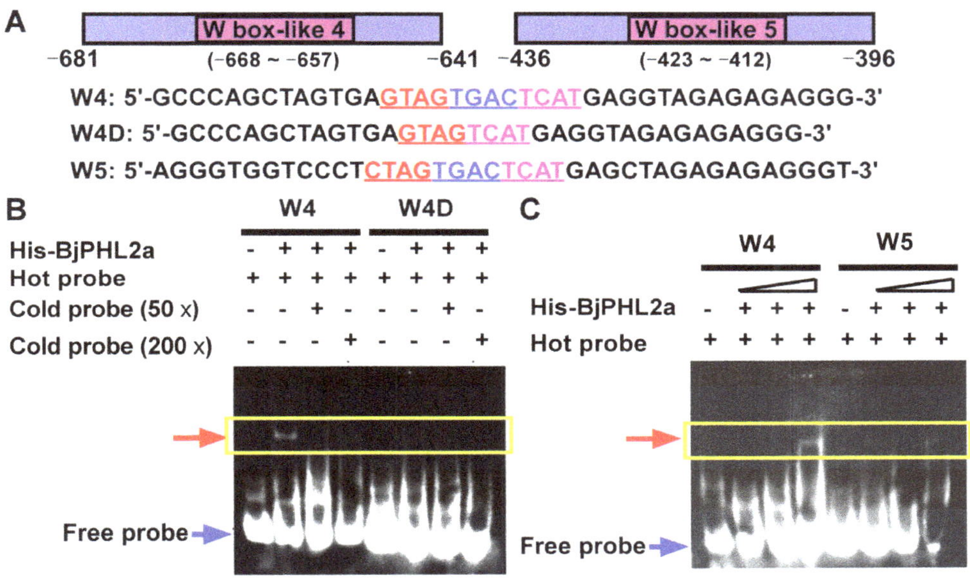

Figure 9. *BjPHL2a* binds to the Wbl-4 element in EMSA. (**A**) Diagrams of probes W4, W4D (W4 mutant), W5, and their nucleotide sequences. W4 contains Wbl-4 element. W4D is a W4 mutant lacking four core base of Wbl-4 element. W5 contains Wbl-5 element. Numbers indicate the nucleotide positions in *BjCHI1* promoter. (**B**,**C**) Electrophoretic mobility shift assays (EMSA) were performed with biotin-labeled (hot probe) or unlabeled probes (cold probe) and recombinant His-BjPHL2a proteins. Specific combinations are shown indicated by a yellow box. An equal amount of *BjPHL2a* protein or hot probe was used in all lanes. Cold probes were added in 50-fold (50×) or 200-fold (200×) excess as binding competitors. The bound and free hot probes are indicated by arrows on the left.

2.10. BjPHL2a Activates a Mini-Promoter Containing a Wbl-4 Element In Vivo

To further analyze whether *BjPHL2a* actually had transactivation activity in vivo, we sought to directly investigate its role in transactivating the Wbl-4 element by transient expression in *Nicotiana benthamiana*. As previously described, we also recruited several key deletion derivatives of the *BjCHI1* promoter, such as P9, P16, P53, P27, and P45 (Figure 9A) as promoters for *GUS* fusion reporter plasmids [14,15]. Consistent with previous reports [14,15], *BjPHL2a* activated the expression of *GUS* driven by P16-, P27-, and P45 promoters containing Wbl-4 box, but not by P53 and P9 (Figure 10B,C). Compared with P16, only the TGAC in the core sequence of the Wbl-4 element was absent in P53 and the Wbl-4 box was removed in P9. Therefore, these observations indicate that *BjPHL2a* can also transactivate the Wbl-4 element in vivo.

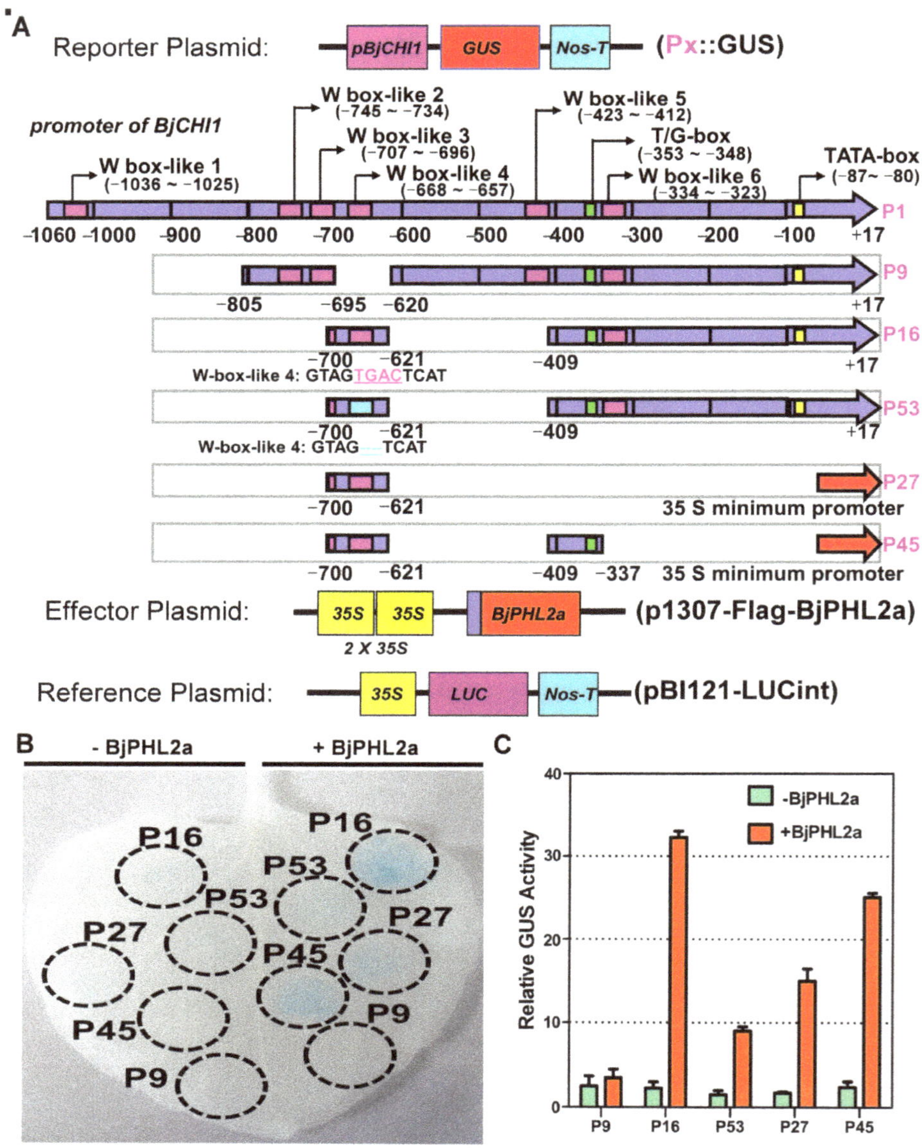

Figure 10. *BjPHL2a* binds to the Wbl-4 element *in vivo*. (**A**) Schematic diagram of the reporter and the effector constructs used in co-transfection experiments. For the reporter construct, the deletion- and mutation-derivatives of the *BjCHI1* promoter (P9, P16, P27, P45, and P53) were fused with GUS. For the effector construct, 2 × 35S is a tandem repeat of the cauliflower mosaic virus (CaMV) 35S promoter that is used to express *BjPHL2a* in tobacco leaves. (**B**) Transactivation activity of the reporter constructs based on *BjPHL2a* expression. Tobacco leaves co-infiltrated with the indicated constructs were stained with X-Gluc for *GUS* expression analysis. The infiltration areas are indicated with dashed cycles. (**C**) Quantization of GUS activity for every construct in transiently transformed tobacco leaves. The GUS activity was normalized to LUC activity. Columns show the ratio of GUS activity induced by *BjPHL2a* to that of no induction. Values represent means ± SD from three replicates.

3. Discussion

The MYB-like superfamily transcription factors are one of the largest transcription factor families in plants and play important roles in a wide range of physiological processes, such as light signaling, the anthocyanin biosynthetic pathway, hypocotyl elongation, phosphate starvation, and resistance to biotic and abiotic stresses [29,34]. Although *MYB* and *MYB-CC* transcription factor genes have been identified in a number of plant species at the genome-wide level [16,17,23–35], there is little information available on this superfamily in the *Brassica juncea* genome. Recently, bioinformatics and publicly released data sets describing the genome and transcriptome of *Brassica juncea* have paved the way for systemically exploring this important crop species [1,37,38]. In the present study, a total of 502 *BjMYB* genes were identified in *Brassica juncea*, which were divided into six subfamilies (1R-, R2R3-, 3R-, 4R-, atypical MYBs, and MYB-CCs) and clustered into 15 groups (Figure 1 and Table S1). The number of *BjMYB* genes was greater than those of *Arabidopsis thaliana* (212) [24,39], *Oryza sativa* (213) [24,40], *Zea mays* (169) [17,23], *Brassica rapa* (354), and *Brassica oleracea* (330), but fewer than that in *Brassica napus* (556) [24,26,27]. BjMYB proteins were not equally distributed into 15 different groups in the phylogenetic tree (Figure 1). Among them, the top largest groups (C1, C9, C13, C14, and C15) represented around 70% of the total BjMYB members (Figure 1 and Table S1), suggesting that these groups may be highly differentiated in the *Brassica juncea* genome. Moreover, the large range of Mw and pI of BjMYBs suggests their functional diversity in distinct biological processes, which may be attributed to their difference in full-length amino acid sequences and rest domains, in addition to similarity in MYB domains.

As displayed in Figure 2, the 471 (93.82%) identified members of *BjMYB* genes were unevenly distributed across all 18 chromosomes at different densities. Among them, 241 *BjMYB* genes were mapped to the A subgenome and 230 *BjMYB* genes to the B subgenome of *Brassica juncea*. The chromosomes A01 to A10 and B01 to B08 represent the two parental diploid ancestors of *Brassica juncea*. The number of *BjMYB* genes (502) was found to be less than the sum of its two parental lines, *Brassica rapa* (301) and *Brassica nigra* (>300) [26,27], indicating loss of *MYB* genes also happened after allopolyplodization. In general, every AtMYB-CC was expected to have four to six homologs in *Brassica juncea*. Consistently, 15 *AtMYB-CCs* had 64 homologous genes in *Brassica juncea* (Figure 4A and Table S1). For example, *AtPHL2* had four homologs (*BjPHL2a*, *BjPHL2b*, *BjPHL2a1*, and *BjPHL2b1*) and *AtPHR1* had five homologs (*BjPHR1a*, *BjPHR1b*, *BjPHR1a1*, *BjPHR1b1*, and *BjPHR1b2*). In terms of the overall gene structure, significant differences were found in the number of introns among *BjMYB-CC* subfamily members (Figure 3). However, the overall intron/exon distribution within members of the same subclade was highly similar, such as PHL2-, PHL3-, and PHR1-subclade (Figure 3), indicating that the biological significance of these homologous genes may be redundantly conserved to a certain extent.

It is tempting to speculate that gene expression patterns can provide important clues for gene function [33,36]. The expression patterns of 64 *BjMYB-CC* genes from three type of abiotic stresses were determined (Figure 5). Only a few BjMYB-CC genes displayed inducible expression by at least one analyzed treatment (Figure 5 and Table S4). Furthermore, qRT-PCR analysis revealed similar expression patterns after *Botrytis cinerea* infection of the three *AtPHL2* homologs in *Brassica juncea* (*BjPHL2a*, *BjPHL2b*, and *BjPHL2a1*, Figure 6).

Previous work has shown that the MYB-CC transcription factor BjMYB1 can bind to W-box or W-box-like elements, but not to AC elements [15]. As mentioned above, MYB–CC type transcription factors are a distinct subfamily of the plant MYB superfamily, which were originally identified as phosphate starvation response (PHR) protein family transcription factors, such as AtPHR1 and OsPHR2 [20]. It is well known that PHR protein family members can transactivate the expression of a broad range of PSI genes by binding to P1BS *cis*-elements in a phosphate (Pi)-dependent manner [17,20]. The P1BS *cis*-regulatory element is characterized by the nucleotide motif GNATATNC or CATATATG [20,41]. We also observed a putative P1BS-like *cis*-element (CATATATCG) near the fungus-responsive Wbl-4 box in the promoter of *BjCHI1*. Besides serving as master transcriptional regulators of the phosphate sensing pathway [22,39,42], PHR1-like MYB-CC proteins (such as AtPHR1 and OsPHR1/2/3) have been recently reported to play central roles in the balance between nutrient stress response and immune regulation or mycorrhizal symbiosis with respect to root microbiota [19,43,44]. One possibility is that the observed binding activity of MYB-CC-like proteins (BjMYB1 and *BjPHL2a*) to the Wbl-4 box containing mini-promoter may also be partially attributed to the putative P1BS-like *cis*-element near the Wbl-4 box. In other words, PHR1-like MYB-CC proteins may also have roles in fighting against *Botrytis cinerea* infection via modulating the expression of *BjCHI1*. We found four AtPHL2 homologs in *Brassica juncea* and similar expression patterns after *Botrytis cinerea* infection was observed among three paralogs (Figure 6). The previously isolated BjMYB1 showed sequence differences compared with other BjPHL2 s (Figures S1–S3). Therefore, it is possible that *BjPHL2a*, together with the closed paralogs, may be redundantly involved in fighting against fungal attack by interacting with the Wbl-4 element in the promoter of *BjCHI1*.

Recently, hierarchical and dynamic regulation of defense-responsive genes by AtMYB51 and AtWRKY33 has been reported [45]. The cognate nucleotide motif of the Wbl-4 box (GTGACT), also shows a minor difference to the core nucleotide sequence of classical W-boxes (C/TTGACC/T) [15,46]. This may be the major reason that we have not isolated any WRKY-like transcription factors to date. However, another possibility is that *BjPHL2a* or BjMYB1 and BjWRKY transcription factors may function coordinately to control the inducible expression of *BjCHI1*.

4. Materials and Methods

4.1. Identification of the MYB Gene Superfamily in Brassica juncea

Protein sequences of *Brassica juncea* var. *tumida* inbred line (T84−66, GenBank accession number: LFQT00000000, BioProject PRJNA285130) were obtained from the BGD (http://brassicadb.bio2db.com/) [38,47]. HUMMER3.1 software (Linux version) was used to isolate *BjMYBs* from *Brassica juncea* var. *tumida* inbred line T84−66 through the HMM profile corresponding to Myb-like DNA binding domain (PF00249, PF08914, PF13921, PF12776, PF13873, PF15963, PF13837) and MYB-CC domain (PF14379), with the threshold set at E-value $< 1 \times 10^{-10}$. A total of 195 AtMYB [24] and 15 MYB-CC protein sequences were downloaded from TAIR (www.arabidopsis.org) and were used as queries to perform a BLASTP search in the local protein database of *Brassica juncea* var. *tumida* inbred line (T84−66). Then, the putative BjMYBs were obtained via taking the intersection of the HUMMER and BLASTP methods. Finally, these proteins were submitted to the NCBI-CDD server (http://www.ncbi.nlm.nih.gov/Structure/cdd/wrpsb.cgi) and the SMART (Simple Modular Architecture Research Tool, http://smart.embl-heidelberg.de/) database to perform the Myb-like DNA binding domain or MYB-CC domain predictions as previously described [48]. The theoretical molecular weight and isoelectric points of BjMYBs were calculated by DNAstar as previously described [49].

4.2. Phylogenetic, Chromosome Distribution, Gene, and Protein Structure Analysis

Protein sequences of the BjMYB members were first aligned with the MUSCLE tool. The phylogenetic tree was then generated using MEGA 11 with the neighbor-joining (NJ) method (bootstrap replications, n = 1000) as previously described [36,49]. The phylogenetic tree was then visualized using Figtree software (version 1.4.4) [36,48]. The physical positional information of each *BjMYB* gene was downloaded from NCBI (accession number PRJNA285130) [38]. Locations of *BjMYB* genes on *Brassica juncea* chromosomes were then deciphered with the MapChart tool as previously described [36]. The overall intron/exon organization of the *MYB-CC* genes was determined based on GFF annotation files with Gene Structure View (Advanced) from TB tools as previously described [48]. Protein sequences of 15 AtMYB-CCs and 64 BjMYB-CCs were aligned using the MUSCLE tool, and the maximum likelihood trees were generated using MEGA 11, as mentioned above. The conserved motifs and protein architecture were predicted by the MEME (Multiple Em for Motif Elicitation) tool (http://meme-suite.org/tools/meme) as previously described [36].

4.3. Expression Profiles of the *BjMYB-CC* Genes

RNA-seq data from *Brassica juncea* under drought, high temperature, and salt stresses were downloaded from GEO (Gene Expression Omnibus, accession numbers GSE66389 and GSE64242) [37]. All RNA-seq data were mapped to the reference genome of *Brassica juncea* with HISAT2 software. Transcript abundance of the *BjMYB-CC* genes were calculated using TPM (transcripts per million) values generated with the FeatureCounts R package and a histogram was generated using TBtools software as previously described [48].

4.4. Plant Materials and Growth Conditions

Plants of *Nicotiana benthamiana* and *Brassica juncea* were grown in a growth chamber at 25 °C (light)/22 °C (dark) under a 16 h light/8 h dark cycle. *Nicotiana benthamiana* was used to carry out transient expression assays and *Brassica juncea* was used for constructing a cDNA library and endogenous gene expression assays.

4.5. Plasmid Construction

To express *BjPHL2a* in yeast, the full-length CDS of *BjPHL2a* was cloned into the pGADT7 vector between the *Bam*HI and *Sal*I sites. The following primer pairs were used for PCR amplification: *BjPHL2a*-2bp*Bam*F: 5′-CGGGATCCGTATGTACTCGGCGATTCGCTC-3′ and *BjPHL2a*-1bp*SalR*: 5′-ACGCGTCGACATCCCATGGTACTACCCGGCACAG-3′ (*Bam*HI and *Sal*I sites are underlined).

To produce FLAG-tagged *BjPHL2a*, the CDS of *BjPHL2a* was amplified with the following primers: *BjPHL2a-Bam*F: 5′-CGGGATCCATGTACTCGGCGATTCGCTC-3′ and *BjPHL2a-SalR*: 5′-ACGCGTCGACTCCCATGGTACTACCCGGCACAG-3′ (*Bam*HI and *Sal*I sites are underlined). The sequence-confirmed full-length CDS of *BjPHL2a* was then cloned into the *Bam*HI and *Sal*I sites of the binary expression vector p1307-3Flag.

To make YFP tagged *BjPHL2a* (YFP-*BjPHL2a*), the CDS was removed from p1307-3Flag-*BjPHL2a* and inserted into the pCM1205-YFP vector between the *Bam*HI and *Sal*I sites, resulting in an N-terminal fusion to YFP.

To obtain His-tagged *BjPHL2a* recombinant protein, the CDS was removed from p1307-Flag-*BjPHL2a* and inserted into pET-28a (Novagen) between the *Bam*HI and *Sal*I sites.

4.6. Yeast One-Hybrid Assay

Yeast one-hybrid assays were performed as previously described [15]. Briefly, pGADT7-*BjPHL2a* plasmids or pGADT7 empty vector were transformed into the bait-reporter yeast strains Y1Hgold [pBait-AbAi] or Y1Hgold [pBait-m-AbAi] by the lithium acetate/single-stranded carrier DNA/polyethylene glycol method [50,51]. Transformed yeast cells were selected on SD/-Leu (no AbA) medium. DNA binding and transactivation were determined by checking the growth of transformed yeast cells with serial dilutions on SD/-Leu/+AbA^{550} (SD/-Leu media containing 550 ng/mL AbA) agar plates for 2~3 days. Growth was recorded by photography.

4.7. Subcellular Localization of BjPHL2a

The YFP-*BjPHL2a* fusion vector (pCM1205-YFP-*BjPHL2a*) was transiently expressed in *Nicotiana benthamiana* leaves by agroinfiltration as previously described [15,50,51]. Briefly, the *Agrobacterium tumefasciens* EHA105 harboring pCM1205-YFP-*BjPHL2a* was grown on agar-LB containing 100 μg/mL chloramphenicol and 30 μg/mL rifampicin, then suspended in MMA buffer (10 mM $MgCl_2$, 10 mM MES pH 5.5, 100 μM acetosyringone) at OD_{600} 0.5 and incubated at 28 °C for 3 h. The fifth and sixth expanded leaves of *Nicotiana benthamiana* were infiltrated at the blade back with a 1 mL needleless syringe. YFP fluorescence of cells from the lower epidermis was observed with a confocal laser scanning microscope 2~3 days after infiltration [15].

4.8. Protein Purification and Electrophoretic Mobility Shift Assay (EMSA)

Recombinant protein purification and an EMSA assay were performed as previously described [15]. Briefly, the expression of recombinant His-tagged *BjPHL2a* fusion protein was induced with 0.6 mM isopropyl-D-thiogalactoside (IPTG) until the OD_{600} reached 0.5. Recombinant His-BjPHL2a protein was purified with Ni^{2+} affinity resin (Ni-NTA, QIAGEN) according to the manufacturer's instructions. Purified His-BjPHL2a was used for the EMSA assay as previously described [15]. Complementary pairs of non-labeled and 3′-biotin-labeled oligonucleotides of W4 (intact Wbl-4 element), W4D (core sequence TGAC in the Wbl-4 element deleted), and W5 (W-box-like 5 elements) were synthesized and used as probes. The following probe sequences were used: W4: 5′-gcccagctagtgagtagtgactcatgaggtagagagaggg-3′, W4D: 5′-gcccagctagtgagtagtcatgaggtagaga gaggg-3′, and W5: 5′-agggtggtccctctagtgactcatgagctagagagagggt-3′ (the core W box-like elements are underlined).

4.9. Histochemical Staining of GUS and GUS Quantitative Assays

GUS staining and GUS quantitative analysis was performed as described in a previous study [14]. Briefly, the agrobacterium strain EHA105 containing the GUS constructor and its derivatives (P16 and P53) were co-infiltrated with *BjPHL2a*, and with pBI121-LUCint in *Nicotiana benthamiana* leaves, and transiently expressed. For histochemical staining of GUS, the infiltrated *Nicotiana benthamiana* leaves were stained and photographed. For the GUS quantitative assay, the infiltrated *Nicotiana benthamiana* leaf discs were cut and used for GUS and LUC assay as described previously [14].

4.10. Determination of the Expression of BjPHL2a and its Paralogs after Botrytis cinerea Infection

Botrytis cinerea was cultured on potato agar media plus 1.5% dextrose (potato 200 g /L, glucose 20 g/L, agar 15 g/L, pH 6.0) at 22 °C for 14 days. The conidia of well-grown *Botrytis cinerea* were suspended in sterile distilled water, filtered with two layers of gauze, and diluted to 5×10^5 cells per milliliter. To determine the expression level of *BjPHL2a* and its paralogs after *Botrytis cinerea* infection, 16-day-old *Brassica juncea* plants were inoculated with the conidial suspension by spraying and sampled at 0, 18, 24, 36, 48, 72, and 96 hours after infection. qRT-PCR was performed using SYBR Premix Ex TaqTM II (TAKARA BIO Inc., Shiga, Japan) on a 7500 Fast Real-Time PCR System (Applied Biosystems, Foster City, CA, USA) with the following conditions: 95 °C for 1 min, followed by 40 cycles of 95 °C for 10 s and 60 °C for 34 s in 20 μL reaction volumes. A melting curve was generated for each reaction to ensure specific amplification. The relative expression was quantified using the comparative $2^{-\Delta\Delta CT}$ method. Primer pairs BjPHL2aF1/ BjPHL2aR1 (BjPHL2aF1: 5′-GGAACAGAACGAGAGTTACC-3′ and BjPHL2aR1: 5′-CTTCCCTTGTGCTTCTATCC-3′), BjPHL2a1F1/BjPHL2a1R1 (BjPHL2a1F1: 5′-CAGAACGAGGGTTACCAAGTC-3′ and BjPHL2a1R1: 5′-TCCTTGTGCCTCTATCCTCAG-3′), BjPHL2bF1/BjPHL2bR1 (BjPHL2bF1: 5′-AGCAACATCACGACCAACTG-3′ and BjPHL2bR1: 5′-TCGCTTCATACCCAACACTG-3′), and BjPHL2b1F1/BjPHL2b1R1 (BjPHL2b1F1: 5′-AGCAGTCCCGTTCTTAGATG-3′ and BjPHL2b1R1: 5′-CCGAACAGTTGGTTGTGATG-3′) were used for qRT-PCR analysis of *BjPHL2a*, *BjPHL2a1*, *BjPHL2b*, and *BjPHL2b1*, respectively. Primer pairs BjActin-F/ BjActin-R of Actin gene in *Brassica juncea* (BjuB025449, an AtACT2-like gene in *Brassica juncea*) were used as internal references (BjActin-F: 5′-CTTCTTACCGAGGCTCCTCT-3′ and BjActin-R: 5′-AAGGATCTTCATGAGGTAATCAGT-3′) as previously described [15].

5. Conclusions

In the present study, the identification, classification, and expression analysis of the *MYB* gene superfamily was comprehensively evaluated at the whole genome level in *Brassica juncea*. Functional and molecular studies showed that *BjPHL2a* functions as a transcription activator by interacting with the Wbl-4 element in the promoter of *BjCHI1* for targeted gene inducible expression after *Botrytis cinerea* infection.

Supplementary Materials: The following are available online at https://www.mdpi.com/article/10.3390/plants12051011/s1. Table S1. Details of the identified *BjMYB* genes in *Brassica juncea;* Table S2. Sequence information of additional conserved motifs identified from BjMYB proteins in *Brassica juncea;* Table S3. Sequence information of the conserved MYB-CC domain from AtMYB-CC and BjMYB-CC proteins; Table S4. Expression of *BjMYB-CCs* under different stresses; Figure S1. Multiple sequence alignment of BjPHL2 subclade members; Figure S2. Sequence alignment of BjMYB1 and BjPHL2b1 at the CDS level; Figure S3. Sequence alignment of BjMYB1 and BjPHL2b1 at the protein level; Figure S4. Protein architecture of BjMYBs.

Author Contributions: Y.G., J.L. and C.G.X. conceived the project and designed the experiment plans; C.G.X., P.J., J.X., S.L., T.S., R.W., S.J., Z.Z. and W.G. conducted the experiments; Y.G. and C.G.X. analyzed the data and wrote the article; W.H., X.Z. and J.L. reviewed and edited the manuscript. All authors have read and agreed to the published version of the manuscript.

Funding: This work was financially supported by grant from the Innovation Project of the Chinese Academy of Agricultural Sciences in China (2060299-2-22), grant from Operating Expenses for Basic Scientific Research of the Institute of Crop Science by the Chinese Academy of Agricultural Sciences in China (S2022XC01), the State Basic Research Project from the Department of Science and Technology of Shaanxi Province (2021JM-086), the State Key Research and Development Project from the Department of Science and Technology of Shaanxi Province (2023-YBNY-023 and 2018NY-097), and grants from the Science and Innovation Project to Undergraduates by the College of Life Sciences, Northwest A&F University.

Data Availability Statement: Not applicable.

Acknowledgments: We are grateful to Wang Qinhu and Pei Guoliang (Northwest A&F University) for help with bioinformatics analyses.

Conflicts of Interest: The authors declare that they have no competing interests.

References

1. Kang, L.; Qian, L.; Zheng, M.; Chen, L.; Chen, H.; Yang, L.; You, L.; Yang, B.; Yan, M.; Gu, Y.; et al. Genomic insights into the origin, domestication and diversification of *Brassica juncea*. *Nat. Genet.* **2021**, *53*, 1392–1402. [CrossRef]
2. Yang, H.; Bayer, P.E.; Tirnaz, S.; Edwards, D.; Batley, J. Genome-wide identification and evolution of receptor-like kinases (RLKs) and receptor like proteins (RLPs) in *Brassica juncea*. *Biology* **2020**, *10*, 17. [CrossRef] [PubMed]
3. Fisher, M.C.; Henk, D.A.; Briggs, C.J.; Brownstein, J.S.; Madoff, L.C.; McCraw, S.L.; Gurr, S.J. Emerging fungal threats to animal, plant and ecosystem health. *Nature* **2012**, *484*, 186–194. [CrossRef] [PubMed]
4. Lenardon, M.D.; Munro, C.A.; Gow, N.A. Chitin synthesis and fungal pathogenesis. *Curr. Opin. Microbiol.* **2010**, *13*, 416–423. [CrossRef]
5. Eckardt, N.A. Chitin signaling in plants: Insights into the perception of fungal pathogens and rhizobacterial symbionts. *Plant Cell* **2008**, *20*, 241–243. [CrossRef] [PubMed]
6. Wan, J.; Zhang, X.C.; Stacey, G. Chitin signaling and plant disease resistance. *Plant Signal. Behav.* **2008**, *3*, 831–833. [CrossRef]
7. Wang, P.; Hsu, C.C.; Du, Y.; Zhu, P.; Zhao, C.; Fu, X.; Zhang, C.; Paez, J.S.; Macho, A.P.; Tao, W.A.; et al. Mapping proteome-wide targets of protein kinases in plant stress responses. *Proc. Natl. Acad. Sci. USA* **2020**, *117*, 3270–3280. [CrossRef] [PubMed]
8. Fung, K.L.; Zhao, K.J.; He, Z.M.; Chye, M.L. Tobacco-expressed *Brassica juncea* chitinase BjCHI1 shows antifungal activity in vitro. *Plant Mol. Biol.* **2002**, *50*, 283–294. [CrossRef]
9. Wu, X.F.; Wang, C.L.; Xie, E.B.; Gao, Y.; Fan, Y.L.; Liu, P.Q.; Zhao, K.J. Molecular cloning and characterization of the promoter for the multiple stress-inducible gene *BjCHI1* from *Brassica juncea*. *Planta* **2009**, *229*, 1231–1242. [CrossRef]
10. Zhao, K.J.; Chye, M.L. Methyl jasmonate induces expression of a novel *Brassica juncea* chitinase with two chitin-binding domains. *Plant Mol. Biol.* **1999**, *40*, 1009–1018. [CrossRef]
11. Gao, Y.; Zhao, K. Molecular mechanism of BjCHI1-mediated plant defense against *Botrytis cinerea* infection. *Plant Signal. Behav.* **2017**, *12*, e1271859. [CrossRef] [PubMed]
12. Chye, M.L.; Zhao, K.J.; He, Z.M.; Ramalingam, S.; Fung, K.L. An agglutinating chitinase with two chitin-binding domains confers fungal protection in transgenic potato. *Planta* **2005**, *220*, 717–730. [CrossRef] [PubMed]
13. Guan, Y.; Ramalingam, S.; Nagegowda, D.; Taylor, P.W.; Chye, M.L. *Brassica juncea* chitinase BjCHI1 inhibits growth of fungal phytopathogens and agglutinates Gram-negative bacteria. *J. Exp. Bot.* **2008**, *59*, 3475–3484. [CrossRef]
14. Gao, Y.; Zan, X.L.; Wu, X.F.; Yao, L.; Chen, Y.L.; Jia, S.W.; Zhao, K.J. Identification of fungus-responsive *cis*-acting element in the promoter of *Brassica juncea* chitinase gene, *BjCHI1*. *Plant Sci.* **2014**, *215*, 190–198. [CrossRef] [PubMed]
15. Gao, Y.; Jia, S.; Wang, C.; Wang, F.; Wang, F.; Zhao, K. BjMYB1, a transcription factor implicated in plant defence through activating *BjCHI1* chitinase expression by binding to a W-box-like element. *J. Exp. Bot.* **2016**, *67*, 4647–4658. [CrossRef] [PubMed]

16. Dubos, C.; Stracke, R.; Grotewold, E.; Weisshaar, B.; Martin, C.; Lepiniec, L. MYB transcription factors in Arabidopsis. *Trends Plant Sci.* **2010**, *15*, 573–581. [CrossRef]
17. Bai, J.; Sun, F.; Wang, M.; Su, L.; Li, R. Genome-wide analysis of the MYB-CC gene family of maize. *Genetica* **2019**, *147*, 1–9. [CrossRef]
18. Chiou, T.J.; Lin, S.I. Signaling network in sensing phosphate availability in plants. *Annu. Rev. Plant Biol.* **2011**, *62*, 185–206. [CrossRef]
19. Motte, H.; Beeckman, T. PHR1 balances between nutrition and immunity in plants. *Dev. Cell* **2017**, *41*, 5–7. [CrossRef]
20. Sega, P.; Pacak, A. Plant PHR transcription factors: Put on a map. *Genes* **2019**, *10*, 1018. [CrossRef]
21. Wang, F.; Deng, M.; Xu, J.; Zhu, X.; Mao, C. Molecular mechanisms of phosphate transport and signaling in higher plants. *Semin. Cell Dev. Biol.* **2018**, *74*, 114–122. [CrossRef] [PubMed]
22. Wang, Y.; Chen, Y.F.; Wu, W.H. Potassium and phosphorus transport and signaling in plants. *J. Integr. Plant Biol.* **2021**, *63*, 34–52. [CrossRef]
23. Du, H.; Feng, B.R.; Yang, S.S.; Huang, Y.B.; Tang, Y.X. The R2R3-MYB transcription factor gene family in maize. *PLoS ONE* **2012**, *7*, e37463. [CrossRef] [PubMed]
24. Katiyar, A.; Smita, S.; Lenka, S.K.; Rajwanshi, R.; Chinnusamy, V.; Bansal, K.C. Genome-wide classification and expression analysis of MYB transcription factor families in rice and Arabidopsis. *BMC Genom.* **2012**, *13*, 544. [CrossRef] [PubMed]
25. Wang, Z.; Tang, J.; Hu, R.; Wu, P.; Hou, X.L.; Song, X.M.; Xiong, A.S. Genome-wide analysis of the R2R3-MYB transcription factor genes in Chinese cabbage (*Brassica rapa* ssp. pekinensis) reveals their stress and hormone responsive patterns. *BMC Genom.* **2015**, *16*, 17. [CrossRef]
26. Gu, B.J.; Tong, Y.K.; Wang, Y.Y.; Zhang, M.L.; Ma, G.J.; Wu, X.Q.; Zhang, J.F.; Xu, F.; Li, J.; Ren, F. Genome-wide evolution and expression analysis of the *MYB-CC* gene family in *Brassica* spp. *PeerJ* **2022**, *10*, e12882. [CrossRef]
27. Chen, D.; Chen, H.; Dai, G.; Zhang, H.; Liu, Y.; Shen, W.; Zhu, B.; Cui, C.; Tan, C. Genome-wide identification of R2R3-MYB gene family and association with anthocyanin biosynthesis in *Brassica* species. *BMC Genom.* **2022**, *23*, 441. [CrossRef]
28. Hajiebrahimi, A.; Owji, H.; Hemmati, S. Genome-wide identification, functional prediction, and evolutionary analysis of the R2R3-MYB superfamily in *Brassica napus*. *Genome* **2017**, *60*, 797–814. [CrossRef]
29. Abbas, F.; Ke, Y.; Zhou, Y.; Yu, Y.; Waseem, M.; Ashraf, U.; Wang, C.; Wang, X.; Li, X.; Yue, Y.; et al. Genome-wide analysis reveals the potential role of MYB transcription factors in floral scent formation in *Hedychium coronarium*. *Front. Plant Sci.* **2021**, *12*, 623742. [CrossRef]
30. Ke, Y.J.; Zheng, Q.D.; Yao, Y.H.; Ou, Y.; Chen, J.Y.; Wang, M.J.; Lai, H.P.; Yan, L.; Liu, Z.J.; Ai, Y. Genome-wide identification of the MYB gene family in *Cymbidium ensifolium* and its expression analysis in different flower colors. *Int. J. Mol. Sci.* **2021**, *22*, 13245. [CrossRef]
31. Abubakar, A.S.; Feng, X.; Gao, G.; Yu, C.; Chen, J.; Chen, K.; Wang, X.; Mou, P.; Shao, D.; Chen, P.; et al. Genome wide characterization of R2R3 MYB transcription factor from *Apocynum venetum* revealed potential stress tolerance and flavonoid biosynthesis genes. *Genomics* **2022**, *114*, 110275. [CrossRef] [PubMed]
32. Duan, A.Q.; Tan, S.S.; Deng, Y.J.; Xu, Z.S.; Xiong, A.S. Genome-wide identification and evolution analysis of R2R3-MYB gene family reveals S6 subfamily R2R3-MYB transcription factors involved in anthocyanin biosynthesis in carrot. *Int. J. Mol. Sci.* **2022**, *23*, 11859. [CrossRef] [PubMed]
33. Zhang, H.; Liu, Z.; Luo, R.; Sun, Y.; Yang, C.; Li, X.; Gao, A.; Pu, J. Genome-wide characterization, identification and expression profile of MYB transcription factor gene family during abiotic and biotic stresses in mango (*Mangifera indica*). *Plants* **2022**, *11*, 3141. [CrossRef] [PubMed]
34. Liu, Y.; Zhang, Z.; Fang, K.; Shan, Q.; He, L.; Dai, X.; Zou, X.; Liu, F. Genome-wide analysis of the MYB-related transcription factor family in pepper and functional studies of *CaMYB37* involvement in capsaicin biosynthesis. *Int. J. Mol. Sci.* **2022**, *23*, 11667. [CrossRef] [PubMed]
35. Sabir, I.A.; Manzoor, M.A.; Shah, I.H.; Liu, X.; Zahid, M.S.; Jiu, S.; Wang, J.; Abdullah, M.; Zhang, C. MYB transcription factor family in sweet cherry (*Prunus avium* L.): Genome-wide investigation, evolution, structure, characterization and expression patterns. *BMC Plant Biol.* **2022**, *22*, 2. [CrossRef]
36. Bai, Y.; Zhu, W.; Hu, X.; Sun, C.; Li, Y.; Wang, D.; Wang, Q.; Pei, G.; Zhang, Y.; Guo, A.; et al. Genome-wide analysis of the bZIP gene family identifies two ABI5-like bZIP transcription factors, BrABI5a and BrABI5b, as positive modulators of ABA signalling in Chinese cabbage. *PLoS ONE* **2016**, *11*, e0158966. [CrossRef]
37. Bhardwaj, A.R.; Joshi, G.; Kukreja, B.; Malik, V.; Arora, P.; Pandey, R.; Shukla, R.N.; Bankar, K.G.; Katiyar-Agarwal, S.; Goel, S.; et al. Global insights into high temperature and drought stress regulated genes by RNA-Seq in economically important oilseed crop *Brassica juncea*. *BMC Plant Biol.* **2015**, *15*, 9. [CrossRef]
38. Yang, J.; Liu, D.; Wang, X.; Ji, C.; Cheng, F.; Liu, B.; Hu, Z.; Chen, S.; Pental, D.; Ju, Y.; et al. The genome sequence of allopolyploid *Brassica juncea* and analysis of differential homoeolog gene expression influencing selection. *Nat. Genet.* **2016**, *48*, 1225–1232. [CrossRef]
39. Sun, L.; Song, L.; Zhang, Y.; Zheng, Z.; Liu, D. Arabidopsis PHL2 and PHR1 act redundantly as the key components of the central regulatory system controlling transcriptional responses to phosphate starvation. *Plant Physiol.* **2016**, *170*, 499–514. [CrossRef]
40. Ruan, W.; Guo, M.; Wu, P.; Yi, K. Phosphate starvation induced OsPHR4 mediates Pi-signaling and homeostasis in rice. *Plant Mol. Biol.* **2017**, *93*, 327–340. [CrossRef]

41. Jiang, M.; Sun, L.; Isupov, M.N.; Littlechild, J.A.; Wu, X.; Wang, Q.; Wang, Q.; Yang, W.; Wu, Y. Structural basis for the target DNA recognition and binding by the MYB domain of phosphate starvation response 1. *FEBS J.* **2019**, *286*, 2809–2821. [CrossRef]
42. Bustos, R.; Castrillo, G.; Linhares, F.; Puga, M.I.; Rubio, V.; Perez-Perez, J.; Solano, R.; Leyva, A.; Paz-Ares, J. A central regulatory system largely controls transcriptional activation and repression responses to phosphate starvation in Arabidopsis. *PLoS Genet.* **2010**, *6*, e1001102. [CrossRef]
43. Castrillo, G.; Teixeira, P.J.; Paredes, S.H.; Law, T.F.; de Lorenzo, L.; Feltcher, M.E.; Finkel, O.M.; Breakfield, N.W.; Mieczkowski, P.; Jones, C.D.; et al. Root microbiota drive direct integration of phosphate stress and immunity. *Nature* **2017**, *543*, 513–518. [CrossRef]
44. Shi, J.; Zhao, B.; Zheng, S.; Zhang, X.; Wang, X.; Dong, W.; Xie, Q.; Wang, G.; Xiao, Y.; Chen, F.; et al. A phosphate starvation response-centered network regulates mycorrhizal symbiosis. *Cell* **2021**, *184*, 5527–5540.e5518. [CrossRef]
45. Barco, B.; Clay, N.K. Hierarchical and dynamic regulation of defense-responsive specialized metabolism by WRKY and MYB transcription factors. *Front. Plant Sci.* **2019**, *10*, 1775. [CrossRef]
46. Eulgem, T.; Somssich, I.E. Networks of WRKY transcription factors in defense signaling. *Curr. Opin. Plant Biol.* **2007**, *10*, 366–371. [CrossRef] [PubMed]
47. Chen, H.; Wang, T.; He, X.; Cai, X.; Lin, R.; Liang, J.; Wu, J.; King, G.; Wang, X. BRAD V3.0: An upgraded Brassicaceae database. *Nucleic Acids Res.* **2022**, *50*, D1432–D1441. [CrossRef]
48. Song, M.; Linghu, B.; Huang, S.; Li, F.; An, R.; Xie, C.; Zhu, Y.; Hu, S.; Mu, J.; Zhang, Y. Genome-wide survey of leucine-rich repeat receptor-like protein kinase genes and CRISPR/Cas9-targeted mutagenesis *BnBRI1* in *Brassica napus*. *Front. Plant Sci.* **2022**, *13*, 865132. [CrossRef] [PubMed]
49. Li, G.; Hu, X.; Hou, L.; Cao, L.; Wang, Q.; Wang, D.; Mu, X.; Zhang, Y.; Zhou, X.; Zhao, Y.; et al. Molecular identification of BrHAB2a, one of the two AtHAB2-like proteins in *Brassica rapa*, is an important component of ABA signaling. *Biochem. Biophys. Res. Commun.* **2018**, *503*, 495–500. [CrossRef] [PubMed]
50. Zhou, X.; Hao, H.; Zhang, Y.; Bai, Y.; Zhu, W.; Qin, Y.; Yuan, F.; Zhao, F.; Wang, M.; Hu, J.; et al. SOS2-LIKE PROTEIN KINASE5, an SNF1-RELATED PROTEIN KINASE3-type protein kinase, is important for abscisic acid responses in Arabidopsis through phosphorylation of ABSCISIC ACID-INSENSITIVE5. *Plant Physiol.* **2015**, *168*, 659–676. [CrossRef] [PubMed]
51. Zhou, X.; Yuan, F.; Wang, M.; Guo, A.; Zhang, Y.; Xie, C.G. Molecular characterization of an *ABA insensitive 5* orthologue in *Brassica oleracea*. *Biochem. Biophys. Res. Commun.* **2013**, *430*, 1140–1146. [CrossRef] [PubMed]

plants

Review

Recent Advancements and Biotechnological Implications of Carotenoid Metabolism of *Brassica*

Lichun Shi [1,2,3,4,5,†], Lin Chang [6,†], Yangjun Yu [2,3,4,5], Deshuang Zhang [2,3,4,5], Xiuyun Zhao [2,3,4,5], Weihong Wang [2,3,4,5], Peirong Li [2,3,4,5], Xiaoyun Xin [2,3,4,5], Fenglan Zhang [2,3,4,5], Shuancang Yu [2,3,4,5], Tongbing Su [2,3,4,5,*], Yang Dong [7,8,*] and Fumei Shi [1,*]

1 School of Life Sciences, Liaocheng University, Liaocheng 252059, China
2 Beijing Vegetable Research Center (BVRC), Beijing Academy of Agriculture and Forestry Science (BAAFS), Beijing 100097, China
3 National Engineering Research Center for Vegetables, Beijing 100097, China
4 Key Laboratory of Biology and Genetic Improvement of Horticultural Crops (North China), Ministry of Agriculture, Beijing 100097, China
5 Beijing Key Laboratory of Vegetable Germplasm Improvement, Beijing 100097, China
6 Marine Science Research Institute of Shandong Province, Qingdao 266104, China
7 State Key Laboratory of Systematic and Evolutionary Botany, Institute of Botany, The Chinese Academy of Sciences, Beijing 100093, China
8 China National Botanical Garden, Beijing 100093, China
* Correspondence: sutongbing@nercv.org (T.S.); yang.dong@ibcas.ac.cn (Y.D.); shifumei@lcu.edu.cn (F.S.); Tel.: +86-13676350008 (F.S.)
† These authors contributed equally to this work.

Abstract: Carotenoids were synthesized in the plant cells involved in photosynthesis and photoprotection. In humans, carotenoids are essential as dietary antioxidants and vitamin A precursors. *Brassica* crops are the major sources of nutritionally important dietary carotenoids. Recent studies have unraveled the major genetic components in the carotenoid metabolic pathway in *Brassica*, including the identification of key factors that directly participate or regulate carotenoid biosynthesis. However, recent genetic advances and the complexity of the mechanism and regulation of *Brassica* carotenoid accumulation have not been reviewed. Herein, we reviewed the recent progress regarding *Brassica* carotenoids from the perspective of forward genetics, discussed biotechnological implications and provided new perspectives on how to transfer the knowledge of carotenoid research in *Brassica* to the crop breeding process.

Keywords: *Brassica*; carotenoid; QTL; genetics; biotechnological implications

Citation: Shi, L.; Chang, L.; Yu, Y.; Zhang, D.; Zhao, X.; Wang, W.; Li, P.; Xin, X.; Zhang, F.; Yu, S.; et al. Recent Advancements and Biotechnological Implications of Carotenoid Metabolism of *Brassica*. *Plants* **2023**, *12*, 1117. https://doi.org/10.3390/plants12051117

Academic Editors: Zanmin Hu, Han Xiao, Yi Ren and Chengming Fan

Received: 23 January 2023
Revised: 26 February 2023
Accepted: 27 February 2023
Published: 2 March 2023

1. Introduction

Carotenoids are red, orange, and yellow pigments that are widely distributed in nature, mainly comprising C40 isoprenes. Studies have shown that lutein and β-carotene are the main components of carotenoids, which also include zeaxanthin, cryptoxanthin, astaxanthin, and lycopene [1]. Carotenoids have important functions in plant photosynthesis and are auxiliary pigments in light absorption. Carotenoid molecules contain multiple conjugated double bonds that can absorb energy and protect plants from reactive oxygen species [2]. Carotenoids endow fruit and flowers with bright colors [3], enabling them to attract the attention of pollinating insects and animals. In addition to their roles in plants, carotenoids are vital for the health and nutrition of humans. People take carotenoids to supplement vitamin A, which is used to treat night blindness [4]. Lutein, as a main component of carotenoids, can protect human vision to a certain extent, prevent vision deterioration, and prevent cataracts, and other eye diseases [5]. Astaxanthin also helps the body fight inflammation and boosts immunity [6]. There have been significant efforts to increase the carotenoid content in agricultural crops to improve their nutritional value and health benefits. Manipulating carotenoid metabolism using biotechnology and genetic engineering has been successfully implemented in many crops, and golden rice is the most relevant example of improving β-carotene in food [7].

The metabolism of carotenoids in plants has been the subject of extensive research because of the significance of carotenoids to both plants and humans. The genus *Brassica* includes many vegetable crops, such as *B. rapa* (AA genome), *B. nigra* (BB genome), *B. oleracea* (CC genome), *B. juncea* (AABB genome), *B. napus* (AACC genome), and *B. carinata* (BBCC genome) [8]. These species contain diverse carotenoid metabolites and high levels of nutritionally significant components. Genus and species, as well as genotype and agricultural conditions, affect the carotenoid composition and content of these organisms. The present review focuses on current situation and recent advances in carotenoid metabolism of *Brassica* crops from the perspective of forward genetics. Moreover, we discuss the evolution of carotenoid composition and certain key carotenoid metabolism genes during the formation of the *Brassica* crops. Lastly, we review the agricultural application and the prospects of carotenoid metabolism and its genetic manipulation in *Brassica*.

2. Overview of the Metabolic Pathway of Carotenoids

2.1. Carotenoid Biosynthesis

Carotenoid biosynthesis pathway components in higher plants have been gradually clarified through biochemical analysis, classical genetics, and molecular genetics. In plants, the methylerythritol phosphate (MEP) pathway is the major carotenoid production route (Figure 1). It can produce the isopentenyl diphosphate (IPP) and its allyl isomer dimethylallyl diphosphate (DMAPP) [9], which are used as substrates to synthesize the C20 geranylgeranyl diphosphate (GGPP) [10]. GGPP is the most direct precursor of plant carotenoids and participates in the synthesis of the earliest carotenoid in plants [11]. The first compound in the carotenoid biosynthesis pathway, 15-cis-phytoene, is synthesized by the condensation of two GGPP molecules, catalyzed by phytoene synthase (PSY) [12]. Phytoene desaturase (PDS) and ζ-carotene desaturase (ZDS) catalyze a four-step dehydrogenation reaction that produces 9,15,9′-cis-ζ-carotene, which is then isomerized into yellow 9,9′-cis-ζ-carotene by ζ-isomerase (ZISO) [13]. Subsequently, carotenoid isomerase (CRTISO) converts yellow prolycopene into red all-trans lycopene [14]. If lycopene β-cyclase (LCYB) acts on both ends of the lycopene molecule, β-carotene is formed. If one of the ends of the lycopene molecule is affected by lycopene ε-cyclase (LCYE), δ-carotene is formed [15]. Then, δ-carotene can be catalyzed by LYCB into α-carotene. Cytochrome P450 carotene β-hydroxylase (CYP97A) and cytochrome P450 carotene ε-hydroxylase (CYP97C) then catalyze α-carotene to create lutein [16]. In addition, β-carotene hydroxylase (BCH) catalyzes β-carotene to produce β-cryptoxanthin, which is converted into zeaxanthin. Zeaxanthin can be further converted to antherxanthin and then to violaxanthin, both of which are catalyzed by zeaxanthin epoxidase (ZEP) [17]. Violaxanthin can also be reversed to form zeaxanthin under the catalysis of violaxanthin de-epoxidase (VDE) [18], via a process termed the xanthophyll cycle [19], which protects plants from light damage. Neoxanthin is then produced from violaxanthin under the catalysis of neoxanthin synthase (NXS) [20].

2.2. Degradation of Carotenoids

Carotenoid cleavage dioxygenases (CCDs) lyse carotenoids to form products that provide leaves, flowers, and fruit with specific colors and flavors, and produces abscisic acid (ABA) and other plant hormones. A group of enzymes known as carotenoid cleavage oxygenases (CCOs) catalyze the particular enzymatic oxidative degradation of carotenoids. The varieties of apocarotenoid breakdown products are determined by the precise bonds at which CCOs cleave the carotenoids, with substrate specificity. The 9-cis-epoxycarotenoid dioxygenases (NCEDs) and carotenoid cleavage dioxygenases make up the plant CCO family (which are CCDs) [21]. CCDs are divided into a number of subfamilies, including CCD1, CCD2, CCD4, CCD7, and CCD8. Among them, CCD1 is mainly involved in the formation of plant aromatic substances, and can split β-carotene into β-ionone, an important aroma component [22]. CCD4 is mainly involved in the formation of pigment substances in flowers and fruit. The enzymes encoded by CCD7 and CCD8 are located in plastids and are mainly involved in the synthesis of strigolactone. NECDs are rate-limiting enzymes controlling the production of ABA from carotenoids, which mainly comprises cleaving violaxanthin or neoxanthin, thereby forming precursors of ABA [23]. In addition to the precise cleavage mediated by CCDs, nonspecific enzymes (lipoxygenases and peroxidases) and photochemical oxidation contribute to carotenoid degradation. Carotenoid nonspecific oxidation produces unspecific apocarotenoid factors via random cleavage.

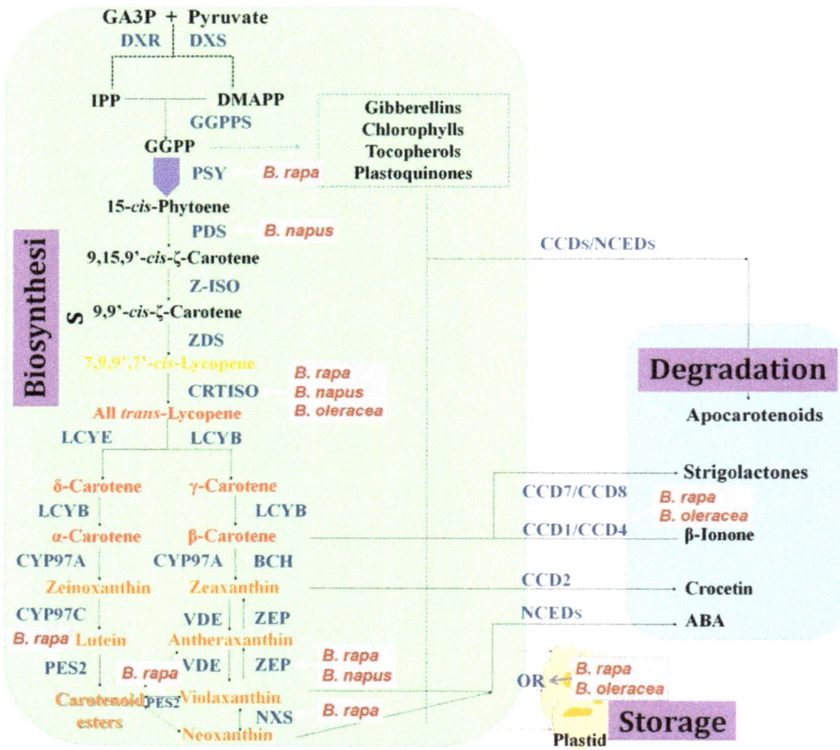

Figure 1. Overview of carotenoid metabolism in *Brassica*.

Genes identified by forward genetic analysis which are involved in carotenoid metabolism in *Brassica* were labelled with grey boxes. Enzymes in the carotenoid metabolic pathway were in dark blue. GA3P, Glyceraldehyde 3-phosphate; IPP, isopentenyl diphosphate; DMAPP, dimethylallyl diphosphate; GGPP, geranylgeranyl diphosphate; DXS, 1-deoxy- D –xylulose 5-phosphate synthase; DXR, 1-deoxy- D -xylulose 5-phosphate reductoisomerase; GGPPS, GGPP synthase; PSY, phytoene synthase; PDS, phytoene desaturase; Z-ISO, ζ-carotene isomerase; ZDS, ζ-carotene desaturase; CrtISO, carotenoid isomerase; LCYE, lycopene ε-cyclase; LCYB, lycopene β-cyclase; BCH, β-carotene hydroxylase; CYP97A, cytochrome P450 carotene β-hydroxylase; CYP97C, cytochrome P450 carotene ε-hydroxylase; ZEP, zeaxanthin epoxidase; VDE, violaxanthin de-epoxidase; NXS, neoxanthin synthase; CCD, carotenoid cleavage dioxygenase; NCED, 9-cis-epoxycarotenoid dioxygenase; OR, ORANGE protein.

3. Genetic Study of Carotenoid Accumulation in *Brassica* Crops

Brassica carotenoid metabolism is important for the development of flower color, which is exploited for decorative and landscaping use. *Brassica* flowers are generally yellow, but can be dark yellow, orange, milky white, or white. Here, we used *B. rapa* (Chinese cabbage) as an example. The main carotenoids in the yellow petals of Chinese cabbage are violaxanthin and lutein [24], while there is more lutein and β-carotene in the orange petals. β-carotene and lutein contents are regarded as key elements related to the yellow pigmentation in the leaves of Chinese cabbage [25]. The lutein content in the internal leaves of yellow cultivars is greater than that in the internal leaves of orange cultivars. Similar observations were made for the inner yellow leaves compared with the white internal leaves. Collectively, these observations demonstrated that carotenoid profiles are distinctive in petals and leaves, and the carotenoid contents of Chinese cabbage with yellow or orange inner leaves are markedly higher than, and different from, those of common white leaf varieties.

CRTISO catalyzes the isomerization of poly-cis-carotenoids to all-trans-carotenoids. Together with PDS and ZDS, CRTISO is required to synthesize lycopene from phytoene. In *B. rapa*, the orange pigmentation of flowers and the inner leaves is under the control of the *BrCRTISO* gene. Using restriction fragment length polymorphism markers, the *orange–yellow* pigmentation gene (*Oy*) in Chinese cabbage was first mapped to linkage group 1 [26]. Linkage analysis finally located the candidate gene to the end of A09. Comparisons of the promoter regions of *CRTISO* promoter revealed insertions/deletions between the two parents, which identified *CRTISO* as the most likely candidate gene for *Br-Oy*. After using BC1 backcross population for high-quality mapping, three SNP markers delimited the *Br-Oy* locus was delimited to a 9.47 kb using three single nucleotide polymorphism (SNP) markers, which contained one functional gene, Bra031539 (*BrCRTISO*) [27]. The *Br-Oy* gene has many sequence variations, such as a 90-bp promoter deletion and a 501-bp 3′ end insertion. Meanwhile, a *CRTISO* mutation was found in an orange Chinese cabbage, which were used to develop molecular markers to differentiate orange from white genotypes [28]. Further study showed that these mutations in *BrCRTISO* reduced the flux of carotenoid synthesis, leading to the formation of orange leaves in Chinese cabbage. Genes involved in carotenoid metabolism were identified using white flowered varieties. According to genetic research, recessive loci *Brwf1* and *Brwf2* regulate Chinese cabbage's white flowers. Insertion/deletion (InDel) and SNP tombstone analysis located *Brwf1* in a 49.6-kb region of chromosome A01 comprising nine annotated genes. *BrCRTISO* (Bra031539) was mapped to a 59.3-kb gap on chromosome A09 comprising 12 known genes and *Brwf2* [29]. To fine map the white flower gene *BrWF3* in Chinese cabbage, an F2 population was created from the F1 plants of a cross between a white flowered line and a yellow flowered line. *BrWF3* was precisely mapped to a 105.6 kb gap. Sequence variation analysis, functional elucidation, and expression profiling demonstrated that the *BrWF3* gene was most likely Bra032957, an *AtPES2* homolog. Carotenoid compound analysis and transmission electron microscopy revealed that *BrWF3* might produce xanthophyll esters, primarily violaxanthin esters, which interfere with chloroplast formation and plastoglobule (PG) generation. The third exon of *BrWF3* had an SNP deletion that prevented the protein from functioning properly and prevented PG assembly, accompanied by decreased expression of carotenoid metabolism-related genes [30]. In another study, a natural mutant of flowering Chinese cabbage (*B. rapa* ssp. *chinensis* var. *parachinensis*) with visually distinguishable pale-yellow petals was obtained in farmland. The pale-yellow petal was controlled by a single recessive gene *BrPYP* [31]. Further study showed that *BrPYP* was mapped to the same locus of *BrWF3*; however, different variations were identified. A functional 1148 bp deletion in the promoter region of *BrPYP* that reduces promoter activity and expression level was identified. In *B. napus*, mutation of *BnaA09.CRTISO* and *BnaC08.CRTISO* using the clustered regularly interspaced short palindromic repeats (CRISPR)/CRISPR associated protein 9 (CAS9) system caused the petals to turn milky white and the leaves to turn pale yellow. Thus, petal and leaf coloration in *B. napus* are regulated crucially and redundantly by

BnaA09.CRTISO and *BnaC08.CRTISO*. The carotenoid concentrations in the petals and leaves of the *BnaCRTISO* double mutant were significantly decreased, according to subsequent observation. The levels of chalcone, a vital component of yellow color, were reduced dramatically in the mutant flower's petals, yet its levels of lycopene, β-carotene, and α-carotene increased slightly [32]. These findings help explain how carotenoids are produced and how *B. napus*'s color variation is controlled. Chinese kale (*B. oleracea* var. *alboglabra*) contains abundant carotenoids, among which neoxanthin is one of the most important. When *BoaCRTISO* expression was downregulated in Chinese kale, virtually all carotenoid biosynthesis genes were downregulated, and the leaves turned yellow. The authors found that *BoaCRTISO* was a photoinduced gene, and a mixture of red, blue, and white light could increase the carotenoid content in plants [33]. Consequently, the CRISPR/Cas9 method was used to target and edit the Chinese kale *BoaCRTISO* gene. Decreases in the overall and individual carotenoid and chlorophyll levels were observed in the mutants. The color of the mutant changed from green to yellow, suggesting a reduced protective effect of carotenoids toward chlorophyll [34].

The hydroxylated β-rings of zeaxanthin were subjected to subsequent epoxidation by ZEAXANTHIN EPOXIDASE (ZEP) to produce antheraxanthin and later, violaxanthin. In addition to CRTISO, ZEPs were also identified to be widely involved in the coloration of flowers or leaves in different *Brassica* crops. In a study of the dark yellow petals of Chinese cabbage, bulked segregant RNA (BSR) sequencing combined with competitive allele-specific PCR (KASP) assays fine mapped *Br-dyp1* to a 53.6 kb region on chromosome A09. Further expression analysis, functional annotation, and sequence variation assessment identified Bra037130 (*BraA09.ZEP*) as a potential candidate gene for *Br-dyp1*. Plants with dark yellow petals had a 679 bp insertion in *BraA09.ZEP* that created a premature stop codon, leading to ZEP loss-of-function. Loss of ZEP activity destroyed the carotenoid metabolism, and increased accumulation of total carotenoid, and finally changed the petals from yellow to dark yellow of Chinese cabbage. Large amounts of violaxanthin lead to yellow petals in *B. napus*, while the orange flowers were caused by the accumulation of lutein. In *B. napus*, the ratio of yellow- to orange-flowered plants was 15:1 in the F2 population and 3:1 in the BC1 population, indicating that two dominant nuclear genes controlled the yellow-flower phenotype [35]. Map-based cloning was used to isolate key genes to reveal the underlying molecular mechanism of the orange-flowered phenotype. Quantitative trait locus (QTL) cloning assigned the change in color from yellow to orange to the loss of *BnaC09.ZEP* and the deletion of a 1695 bp fragment of *BnaA09.ZEP*. Further CRISPR/Cas9 and genetic complementation analyses showed that the nullification of *BnaA09.ZEP* and *BnaC09.ZEP* contributed to markedly increased lutein levels and a large decrease in violaxanthin levels in petals rather than in leaves [36].

CCDs encompass a superfamily of mononuclear non-heme iron proteins that catalyze the oxygenolytic splitting of alkene bonds in carotenoids to develop apocarotenoid products. β-carotenoid cleavage dioxygenase 4 (CCD4) was reported to be the primary negative switch for seed carotenoids, particularly β-carotene, according to linkage mapping and genome-wide association studies of *Arabidopsis* carotenoids [37]. In *B. napus*, positional cloning identified a carotenoid cleavage dioxygenase 4 gene, *BnaC3.CCD4*, as being responsible for the composition of white or yellow flower color, with white-bloomed *B. napus* lines having higher expression of *BnaC3.CCD4* in their petals. In yellow-flowered varieties, a CACTA-like transposable element 1 (TE1) is embedded in the coding region of *BnaC3.CCD4*, which disrupts *BnaC3.CCD4* expression. Further investigation uncovered that this TE insertion occurred frequently in the *BnaC3.CCD4* gene of yellow-flowered *B. napus* [38]. In *B. oleracea*, analyses demonstrated that this yellow-white petal characteristic was dependent a single locus on C03, and in 2019, Han mapped the gene to a 207-kb locus, suggesting that the candidate gene was *BoCCD4*. Further sequence analysis, expression pattern assessment, and functional complementation analyses in *B. oleracea* accessions showed that *BoCCD4* functional failure resulted the yellow petal phenotype. Overexpression of *BoCCD4* changed the color of the petals from yellow to white or pale yellow [39]. In cauliflower, the yellow-flower locus was fine-mapped, which identified *BoCCD4* as the most likely candidate gene. Further investigation revealed the presence of a novel 10,608 bp CACTA-like transposon that inhibited the function of *BoCCD4* [40]. The yellow-petal feature could be induced in a white-petal natural line by *BoCCD4* functional complementation. *BoCCD4* was observed to be particularly expressed in the petal tissue of white-petal plants, and a genetic study revealed that in carotenoid metabolism, CCD4 homologs might share evolutionarily conserved functions. BoAAO3 is a key enzyme that interacts with *BoCCD4* to regulate petal carotenoid deterioration. Likewise, *BoCCD4* co-regulates carotenoid metabolism accompanied by two key transcription components, Bo2g151880 (WRKY) and Bo3g024180 (SBP). Together, they regulate carotenoid biosynthesis in petals, which in turn adjusts whether petals are white or yellow [41].

The bioavailability of carotenoids from new and processed snacks heavily depends on their natural deposition form. In cauliflower (*B. oleracea* var *botrytis*), a spontaneous semidominant *Orange* (*OR*) mutant has a fascinating genetic mutation that results in the accumulation of β-carotene in commonly unpigmented tissues. Using positional cloning, the gene responsible for *OR* was determined, and functional complementation confirmed this identification in wild-type cauliflower. *OR* encodes a DnaJ Cys-rich domain-containing plastid-associated protein. The *OR* allele contains an inserted long terminal repeat retrotransposon, resulting in the *OR* gene mutation. Analyses of the gene, its output, and the effects of an *OR* transgene on cells indicated that *OR*'s function is associated with a biological process that promotes the differentiation of proplastids or other noncolored plastids into chromoplasts for carotene accumulation. Additionally, the study demonstrated that that regulating chromoplast formation is essential for the control of carotenoid production in plants [42]. Regarding leafy *Brassica* crops, cultivars with golden leafy heads are becoming increasingly recognized. The golden cultivars are abundant in β-carotene and lutein. In comparison with the white line, the β-carotene level was increased by 13.6—fold. Bulked-segregant study sequencing identified *BraA09g007080.3C* (encoding the ORANGE protein) as the candidate gene. A 4.67 kb long terminal repeat was observed to be inserted in exon three of *BrGOLDEN*, which resulted in the expression of three alternatively spliced transcripts. Spatiotemporal expression analysis showed that *BrGOLDEN* might affect the expression levels of carotenoid-related genes [43]. Further sequence analysis revealed that *BrGOLDEN* was probably transferred into *B. rapa* through distant hybridization between *B. rapa* and *B. oleracea*. In addition, in the first step of carotenoid biosynthesis, orange protein (OR) is known to interact with phytoene synthase (PSY) and is a major post-transcriptional regulator on PSY in *Arabidopsis* [44].

Additionally, turnip (*B. rapa* ssp. rapa) is a nutritious and fitness-promoting vegetable, with yellow and white flesh, among which yellow-fleshed turnips have a higher nutritional value. Combined transcriptomic and metabolomic investigations identified that *PSY* is the important gene that affects carotenoid creation in turnip. High expression of *PSY* results in yellow turnips rather than mutations in *PSY*. This suggested that carotenoids might be produced via a post-transcriptional regulatory mechanism.

In addition to *CRTISO*, *ZEP*, *CCD*, and *OR*, certain other genes were reported to be involved in carotenoid absorption. A mutant with *yellowish-white* flowers (*ywf*), which was developed from Zhongshuang 9 (ZS9) using ethyl methane sulfonate mutagenesis was analyzed. The *ywf* locus displayed a lower petal carotenoid content. A genetic study revealed that a single recessive gene regulated the yellowish-white trait. Bulked segregant analysis sequencing mapped the *ywf* locus to *YWF*, encoding phytoene desaturase 3 (PDS3). Moreover, *ywf* accommodated a C-to-T replacement in the coding region, which caused premature translation termination. RNA-seq and carotenoid constituent investigation showed that in *ywf* petals, the truncated *BnaA08.PDS3* disrupted carotenoid biosynthesis [45]. In kale, the neoxanthin synthase gene (*BoaNXS*) functions to adjust leaf color, and in *BoaNXS* overexpressing plants, the color alternated from yellow-green to green, and the total carotenoids and individual carotenoid contents were considerably elevated [46].

4. Evolution of Carotenoid Biosynthesis and Some Key Carotenoid Genes in *Brassica*

The consumption of *Brassica* species confers unique health attributes, and they contain high carotenoid levels. The six genetically linked *Brassica* species share a common evolutionary history and are currently being bred using interspecific hybridization for species improvement [47]. In this background, to increase the nutritional value of *Brassica*, identifying the genetic connections among *Brassica* species related to carotenoid accumulation would promote their genetic improvement. The main carotenoids that accumulate in genetically-related *Brassica* species have been identified. *B. rapa* was reported to accumulate the highest concentrations of antheraxanthin, lutein, and zeaxanthin. The maximum concentrations of β-carotene and total chlorophyll were detected in *B. juncea*. *B. nigra* contained the highest levels of 5,6-epoxylutein and violaxanthin, while *B. oleracea* had the highest neoxanthin levels. Interestingly, the amphidiploids *B. carinata* and *B. napus* were found to contain significantly reduced levels of carotenoids compared with those in the diploid species and *B. juncea* [8].

Plant whole-genome triplication (WGT) resulted in multiple duplicates of carotenoid biosynthetic genes, most of which retain their syntenic relationships with their *A. thaliana* orthologs. Taking *B. rapa* as an example, *B. rapa* diverged from *A. thaliana* and its nucleus contains three subgenomes [48]. Gene loss events in the three subgenomes of *B. rapa* show bias. However, the proportions of carotenoid biosynthetic genes in the respective *B. rapa* subgenomes were not considerably different from those in the whole genome. Flower color is mostly determined by the presence or deficiency of carotenoid pigments. The production of carotenoids in chromoplasts of petal cells leads to the yellow color of petals. The prevailing flower color of *Brassica* spp. according to the triangle of U hypothesis is yellow. However, among the subspecies of the *B. oleracea* cytodeme, some white-flowered varieties are observed. Moreover, white flowers were reported in rapeseed lines created via interspecific hybridization between white-flowered *B. oleracea* and *B. rapa*. Herein, we used *B. napus* as an example to explain the evolution of carotenoids in *Brassica*. *B. napus* (2n = 4 = 38, AACC) is an allopolyploid crop plant formed by an interspecific crosses between *B. rapa* (2n = 2 = 20, AA) and *B. oleracea* (2n = 2 = 18, CC). Flower pigment divergence was evident in the selfed offspring from a single *B. napus* parent, varying from white to bright yellow. This variation was assumed to be genetically related because all offspring were identified under controlled environmental conditions. In *B. napus* and *B. oleracea*, the *carotenoid cleavage dioxygenase 4* gene (*CCD4*) exhibits the white flower trait. Among *B. carinata*, *B. rapa*, and *B. oleracea*, five distinct alleles of *C3.CCD4* were identified. The wild-type (WT) allele was designated as that contained by white-flowered lines of *B. rapa* and *B. oleracea*. The four variant alleles comprised two InDels (M2 and M3) and two with TE insertions (M1 and M4), and were contained in yellow-flowered *Brassicas*: *B. oleracea* and the allotetraploids *B. napus* and *B. carinata*. The color of the petals changed from white to yellow as a result of these variant alleles disrupting *C3.CCD4*'s function. Interestingly, all yellow-flowered *B. napus* and *B. carinata* plants and some yellow-flowered *B. oleracea* plants are homozygous for M1 or M4, suggesting that the CACTA-like TE insertion into *C3.CCD4* occurred before *B. napus* and *B. carinata* allopolyploidization. The two InDels were only detected in *B. oleracea BolC3.CCD4*, but not in the *CCD4* genes of *B. napus* or *B. carinata*. This suggested that mutations M2 and M3 occurred outside of the centers of origin of *B. napus* and *B. carinata*, making no contribution to their speciation. Using *AtCCD4* as the outgroup, a phylogenetic tree of the five *Brassica C3.CCD4* alleles was constructed, which suggested that the WT allele was the ancestral and functional type, and the others were loss-of-function types. This prompted the conclusion that the *B. oleracea* flower color diversified into yellow and white flowers prior to the emergence of the amphidiploids *B. napus* and *B. carinata*, resulting from CACTA-like TE insertions in the *B. oleracea BolC3.CCD4* gene. Thus, we presumed that the development of blossom color in *B. carinata*, *B. napus*, and *B. oleracea* is governed by the evolution of *CCD4* in the *Brassica* C genome [38].

In addition, the dynamic expression pattern of genes played a substantial role in the development of carotenoid biosynthesis during evolution. *Brassica* diploids were found to contain two ZEP homologs, while *Brassica* allotetraploids were found to have four. Genetic analyses allowed the assignment of *Brassica* ZEPs to subclades based on their localization to the A or C genome. ZEPs translated from genes located on chromosomeA07/C07 formed a subclade that was most closely related to *Arabidopsis* ZEP proteins, forming a sister cluster to the ZEPs located on *Brassica* chromosomes A09/C09. These results suggested that *Brassica* ZEPs diverged prior to the allopolyploidization event. Furthermore, *BnaA09.ZEP* and *BnaC09.ZEP* were reported to be mostly expressed in floral tissues, while homologous *BnaA07.ZEP* and *BnaC07.ZEP* were mainly expressed in leaves. These observations revealed that *BnaZEPs* were redundant and experienced tissue-specific diversification [36]. These discoveries, taken together with earlier data for CCD4, revealed the central role exerted by gene duplication and tissue-specific expression in the evolution of carotenoid accumulation in *Brassica*. Indeed, accumulating evidence demonstrated that the evolution of tissue-specific expression following gene duplication might be unique to plant carotenoid metabolism. In tomatoes, many carotenoid-metabolic genes, such as GGPPS, PSY, LCYB, and BCH, are present in multiple copies: one copy is preferentially expressed in green tissues, and the other in flowers or fruit.

5. Biotechnological Implications of Carotenoid Genes Identified by QTL-Mapping

Research has sought to develop crops with higher carotenoid contents because of their status as essential phytonutrients. Vegetable yield and quality are affected by cultivars, harvesting dates, climate, location, and conditions. However, conventional methods to improve quality, e.g., chemical spraying and hybridization, are often lengthy and ineffective. Plant disease resistance and yield have been improved using the efficient and convenient CRISPR/Cas9 system [49]. Such technology will drive innovative strategies for metabolic engineering to produce crops with higher nutritional value. Improved crop varieties with modified carotenoid pathways will have a wide application in horticulture.

The breeding of plants with different colors has been facilitated by the cloning and functional analysis of flower pigment regulatory gene. Indeed, certain transgenic ornamental plant varieties with modified flower color have been developed. Agro-transformation of *B. napus* has successfully altered its flower color. For example, *B. napus* with yellow flowers was transformed with *PAP2* (encoding phytochrome-associated protein 2) controlled by a petal-specific promoter, which increased the petal anthocyanin content, generating red flowers [50]. Moreover, both *BnaA09.ZEP* and *BnaC09.ZEP* were mutated using CRISPR/Cas9, which changed the profile of carotenoids in the petals of the mutant plants, without disturbing their growth and development. The CRISPR/Cas9 system was also used to edit *B. napus BnaA09.CRTISO* and *BnaC08.CRTISO*, resulting in a mutant phenotype of yellowish leaves and creamy white petals in the double mutant plants. The carotenoid isomerase gene (*BoaCRTISO*) of Chinese kale was also targeted and edited, which decreased the chlorophyll content and the total and individual carotenoid levels. These studies showed that the manipulation of known genes that are genetically determined by QTL mapping represents an auspicious approach to alter the appearance of flowers and to breed *Brassica* crop varieties with different colored flowers with high ornamental worth.

6. Conclusions and Perspective

Consumption of carotenoids has been associated with various health benefits, including a reduced risk of age-related macular degeneration and cataract, some cancers and coronary heart disease [51]. There is also some evidence of a beneficial effect on cognitive function [52]. The content and composition of carotenoids have vital functions in the nutrition and quality of *Brassica* crops [53]. However, humans cannot synthesize carotenoids and must ingest them in food or via supplementation. In recent times, large numbers of genes and the mechanisms that regulate carotenoid biosynthesis and accumulation have been discovered (Figure 1 and Table 1), facilitating the smooth breeding of *Brassica* crop [54].

Here, we reviewed recent progress regarding carotenoids of *Brassica* from the perspective of forward genetics. We believe that this review provides biotechnological implications and some new perspectives on carotenoid research in *Brassica*, as well as other horticultural crops. However, the regulation of carotenoids in higher plants is complex and multi-faceted; thus, more effort should be made to elucidate the mechanisms of carotenoid function.

Table 1. Enzymes and genes that regulate carotenoid accumulation in *Brassica* crops.

Regulated Genes	Species	Major Changes	Genes (Gene Accession)	Color Change/Tissue	Reference
CRTISO	*B. rapa*	The mutant Br-oy protein cannot convert prolycopene to all- trans -lycopene.	*BrOy* (*Bra031539*)	white/yellow → orange inner leaf	[26,27]
		The loss of BrCRTISO function leads to the accumulation of prolycopene.	*Br-oy or BrCRTISO* (*Bra031539*)		
		Loss of BrWF3 function interferes with plastoglobules assembly and decreases expression levels of genes associated with carotenoid metabolism.	*Brwf3* (*Bra032957*)	yellow → white/flower petal	[30]
		The key factor for the pale-yellow color of petals was the decrease in esterified carotenoid content due to the loss of PYP function.	*BrPYP* (*BraA02g037170.3C*)	yellow → pale-yellow/flower petal	[31]
	B. napus	The contents of carotenoids in petals and leaves of BnaCRTISO double mutant were reduced. In petals, the content of chalcone decreased, the content of some carotene (lycopene, α-carotene, γ-carotene) increased.	*BnaA09. CRTISO* (*BnaA09g49740D*) *BnaC08. CRTISO* (*BnaC08g44970D*)	yellow → milky white flower petals yellow → pale yellow leaves	[32]
	B.oleracea	Carotenoid and chlorophyll levels were reduced in the mutant of BoaCRTISO.	*BoaCRTISO* (*GenBank accession MN810158*)	green → yellowing leaves	[33,34]
ZEP	*B. rapa*	The loss of function of ZEP disrupts the metabolism of carotenoids and leads to the increase in total carotenoid accumulation.	*Br-dyp1* (*Bra037130*)	yellow → dark yellow flower petal	[35]
	B. napus	The abolishment of both genes led to a substantial increase in lutein content and a sharp decline in violaxanthin content in petals.	*BnaA09. ZEP* (*BnaA09g07610D*) *BnaC09. ZEP* (*BnaC07g16350D*)	yellow → orange flower petal	[36]
CCD4	*B. napus*	In yellow petals, a large amount of α-carotene, α-cryptoxanthin, β-cryptoxanthin, violaxanthin, 9-*cis*-violaxanthin, lutein, and *cis*-neoxanthinwere accumulated.	*BnaC3.CCD4* (*Bol029878*)	white → yellow flower petal	[38]
	B.oleracea	Not available.	*BoCCD4* (*Bol029878*)	white/pale yellow → yellow flower petal	[39]
		These key genes may interact with *BoCCD4* to jointly regulate carotenoid biosynthesis in petals.	*WRKY* (*Bo2g151880*) *SBP* (*Bo3g024180*)		[41]

Table 1. *Cont.*

Regulated Genes	Species	Major Changes	Genes (Gene Accession)	Color Change/Tissue	Reference
OR	*B.oleracea*	The *OR* gene mutation confers the accumulation of high levels of β-carotene in various tissues normally devoid of carotenoids.	*OR (GenBank accession DQ482460)*	white → orange	[42]
	B. rapa	The *BrGOLDEN* lines are rich in β-carotene and lutein.	*BrGOLDEN* (*BraA09g007080.3C*)	golden → light yellow inner leaf	[43]

Author Contributions: F.S., L.C. and L.S. conceived the study and wrote the manuscript. T.S., F.Z., S.Y. and Y.D. discussed the focus, structure, and content of the review. Y.Y., D.Z., X.Z., W.W., P.L. and X.X. discussed the results and commented on the manuscript. All authors have read and agreed to the published version of the manuscript.

Funding: This research was funded by grants from the Innovation and Capacity-Building Project of BAAFS (KJCX20200204 and KYCX20210427), Beijing Joint Research Program for Germplasm Innovation and New Variety Breeding (G20220628003-01), and the Outstanding Scientists Training Program of Beijing Academy of Agriculture Forestry Sciences (JKZX201906).

Institutional Review Board Statement: Not applicable.

Informed Consent Statement: Not applicable.

Data Availability Statement: The data used to support the study are available from the corresponding author upon request.

Acknowledgments: We gratefully acknowledge the support of our colleagues for critical review of the manuscript. We apologize to the colleagues whose work could not be discussed due to space limitations.

Conflicts of Interest: The authors declare that the research was conducted in the absence of any commercial or financial relationships that could be construed as a potential conflict of interest.

References

1. George, B. Carotenoid research: History and new perspectives for chemistry in biological systems. *Biochim. Biophys. Acta Mol. Cell Biol. Lipids* **2020**, *1865*, 158699.
2. Ho, J.; Kish, E.; Méndez-Hernández, D.D.; WongCarter, K.; Pillai, S.; Kodis, G.; Niklas, J.; Poluektov, O.G.; Gust, D.; Moore, T.A.; et al. Triplet-triplet energy transfer in artificial and natural photosynthetic antennas. *Proc. Natl. Acad. Sci. USA* **2017**, *114*, 5513–5521. [CrossRef]
3. Cazzonelli, C.I. Carotenoids in nature: Insights from plants and beyond. *Funct. Plant Biol.* **2011**, *38*, 833–847. [CrossRef]
4. Palmer, A.C.; Healy, K.; Barffour, M.A.; Siamusantu, W.; Chileshe, J.; Schulze, K.J.; West, K.P.; Labrique, A.B. Provitamin A Carotenoid-Biofortified Maize Consumption Increases Pupillary Responsiveness among Zambian Children in a Randomized Controlled Trial. *J. Nutr.* **2016**, *146*, 2551–2558. [CrossRef]
5. Xiao, H.L.; Rong, B.Y.; Rong, L.; Zhen, X.H.; Cheng, C.H.; Zhong, H.Z.; Le, M. Association between Lutein and Zeaxanthin Status and the Risk of Cataract: A Meta-Analysis. *Nutrients* **2014**, *6*, 452–465. [CrossRef]
6. Chang, M.X.; Xiong, F. Astaxanthin and Its Effects in Inflammatory Responses and Inflammation-Associated Diseases: Recent Advances and Future Directions. *Molecules* **2020**, *25*, 5342. [CrossRef]
7. Alós, E.; Rodrigo, M.J.; Zacarias, L. Manipulation of Carotenoid Content in Plants to Improve Human Health. *Sub-Cell. Biochem.* **2016**, *79*, 311–343. [CrossRef]
8. Carl, E.S.; David, E.K.; Dean, A.K.; Scott, M.J. Genetic variation in carotenoid concentrations among diploid and amphidiploid rapid-cycling *Brassica* species. *HortScience* **2007**, *42*, 461–465. [CrossRef]
9. Cheng, L.; Charles, A.S.; Anthony, J.S. Modular engineering for microbial production of carotenoids. *Metab. Eng. Commun.* **2020**, *10*, e00118. [CrossRef]
10. Bramley, P.M. Regulation of carotenoid formation during tomato fruit ripening and development. *J. Exp. Bot.* **2002**, *53*, 2107–2113. [CrossRef]
11. Ruiz-Sola, M.Á.; Rodríguez-Concepción, M. Carotenoid Biosynthesis in *Arabidopsis*: A Colorful Pathway. *Arab. Book* **2012**, *10*, e0158. [CrossRef]

12. Barj, M.V.; Ezquerro, M.; Beretta, S.; Diretto, G.; FlorezSarasa, I.; Feixes, E.; Fiore, A.; Karlova, R.; Fernie, A.R.; Beekwilder, J.; et al. Several geranylgeranyl diphosphate synthase isoforms supply metabolic substrates for carotenoid biosynthesis in tomato. *New Phytol.* **2021**, *231*, 255–272. [CrossRef]
13. Hermanns, A.S.; Zhou, X.S.; Xu, Q.; Tadmor, Y.; Li, L. Carotenoid Pigment Accumulation in Horticultural Plants. *Hortic. Plant J.* **2020**, *6*, 343–360. [CrossRef]
14. Breitenbach, J.; Sandmann, G. ζ-Carotene *cis* isomers as products and substrates in the plant poly-*cis* carotenoid biosynthetic pathway to lycopene. *Planta* **2005**, *220*, 785–793. [CrossRef]
15. Francis, X.C. Regulation of carotenoid synthesis and accumulation in plants. *Pure Appl. Chem.* **2013**, *74*, 1409–1417. [CrossRef]
16. Gupta, P.; Hirschberg, J. The Genetic Components of a Natural Color Palette: A Comprehensive List of Carotenoid Pathway Mutations in Plants. *Front. Plant Sci.* **2022**, *12*, 806184. [CrossRef]
17. Luan, Y.T.; Fu, X.M.; Lu, P.J.; Grierson, D.; Xu, C.J. Molecular Mechanisms Determining the Differential Accumulation of Carotenoids in Plant Species and Varieties. *Crit. Rev. Plant Sci.* **2020**, *39*, 125–139. [CrossRef]
18. Demmig, A.B.; Gilmore, A.M.; Adams, W.W. Carotenoids 3: In vivo function of carotenoids in higher plants. *FASEB J.* **1996**, *10*, 403–412. [CrossRef]
19. Beatrycze, N.; Wojciech, S.; Kazimierz, S. New transgenic line of *Arabidopsis thaliana* with partly disabled zeaxanthin epoxidase activity displays changed carotenoid composition, xanthophyll cycle activity and non-photochemical quenching kinetics. *J. Plant Physiol.* **2009**, *166*, 1045–1056. [CrossRef]
20. Wei, Z.; Arazi, T.; Hod, N.; Zohar, M.; Isaacson, T.; Doron-Faigenboim, A.; Reznik, N.; Yedidia, I. Transcriptome Profiling of *Ornithogalum dubium* Leaves and Flowers to Identify Key Carotenoid Genes for CRISPR Gene Editing. *Plants* **2020**, *9*, 540. [CrossRef]
21. Zhao, J.; Li, J.; Zhang, J.; Chen, D.; Zhang, H.; Liu, C.; Qin, G. Genome-Wide Identification and Expression Analysis of the Carotenoid Cleavage Oxygenase Gene Family in Five *Rosaceae* Species. *Plant Mol. Biol. Rep.* **2021**, *39*, 739–751. [CrossRef]
22. Baldermann, S.; Kato, M.; Kurosawa, M.; Kurobayashi, Y.; Fujita, A.; Fleischmann, P.; Watanabe, N. Functional characterization of a carotenoid cleavage dioxygenase 1 and its relation to the carotenoid accumulation and volatile emission during the floral development of *Osmanthus fragrans* Lour. *J. Exp. Bot.* **2010**, *61*, 2967–2977. [CrossRef]
23. Song, H.X.; Lu, Q.; Hou, L.P.; Li, M.L. The genes crucial to carotenoid metabolism under elevated CO_2 levels in carrot (*Daucus carota* L.). *Sci. Rep.* **2021**, *11*, 12073. [CrossRef]
24. Zhang, N.; Ma, X.M.; Li, R.; Xue, Y.H.; Sun, Y.S.; Nie, S.S.; Zhang, L.G. Transcriptome-based analysis of carotenoid accumulation-related gene expression in petals of Chinese cabbage (*Brassica rapa* L.). *3 Biotech* **2019**, *9*, 274. [CrossRef]
25. Jung, H.J.; Manoharan, R.K.; Park, J.I.; Chung, M.Y.; Lee, J.; Lim, Y.P.; Hur, Y.; Nou, I.S. Identification of Yellow Pigmentation Genes in *Brassica rapa* ssp. *pekinensis* Using Br300 Microarray. *Int. J. Genom.* **2014**, *2014*, 204969. [CrossRef]
26. Etsuo, M.; Chika, Y.; Michio, O.; Motohisa, T. Linkage analysis of RFLP markers for clubroot resistance and pigmentation in Chinese cabbage (*Brassica rapa* ssp. *pekinensis*). *Euphytica* **1998**, *104*, 79–86. [CrossRef]
27. Su, T.B.; Yu, S.C.; Wang, J.; Zhang, F.L.; Yu, Y.J.; Zhang, D.S.; Zhao, X.Y.; Wang, W.H. Loss of Function of the Carotenoid Isomerase Gene BrCRTISO Confers Orange Color to the Inner Leaves of Chinese Cabbage (*Brassica rapa* L. ssp. *pekinensis*). *Plant Mol. Biol. Report.* **2015**, *33*, 648–659. [CrossRef]
28. Seohee, L.; Sangchoon, L.; Donghae, B.; Dongyoung, L.; Jeeyoung, P.; Jonghoon, L.; Hyunoh, L.; Sanghyun, S.; Taejin, Y. Association of molecular markers derived from the BrCRISTO1 gene with prolycopene-enriched orange-colored leaves in *Brassica rapa*. *Theor. Appl. Genet.* **2014**, *127*, 179–191. [CrossRef]
29. Zhang, N.; Chen, L.; Ma, S.; Wang, R.F.; He, Q.; Tian, M.; Zhang, L.G. Fine mapping and candidate gene analysis of the white flower gene *Brwf* in Chinese cabbage (*Brassica rapa* L.). *Sci. Rep.* **2020**, *10*, 6080. [CrossRef]
30. Yang, S.J.; Tian, X.X.; Wang, Z.Y.; Wei, X.C.; Zhao, Y.Y.; Su, H.N.; Zhao, X.B.; Tian, B.M.; Yuan, Y.X.; Zhang, X.W. Fine Mapping and Candidate Gene Identification of a White Flower Gene *BrWF3* in Chinese Cabbage (*Brassica rapa* L. ssp. *pekinensis*). *Front. Plant Sci.* **2021**, *12*, 646222. [CrossRef]
31. Li, P.; Lv, S.; Zhang, D.; Su, T.; Xin, X.; Wang, W.; Zhao, X.; Yu, Y.; Zhang, Y.; Yu, S.; et al. The Carotenoid Esterification Gene *BrPYP* Controls Pale-Yellow Petal Color in Flowering Chinese Cabbage (*Brassica rapa* L. subsp. *parachinensis*). *Front. Plant Sci.* **2022**, *13*, 844140. [CrossRef]
32. Li, H.L.; Yu, K.D.; Amoo, O.; Yu, Y.L.; Guo, M.X.; Deng, S.Y.; Li, M.T.; Hu, L.M.; Wang, J.Z.; Fan, C.C.; et al. Site-Directed Mutagenesis of the Carotenoid Isomerase Gene *BnaCRTISO* Alters the Color of Petals and Leaves in *Brassica napus* L. *Front. Plant Sci.* **2022**, *13*, 801456. [CrossRef]
33. Jiang, M.; Zhang, F.; Yuan, Q.; Lin, P.X.; Zheng, H.; Liang, S.; Jian, Y.; Miao, H.Y.; Li, H.X.; Wang, Q.M.; et al. Characterization of *BoaCRTISO* Reveals Its Role in Carotenoid Biosynthesis in Chinese Kale. *Front. Plant Sci.* **2021**, *12*, 662684. [CrossRef] [PubMed]
34. Bo, S.; Min, J.; Hao, Z.; Yue, J.; Wen, L.H.; Qiao, Y.; Ai, H.Z.; Qing, C.; Yun, T.Z.; Yuan, X.L.; et al. Color-related chlorophyll and carotenoid concentrations of Chinese kale can be altered through CRISPR/Cas9 targeted editing of the carotenoid isomerase gene *BoaCRTISO*. *Hortic. Res.* **2020**, *7*, 94–102. [CrossRef]
35. Yang, S.J.; Liu, H.L.; Zhao, Y.Y.; Su, H.N.; Wei, X.C.; Wang, Z.Y.; Zhao, X.B.; Zhang, X.W.; Yuan, Y.X. Map-Based Cloning and Characterization of Br-dyp1, a Gene Conferring Dark Yellow Petal Color Trait in Chinese Cabbage (*Brassica rapa* L. ssp. *pekinensis*). *Front. Plant Sci.* **2022**, *13*, 841328. [CrossRef]

36. Liu, Y.J.; Ye, S.H.; Yuan, G.G.; Ma, X.W.; Heng, S.P.; Yi, B.; Ma, C.Z.; Shen, J.X.; Tu, J.X.; Fu, T.D.; et al. Gene silencing of BnaA09.ZEP and BnaC09.ZEP confers orange color in *Brassica napus* flowers. *Plant J.* **2020**, *104*, 932–949. [CrossRef]
37. Gonzalez, J.S.; Ha, S.H.; Magallanes, L.M.; Gilliland, L.U.; Zhou, A.; Lipka, A.E.; Nguyen, Y.N.; Angelovici, R.; Lin, H.N.; Cepela, J.; et al. Carotenoid cleavage dioxygenase4 is a negative regulator of β-carotene content in *Arabidopsis* seeds. *Plant Cell* **2013**, *25*, 4812–4826. [CrossRef]
38. Zhang, B.; Liu, C.; Wang, Y.Q.; Yao, X.; Wang, F.; Wu, J.S.; King, G.J.; Liu, K. Disruption of a CAROTENOID CLEAVAGE DIOXYGENASE 4 gene converts flower colour from white to yellow in *Brassica* species. *New Phytol.* **2015**, *206*, 1513–1526. [CrossRef]
39. Han, F.; Cui, H.; Zhang, B.; Liu, X.; Yang, L.; Zhuang, M.; Lv, H.; Li, Z.; Wang, Y.; Fang, Z.; et al. Map-based cloning and characterization of BoCCD4, a gene responsible for white/yellow petal color in *B. oleracea*. *BMC Genom.* **2019**, *20*, 242. [CrossRef]
40. Yan, C.; Huang, Y.; Liu, Z.; Guo, F.; Jiao, Z.; Yang, W.; Zhu, F.; Qiu, Z. Rapid identification of yellow-flowered gene Bofc in cauliflower (*Brassica oleracea* var. *botrytis*) by bulked segregant analysis and whole-genome resequencing. *Euphytica* **2020**, *216*, 1348–1351. [CrossRef]
41. Zhang, B.; Wang, J.; Chen, L.; Ren, W.J.; Han, F.Q.; Fang, Z.Y.; Yang, L.M.; Zhuang, M.; Lv, H.H.; Wang, Y.; et al. Transcriptome Analysis Reveals Key Genes and Pathways Associated with the Petal Color Formation in Cabbage (*Brassica oleracea* L. var. *capitata*). *Int. J. Mol. Sci.* **2022**, *23*, 6656. [CrossRef]
42. Lu, S.; Van, E.J.; Zhou, X.J.; Lopez, A.B.; O'Halloran, D.M.; Cosman, K.M.; Conlin, B.J.; Paolillo, D.J.; Garvin, D.F.; Vrebalov, J.; et al. The cauliflower Or gene encodes a DnaJ cysteine-rich domain-containing protein that mediates high levels of beta-carotene accumulation. *Plant Cell* **2006**, *18*, 3594–3605. [CrossRef]
43. Zhang, L.; Zhang, S.F.; Dai, Y.; Wang, S.X.; Wang, C.G.; Li, F.; Zhang, H.; Chen, G.H.; Yuan, L.Y.; Hou, J.F.; et al. Mapping and Validation of *BrGOLDEN*: A Dominant Gene Regulating Carotenoid Accumulation in *Brassica rapa*. *Int. J. Mol. Sci.* **2022**, *23*, 12442. [CrossRef]
44. Yuto, O.; Miho, T.; Atsushi, S.; Norihiko, M.; Hiroshi, S. Orange protein, phytoene synthase regulator, has protein disulfide reductase activity. *Plant Signal. Behav.* **2022**, *17*, 2072094. [CrossRef]
45. Zhao, C.J.; Bin, S.L.; Xie, M.L.; Shi, M.J.; Dong, Z.X.; Yang, L.; Cheng, X.H.; Liu, Y.Y.; Bai, Z.T.; Xiang, Y.; et al. Mutation of the PHYTOENE DESATURASE 3 gene causes yellowish-white petals in *Brassica napus*. *Crop J.* **2021**, *9*, 1124–1134. [CrossRef]
46. Jian, Y.; Zhang, C.L.; Wang, Y.T.; Li, Z.Q.; Chen, J.; Zhou, W.T.; Huang, W.L.; Jiang, M.; Zheng, H.; Li, M.Y.; et al. Characterization of the Role of the Neoxanthin Synthase Gene *BoaNXS* in Carotenoid Biosynthesis in Chinese Kale. *Genes* **2021**, *12*, 1122. [CrossRef]
47. Elvis, K.; Daniela, Q.M.; Elizabeth, I.K.; Paula, V.T.; Annaliese, S.M. Interspecific Hybridization for *Brassica* Crop Improvement. *Crop Breed. Genet. Genom.* **2019**, *1*, e190007. [CrossRef]
48. Wang, X.W.; Wang, H.Z.; Wang, J.; Sun, R.F.; Wu, J.; Liu, S.Y.; Bai, Y.Q.; Mun, J.H.; Bancroft, I.; Cheng, F.; et al. The genome of the mesopolyploid crop species *Brassica rapa*. *Nat. Genet.* **2011**, *43*, 1035–1039. [CrossRef]
49. Raheeba, T.N.; Shaheen, K.J.; Farooq, A.B.; Tariq, R.R.; Asha, N.; Altaf, A.W.; Rehana, A.; Nahida, A.; Moneesa, B.; Ishfaq, M.S.; et al. Review on "Crispr-CAS9—A Genome editing tools for plant disease management". *Plant Cell Biotechnol. Mol. Biol.* **2022**, *23*, 1–14.
50. Fu, H.; Chao, H.B.; Zhao, X.J.; Wang, H.Y.; Li, H.X.; Zhao, W.G.; Sun, T.; Li, M.T.; Huang, J.Y. Anthocyanins identification and transcriptional regulation of anthocyanin biosynthesis in purple *Brassica napus*. *Plant Mol. Biol.* **2022**, *110*, 53–68. [CrossRef]
51. Milani, A.; Basirnejad, M.; Shahbazi, S.; Bolhassani, A. Carotenoids: Biochemistry, pharmacology and treatment. *Br. J. Pharmacol.* **2017**, *174*, 1290–1324. [CrossRef]
52. Johnson, E.J. A possible role for lutein and zeaxanthin in cognitive function in the elderly. *Am. J. Clin. Nutr.* **2012**, *96*, 1161S–1165S. [CrossRef]
53. Cao, W.X.; Wang, P.; Yang, L.M.; Fang, Z.Y.; Zhang, Y.Y.; Zhuang, M.; Lv, H.H.; Wang, Y.; Ji, J.L. Carotenoid Biosynthetic Genes in Cabbage: Genome-Wide Identification, Evolution, and Expression Analysis. *Genes* **2021**, *12*, 2027. [CrossRef]
54. Pham, A.T.; Jae, K.K.; Jeongyeo, L.; Woo, T.P.; Do, Y.K.; Yeon, B.K.; Haeng, H.K.; Hye, R.K. Analysis of carotenoid accumulation and expression of carotenoid biosynthesis genes in different organs of Chinese cabbage (*Brassica rapa* subsp. *pekinensis*). *EXCLI J.* **2012**, *11*, 508–516.

plants

MDPI

Article

An Efficient *Agrobacterium*-Mediated Genetic Transformation Method for *Solanum betaceum* Cav. Embryogenic Callus

Daniela Cordeiro [1,*], Ana Alves [2], Ricardo Ferraz [1], Bruno Casimiro [1], Jorge Canhoto [1] and Sandra Correia [1,3,*]

1 Centre for Functional Ecology, TERRA Associate Laboratory, Department of Life Sciences, University of Coimbra, Calçada Martim de Freitas, 3000-456 Coimbra, Portugal
2 BioISI—Biosystems & Integrative Sciences Institute, Faculty of Sciences, University of Lisbon, Campo Grande, 1749-016 Lisbon, Portugal
3 InnovPlantProtect CoLab, Estrada de Gil Vaz, 7350-478 Elvas, Portugal
* Correspondence: danielacordeiro@outlook.pt (D.C.); sandraimc@uc.pt (S.C.)

Abstract: Somatic embryogenesis in *Solanum betaceum* (tamarillo) has proven to be an effective model system for studying morphogenesis, since optimized plant regeneration protocols are available, and embryogenic competent cell lines can be induced from different explants. Nevertheless, an efficient genetic transformation system for embryogenic callus (EC) has not yet been implemented for this species. Here, an optimized faster protocol of genetic transformation using *Agrobacterium tumefaciens* is described for EC. The sensitivity of EC to three antibiotics was determined, and kanamycin proved to be the best selective agent for tamarillo callus. Two *Agrobacterium* strains, EHA105 and LBA4404, both harboring the p35SGUSINT plasmid, carrying the reporter gene for β-glucuronidase (*gus*) and the marker gene neomycin phosphotransferase (*nptII*), were used to test the efficiency of the process. To increase the success of the genetic transformation, a cold-shock treatment, coconut water, polyvinylpyrrolidone and an appropriate selection schedule based on antibiotic resistance were employed. The genetic transformation was evaluated by GUS assay and PCR-based techniques, and a 100% efficiency rate was confirmed in the kanamycin-resistant EC clumps. Genetic transformation with the EHA105 strain resulted in higher values for *gus* insertion in the genome. The protocol presented provides a useful tool for functional gene analysis and biotechnology approaches.

Keywords: antibiotic resistance; functional genomics; plant cell culture; somatic embryogenesis; tree tomato

Citation: Cordeiro, D.; Alves, A.; Ferraz, R.; Casimiro, B.; Canhoto, J.; Correia, S. An Efficient *Agrobacterium*-Mediated Genetic Transformation Method for *Solanum betaceum* Cav. Embryogenic Callus. *Plants* **2023**, *12*, 1202. https://doi.org/10.3390/plants12051202

Academic Editors: Zanmin Hu, Han Xiao, Yi Ren and Chengming Fan

Received: 17 January 2023
Revised: 2 March 2023
Accepted: 3 March 2023
Published: 6 March 2023

1. Introduction

Besides the application to obtain cultivars with new characteristics, genetic engineering is also a fundamental tool to discover gene function. Nowadays, several approaches can be applied to modulate gene expression, either to overexpress or to knock down, such as RNA interference and CRISPR-Cas approaches [1–3]. Though, an efficient transformation method is an indispensable tool for gene functional analysis. *Agrobacterium*-mediated genetic transformation is the most widely used biotechnological method in plant gene function analysis and crop improvement, mainly in species with no available genome sequencing data [1,4,5]. Indeed, compared with particle bombardment, the *Agrobacterium*-mediated method gives more advantages for genetic transformation, such as lower copy number, less complex insertion sites, more stable transgene expression and the chance to segregate away marker genes [6]. Through this method, the transgene is introduced into plant cells and, in the case of a stable expression, integrated into the nuclear genome [7]. Transformed cells are further selected and induced to differentiate into shoot meristems or somatic embryos. *Agrobacterium*-mediated methods are low-cost procedures that allow for constitutive or tissue-specific expression [8]. Nevertheless, transformation efficiency varies depending on the species, genotype, explant type, *Agrobacterium* strain used, co-cultivation period and cells' sensitivity to antibiotics [9].

Solanum betaceum Cav. is an Andean fruit tree commonly known as tamarillo or tree tomato. Due to the highly nutritious fruits with a distinctive sweet and tart flavor, the economic importance of this species has been increasing [10,11]. Furthermore, the discovery of several beneficial health effects of its fruits is rescuing this species from the so-called orphan species to a true fruit crop [12]. In vitro, this species is easily manipulated and regenerated, and a robust set of information with optimized protocols is currently available [13–18]. Somatic embryogenesis (SE) and further plant regeneration have been efficiently achieved, making this species a good model system for different studies in fundamental plant biology research, such as experimental embryology and cell reprogramming.

In *S. betaceum*, embryogenic (EC) and non-embryogenic (NEC) callus can be obtained from several explants through induction in the presence of exogenous auxin and high sucrose levels [19]. These cultures can be maintained for several subcultures and be used in functional genomic analysis. Indeed, embryogenic cultures have been used as target material for *Agrobacterium*-mediated genetic transformation in different species [20–23]. Using EC as a source of explants for transformation is rapid, easily scalable and less laborious and results in an increased number of transformation events compared with transformation via SE induction [22]. Moreover, Ratjens and colleagues [23] reported the stability of the regenerated plants as an advantage of using EC as the target material.

Successful *Agrobacterium*-mediated transformation has already been reported in Solanaceae species, such as *Datura stramonium* [24], *Nicotiana glauca* [25], *N. tabacum* [26] and *Physalis pruinosa* [27], and more specifically within the genus *Solanum*: *S. chrysotrichum* [28], *S. demissum* [29], *S. dulcamara* [30], *S. hjertingii* [29], *S. lycopersicum* [31], *S. melongena* [32], *S. muricatum* [33], *S. nigrum* [34], *S. papita* [29], *S. phureja* [35], *S. stoloniferum* [29], *S. torvum* [36], *S. trilobatum* [37], *S. tuberosum* [38] and *S. verrucosum* [29]. In *S. betaceum*, the *Agrobacterium*-mediated transformation method was used to obtain transgenic plants regenerated via organogenesis [39,40], and more recently, it was applied in the functional analysis of an rRNA methyltransferase, in which the transgenic plants were regenerated through SE [41]. However, these works revealed low transformation efficiencies and lengthy processes due to the need to obtain callus before regeneration. Thus, the main objective of this work was to develop a reproducible and efficient *Agrobacterium*-mediated transformation protocol for *S. betaceum* using EC as the target explant, enabling further gene function analysis in a rapid and stable way. For this, and aiming to obtain maximum efficiency, several conditions were tested, including the antibiotics resistance, *Agrobacterium* strain, OD, pre-treatment inclusion, co-cultivation period and the use of acetosyringone, glutamine, polyvinylpyrrolidone (PVP) and coconut water.

Several studies have reported different transformation efficiencies depending on the selective agent used. For instance, Aida and colleagues [42] verified higher transformation efficiencies using hygromycin than with kanamycin as the selective agent of transformed cells. Therefore, we tested the effect of three antibiotics on the tamarillo callus. Then, to test the effectiveness of the genetic transformation in this material, two *Agrobacterium tumefaciens* strains, EHA105 [43] and LBA4404 [44], were used. Considered hypervirulent strains, they are part of the most used in plant transformation [2]. Both strains carried the binary vector p35SGUSINT [45], containing the selection marker gene *neomycin phosphotransferase* (*nptII*) and the reporter gene *gus*. *nptII* is the most used selectable marker gene for plant transformation [46]. This gene confers resistance to aminoglycoside antibiotics, such as kanamycin and paromomycin, which grants a negative selection with growth inhibition and death of non-transformed cells by ribosome activity block and protein synthesis inhibition [2]. In p35SGUSINT, *nptII* is under the control of the *T-nos* transcriptional terminator sequence from the *nopaline synthase* gene of *A. tumefaciens*, which is the most successfully used in plants [2]. The widely employed reporter gene *gus* in genetic transformation optimizations was selected since it enables an easy and rapid test to confirm genetic transformation [23,36,47,48].

The protocol presented provides a useful tool for functional gene analysis and represents a biotechnology approach for genetic improvement that can also be applied to other important crop species.

2. Results and Discussion

2.1. Effect of Antibacterial Antibiotics on *Agrobacterium*

As broad-spectrum antibiotics used in plant tissue culture, carbenicillin and cefotaxime were selected to eliminate any remaining *Agrobacterium* after the transformation procedure. Several concentrations of these antibiotics were used to test their inhibitory effect on *Agrobacterium* growth and ensure that the proper antibiotics concentration would be used. From the screening, bacteria growth was only observed in the control treatment (no antibiotics) and in the presence of 50 mg/L carbenicillin or cefotaxime (Table 1). Concentrations higher than 100 mg/L of these antibiotics have shown an effective antibacterial effect in bacteria growing on the surface of the medium. Nevertheless, in callus genetic transformation, bacteria will grow not only on the surface of the medium, but also in the inner parts and recesses of the cell clusters, and therefore, a higher antibiotics concentration could be needed. This requires that a balance in the concentration of antibiotics is achieved to assure that bacteria do not grow, but also that callus proliferation is not affected [49]. In previous tamarillo genetic transformation works, 300 mg/L of cefotaxime [39] and a combination of 250 mg/L each of carbenicillin and cefotaxime [41] were reported to inhibit bacteria growth. Even though much lower concentrations of these antibiotics were effective against *Agrobacterium* (Table 1), the effect of 250 mg/L cefotaxime or carbenicillin alone or in combination at 200 or 250 mg/L on EC proliferation was evaluated, so that more efficient bacteria elimination could be achieved. After one month in the presence of these antibiotics, EC proliferated normally, with no effects observed when compared to the control (see Supplementary Materials, Figure S1). Therefore, carbenicillin and cefotaxime at a 200 mg/L concentration were used in the transformation assay.

Table 1. Antibacterial antibiotic effect screening on *Agrobacterium*.

Antibiotic Concentrations (mg/L)		Antibacterial Effect
Carbenicillin	**Cefotaxime**	
0	0	•
50	0	•
0	50	•
50	50	○
100	0	○
0	100	○
100	100	○
200	0	○
0	200	○
200	200	○
250	250	○
300	300	○

•—means no antibacterial effect, the plate was opaque; ○—means a complete inhibitory effect, the plate was transparent.

2.2. Determination of the Most Selective Antibiotic

The main goal of this work was to establish an effective *Agrobacterium*-mediated transformation protocol for tamarillo EC that could be applied when using different transformation vectors with different selection marker genes, enabling further gene functional analysis. Among the most used selection marker genes for plant transformation are the *neomycin phosphotransferase II* gene (*nptII*) and the *hygromycin-B-phosphotransferase* gene (*hph*) [1]. While *nptII* confers resistance to aminoglycoside antibiotics, such as kanamycin and paromomycin, *hph* confers resistance to hygromycin. Although kanamycin has been the most frequently used selection agent, paromomycin and hygromycin have shown higher effectiveness compared with kanamycin [42,50].

Thus, to determine the most effective antibiotic and the concentration needed for the effective growth inhibition of non-transformed cells, a sensitivity screening assay was performed before *Agrobacterium* transformation. The tolerance of tamarillo callus to increasing concentrations of kanamycin, paromomycin and hygromycin was tested by the incubation of cells in a medium supplemented with these antibiotics. Proliferation in the antibiotic-free control medium resulted in a threefold fresh mass growth after 30 days of culture (Figure 1). For all conditions tested, cell proliferation occurred, with no treatment showing a complete inhibitory effect. In the kanamycin assay, as the antibiotic concentration increased, the callus proliferation rate gradually decreased, with a significant difference ($p \leq 0.0001$) for 100 and 125 mg/L concentrations. These results are similar to the ones reported for Chinese chestnut (*Castanea mollissima*), in which the EC proliferation rate decreased abruptly when incubated in kanamycin concentrations higher than 90 mg/L [51]. Treatments with paromomycin revealed a high tolerance of tamarillo callus to this antibiotic at the concentrations tested, since compared to the control, no significant differences were found in the callus proliferation rate. This antibiotic was chosen to be tested due to its high effectiveness reported in the selection of *Vitis* embryogenic suspension-cultured cells [50]; however, tamarillo cells were almost insensitive to this range of paromomycin concentrations. Therefore, before the use of paromomycin as the selection agent, higher concentrations should be tested. By contrast, for hygromycin, callus showed a reduced proliferation rate for all concentrations tested, with an emphasis on 20 mg/L. Thus, kanamycin and hygromycin showed to be effective selection agents for tamarillo callus transformation assays.

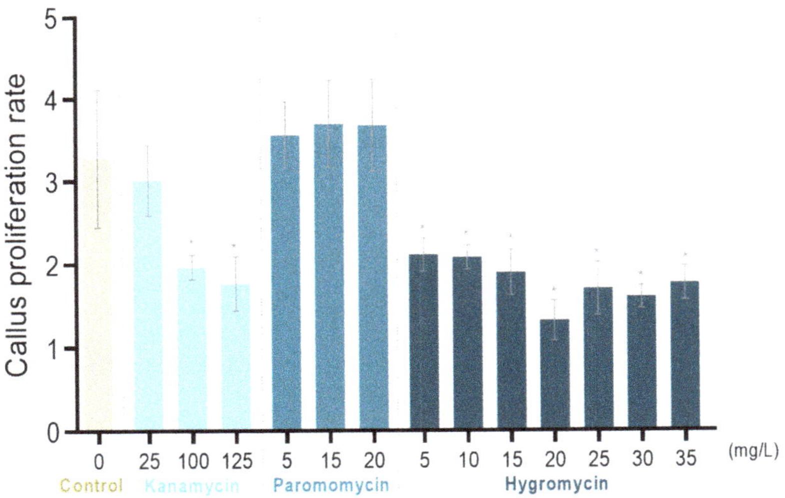

Figure 1. Effect of kanamycin, paromomycin and hygromycin on the proliferation rate of tamarillo callus. Callus proliferation rate is the ratio of fresh weight after thirty days of treatment and the initial one ± SD of at least three biological replicates. * means significant difference (p-value ≤ 0.0001) between the treatment and the control, according to Dunnett's test.

2.3. Agrobacterium-Mediated Tamarillo Callus Transformation

SE-induced EC in this species is usually a cell cluster forming a compact tissue (Figure 2). The difficulty of reaching the inner cells of the clusters by the bacteria compromises the effectiveness of the transformation. Thus, two considered hypervirulent strains, EHA105 and LBA4404 [2], were tested, and their efficiency was compared. Bacteria harboring the p35SGUSINT plasmid were chosen to step up the protocol, since the presence of the *gus* gene allows a rapid verification of the transformation, through a histochemical assay with a small amount of material, and kanamycin could be used as the selection agent.

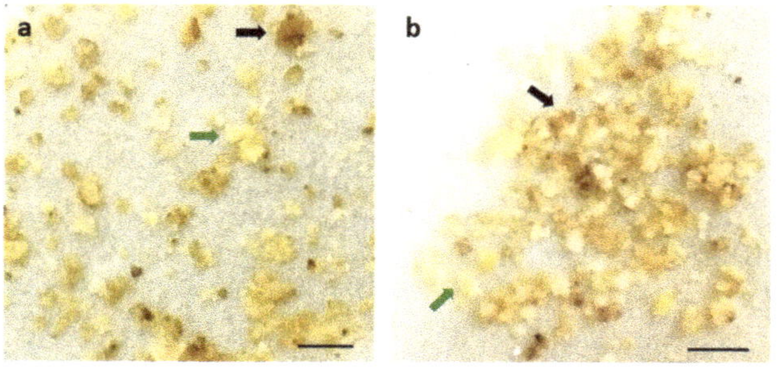

Figure 2. Tamarillo embryogenic callus inoculated with EHA105 (**a**) and LBA4404 (**b**) *Agrobacterium* strains, in selective media for 16 weeks and subcultured in increasing antibiotic concentration every 4 weeks. The black arrows show kanamycin-non-resistant cells, while the green arrows show kanamycin-resistant cells. Bars represent 0.5 cm.

Firstly, *Agrobacterium* culture was prepared in a selective medium with rifampicin and kanamycin, only enabling the growth of the bacteria containing the plasmid. Cultures reaching an OD = 0.8 were selected, as this OD has allowed the highest transformation efficiency in other species when using EC for transformation [51,52], and because it is the one recommended for other Solanaceae species [26,46]. Moreover, lower densities could be ineffective, while higher densities might cause callus oxidation due to *Agrobacterium* overgrowth, as reported by Ma and colleagues [52].

It is noteworthy that the physiological status of the callus influences the transformation efficiency. Therefore, actively growing whitish compact calluses were used for this assay. Even though, EC is very sensitive and tends to oxidize with manipulation (see Supplementary Materials, Figure S2a). In an attempt to reduce callus browning, a cold shock treatment with callus immersion in a 3% (w/v) maltose solution for 20 min was applied before infection, as was reported for *Lolium perenne* [53].

In fact, one of the main obstacles to the application of this protocol was tissue browning, most likely due to oxidation. To reduce browning and subsequently improve the survival rate of cells, PVP and coconut water were included during infection and co-cultivation and in the proliferation medium. With good results demonstrated in rice and sorghum [49,54], the presence of these compounds resulted in infected EC with a healthier appearance and lower necrotic response (Figure 2). PVP and coconut water might reduce the damage of explants by *Agrobacterium* and the phenolic production by the cells, as explained by Priya and colleagues [49]. Together, the pre-treatment and the use of glutamine, PVP and coconut water effectively reduced further callus browning (see Supplementary Materials, Figure S2b).

Infection was performed by a 20 min callus immersion in bacteria. Such an incubation period was also reported as the best inoculation period for tobacco and tomato [26,47]. The use of acetosyringone during infection has also been reported in the vast majority of *Agrobacterium*-mediated protocols. This phenolic compound is released by plant cells during natural *Agrobacterium* infection, activating its virulence genes. Therefore, acetosyringone addition promotes high-efficiency transformation [55]. Together, glutamine was also added to further enhance transformation [48,53]. In addition, vacuum infiltration was also performed to improve the penetration of bacteria into the compact callus cluster (Figure 3a).

To minimize the loss of cells during the infection process, the callus was directly immersed into the *Agrobacterium* culture, and cell strainers were used to collect cells before transfer to co-cultivation. Nevertheless, since a full cell mass recovery is very difficult, starting with a large amount of material, such as 800 mg, is recommended.

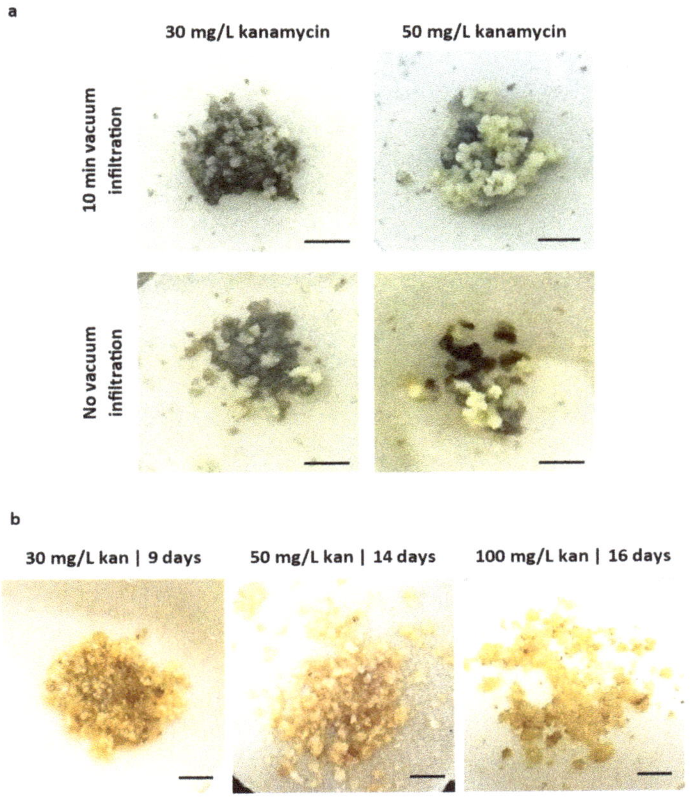

Figure 3. Tamarillo embryogenic callus inoculated with EHA105 *Agrobacterium* strain and incubated with increasing kanamycin concentrations. (**a**) Images represent inoculation with and without a 10 min vacuum, followed by incubation at 80 rpm and 28 °C for another 10 min. (**b**) Incubation in the selective medium with kanamycin at 30 mg/L for 9 days, 50 mg/L for 14 days and 100 mg/L for 16 days. Bars represent 0.5 cm.

Co-cultivation of EC with *Agrobacterium* was performed on filter paper thoroughly soaked in liquid MS containing Gln, PVP and CW. This strategy was preferred since it was reported to result in a higher transformation efficiency and reduces culture browning [20,56]. Co-cultivation took 3 days. Longer periods made bacteria removal impossible afterwards and caused callus browning (see Supplementary Materials, Figure S3).

A high concentration of carbenicillin and cefotaxime (500 mg/L) was used to wash the infected cells and eliminate *Agrobacterium*. A vacuum filtration system was also used to better wash the cells, since a considerable amount of bacteria lodges in callus protuberances. After the washes, incubation for 3 days with no selective agent followed by a gradual increase of selection was performed to avoid escapes while decreasing tissue oxidation.

Although the plasmid used for tamarillo transformation (p35SGUSINT) confers resistance to aminoglycoside antibiotics, such as kanamycin and paromomycin, kanamycin was used for transformed cell selection, since tamarillo callus showed no sensitivity to the paromomycin concentrations tested. During the selection process, an accurate evaluation of subculture frequency is important for effectiveness. As tamarillo EC is sensitive to changes in the medium, less than three weeks is a short period for cells to adapt to the new conditions. However, culture periods longer than four weeks favor tissue browning.

Therefore, transformed cells were subcultured at four-week intervals (Figure 3b). During selection, no bacterial growth was observed.

2.4. Detection of Genetically Transformed Cells

The genetic transformation was first verified by the histochemical GUS assay. Kanamycin-resistant callus was subjected to staining with X-gluc, a substrate cleaved by β-glucuronidase (encoded by the *gus* gene and present in p35SGUSINT). Positive transformants showed typical indigo-blue coloration, proving the efficiency of the protocol, while non-transformed cells did not show any blue color (Figure 4).

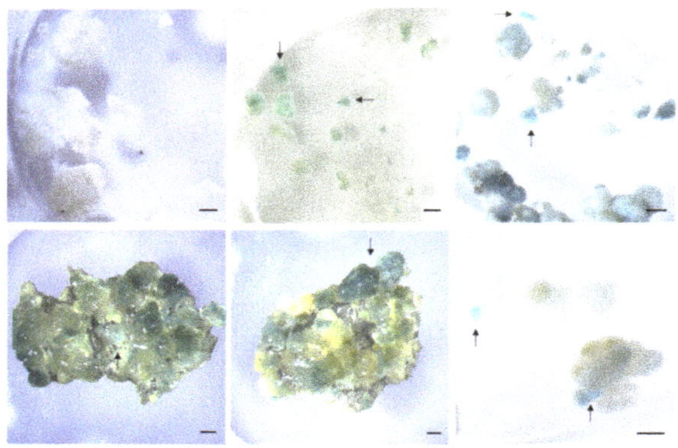

Figure 4. GUS assay in *Agrobacterium*-mediated genetic transformation of tamarillo embryogenic callus (EC). The negative control is non-transformed EC (first image). Arrows indicate indigo-blue coloration from *GUS* expression. Bars represent 1 mm.

Then, to characterize the transformation event, molecular analyses were performed in genomic DNA from 4 biological replicates of kanamycin-resistant callus, which were grown in selective media for 16 weeks and subcultured in increasing antibiotic concentrations every 4 weeks. First, PCR analysis of *gus* fragments was conducted to confirm the presence of the transgene in the tamarillo genome. The results showed amplification of a 636 bp fragment in all samples corresponding to the fragment amplified in the plasmid (Figure 5), while no amplification occurred in non-transformed cells.

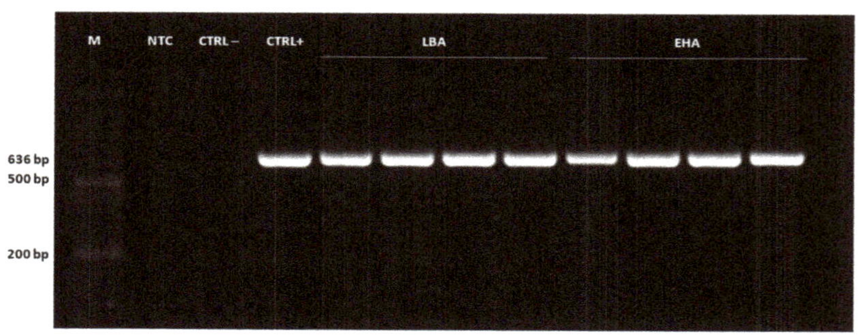

Figure 5. Confirmation of the presence of the transgene in gDNA from tamarillo embryogenic samples by PCR amplification of *gus* gene fragment (636 bp). M = NZYDNA Ladder VI (NZYtech, Lisboa, Portugal) ranging from 50 to 1500 bp, NTC = no template control (H_2O), CTRL− = non-transformed sample, CTRL+ = plasmid (p35SGUSINT), lanes 5–8 = samples transformed with LBA4404 strain, lanes 9–12 = samples transformed with EHA105 strain.

Secondly, qPCR was conducted to confirm the transgene integration. This rapid and high throughput approach requires a lower amount of DNA and is a valid alternative to the conventional, highly laborious and time-consuming southern blot analysis in transgenic woody plants [57,58]. Indeed, it has been applied in several plant genetic transformation works [59–61]. In gene functional analysis, a RT-qPCR could be applied to quantify the expression of the gene of interest and, in this way, also confirm the gene insertion [52].

qPCR analysis showed the transgene integration quantification relative to the *ACT* gene (Figure 6). Although both strains showed an effective tamarillo EC transformation, strain EHA105 showed significantly higher values of *gus* insertion compared with LBA4404. While transformation with LBA4404 resulted in an insertion of 2.04×10^4 relative to ACT, with EHA105, it reached 7.27×10^4. Indeed, van Eck and colleagues [31] noticed that transformation with LBA4404 resulted in a low copy number of the introduced transgene compared to other *Agrobacterium* strains. Further, EHA105 was reported as the strain allowing high transformation efficiencies in banana, when compared with EHA101 and LBA4404 [62]. For tobacco, LBA4404 was reported as the most suitable strain for transformation, but when compared with AGL1 and GV3101 [26]. These results can also relate to the low efficiencies previously reported for other tamarillo explants [39–41], in which *Agrobacterium* LBA4404 was used.

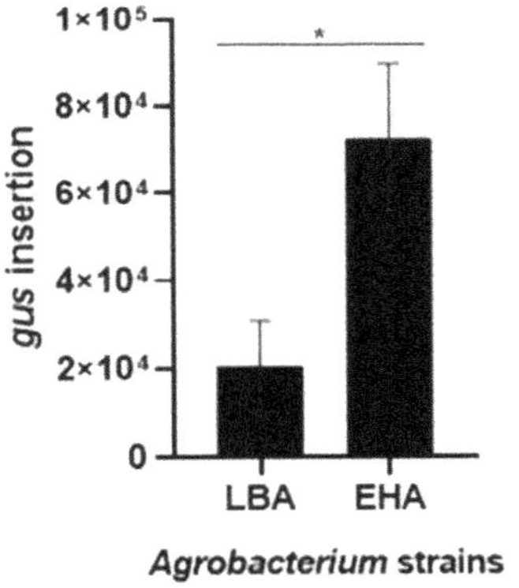

Figure 6. Quantification by qPCR of the *gus* insertion in tamarillo embryogenic callus genetically transformed by the *Agrobacterium*-mediated method using LBA4404 (LBA) and EHA105 (EHA) strains. Data were normalized using *ACT* as the reference gene. Results are presented as the mean ± SEM of four biological replicates. * indicates significant differences by *t*-test at $p < 0.05$.

Along with the GUS assay results, this confirms that tamarillo EC is amenable to *Agrobacterium*-mediated genetic transformation and that the *gus* gene was effectively inserted in the genome. Moreover, the results prove that both strains are effective for tamarillo EC transformation, in line with what was already reported for other Solanaceae [26,31–38].

3. Materials and Methods

3.1. Induction and Proliferation of Embryogenic Callus Cultures

Embryogenic callus (EC) was induced by somatic embryogenesis from tamarillo leaves, obtained from in vitro proliferating shoots, following the methodology previously described in detail [19]. EC proliferation was carried out in Murashige and Skoog (MS) medium [63] (© Duchefa Biochemie, Haarlem, The Netherlands) supplemented with 20 µM Picloram (Sigma-Aldrich®, St. Louis, MO, USA) plus 9% (*w*/*v*) sucrose and gelified with 0.25% (*w*/*v*) Phytagel™ (Sigma-Aldrich®) at pH 5.7 (hereafter designated proliferation medium). Cultures were maintained at 24 ± 1 °C under dark conditions and monthly subcultured.

3.2. *Agrobacterium* Strains and Vector

The transformation protocol was performed using EHA105 [43] and LBA4404 [44] *Agrobacterium tumefaciens* strains, containing a rifampicin resistance gene. Both strains harbor the p35SGUSINT vector [44]. This plasmid carries the selectable marker gene *nptII* for plant selection (conferring kanamycin resistance) under the control of the *nos* promoter and terminator, and the *gusA* reporter gene with a plant intron under the control of the *CaMV35S* promoter, located near the left border.

3.3. *Effect of Antibacterial Antibiotics on Agrobacterium Growth*

The inhibitory effect of antibiotics on *Agrobacterium* growth was tested. A total of 20 µL of bacterial culture (optical density (OD) = 1) were inoculated into plates containing LB (Miller) medium [64] with carbenicillin and cefotaxime alone or in combination at concentrations of 50, 100, 200, 250 and 300 mg/L. Three plates for each treatment were incubated at 28 °C in the dark for 5 days. As a control, *Agrobacterium* was grown in LB medium without antibiotics.

3.4. *Sensitivity Screening of Embryogenic Callus to Antibiotics*

To identify the most suitable antibiotic and its concentrations for the selection of transgenic cells, the tolerance of the tamarillo EC to kanamycin, paromomycin and hygromycin was tested. Different ranges of antibiotic concentrations were evaluated according to the ones described in the literature for several woody and/or Solanaceae species. Accordingly, three clumps of 100 mg of non-transformed EC were cultured on a plate with proliferation medium supplemented with different concentrations of kanamycin (25, 50, 100 and 125 mg/L) [51,65–67], hygromycin (5, 10, 15, 20, 25, 30 and 35 mg/L) [68,69] and paromomycin (5, 10, 15 and 20 mg/L) [50]. Callus growth in the proliferation medium without antibiotics was used as a control. Two plates were used for each treatment (six biological replicates in total). However, due to culture contamination, some data (biological replicates and some antibiotic concentrations) could not be included. Antibiotics were dissolved in sterile deionized water, filter-sterilized (0.22 µm) and added to the autoclaved medium. After incubation at 24 ± 1 °C in the dark for 30 days, the EC final fresh weight was measured. Data were expressed as the callus proliferation rate (ratio between final and initial weight) with ± SD. One-way ANOVA was performed ($p < 0.05$), and each treatment was compared to the control by a Dunnett's multiple comparisons test at $p < 0.05$.

3.5. *Agrobacterium Culture Preparation*

Frozen *Agrobacterium* glycerol stock cultures from both stains were thawed, and 20 µL were grown in solid LB medium supplemented with 50 mg/L rifampicin and 100 mg/L kanamycin at 28 °C in darkness for 2–3 days (Figure 7a). To isolate single colonies, a portion of the culture was streaked out onto a new plate and incubated overnight in the same conditions. A single colony was transferred to a 50 mL centrifuge tube containing 5 mL of liquid LB medium supplemented with 50 mg/L rifampicin and 100 mg/L kanamycin. Bacterial cultures were incubated on a shaker at 220 rpm and 28 °C until the OD at 600 nm reached 0.8 (16–21 h). *Agrobacterium* was centrifugated at 4200× *g* for 8 min, washed with MS plus 3% (*w*/*v*) sucrose and centrifugated again. The bacterial pellet was finally resuspended in 4 mL MS plus 3% (*w*/*v*) sucrose supplemented with 100 µM acetosyringone. Bacterial cultures were incubated on a shaker at 80 rpm and 28 °C for another 4 h before infection.

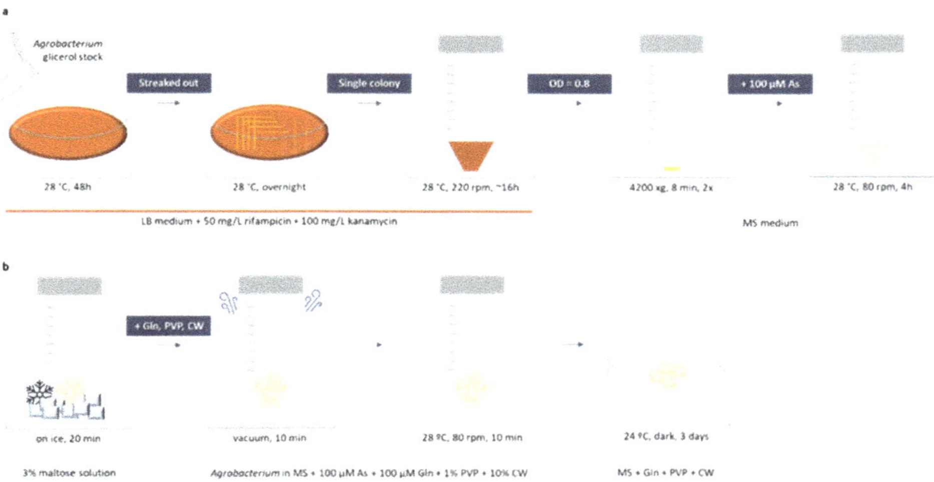

Figure 7. Detailed workflow for *Agrobacterium*-mediated genetic transformation of embryogenic callus from *Solanum betaceum*. (**a**) *Agrobacterium* culture preparation for genetic transformation. (**b**) Tamarillo embryogenic callus transformation procedure. As—acetosyringone, CW—coconut water, Gln—glutamine, MS—Murashige and Skoog medium, OD—optical density, PVP—polyvinylpyrrolidone.

3.6. Agrobacterium-Mediated Callus Transformation and Co-Cultivation

Before *Agrobacterium* infection, a cold-shock treatment was applied to callus explants. In total, 800 mg of tamarillo EC were weighed and immersed in a 3% maltose solution on ice for 20 min (Figure 7b). Callus was then transferred into the bacterial culture (of each *Agrobacterium* strain), and 100 µM glutamine (Gln), 1% (*w*/*v*) polyvinylpyrrolidone 40 (PVP) and 10% (*v/v*) coconut water (CW) were added to the infection solution. Callus was subjected to vacuum infiltration for 10 min to facilitate infection and incubated at 80 rpm and 28 °C for 10 min. Infected cells were poured into a 40 µm cell strainer to remove the excess bacteria and transferred to a sterile filter paper to allow them to dry excessive liquid. Callus was then distributed to four plates containing a sterile Whatman™ grade 1 qualitative filter paper thoroughly soaked with 1 mL MS plus 3% (*w*/*v*) sucrose supplemented with 100 µM Gln, 1% PVP and 10% CW. Co-cultivation was performed at 24 ± 1 °C in the dark for 3 days.

3.7. Selection of Transformed Cells

Following co-cultivation, several washes with MS supplemented with 500 mg/L carbenicillin and cefotaxime were performed in infected EC using a vacuum filtration system to eliminate *Agrobacterium*. Washed callus was transferred to a proliferation medium supplemented with 200 mg/L each of carbenicillin and cefotaxime, 1% PVP and 10% CW with no selective agent, for three days. For selection, cells were transferred to the same medium, now supplemented with 30 mg/L kanamycin. Resistant white EC was subsequently subcultured in gradually increased kanamycin concentrations (50, 75, 100 and 120 mg/L) at 4-week intervals, while concentrations of carbenicillin and cefotaxime were maintained. Suitable controls (untransformed callus) were also proliferated. All cultures were incubated at 24 ± 1 °C in the dark.

3.8. Staining for GUS Activity

β-glucuronidase, encoded by the *gus* gene, catalyzes the cleavage of the X-gluc (5-bromo-4-chloro-3-indolyl-β-D-glucuronide) substrate. The incubation of transformed cells with X-Gluc results in the production of a visible blue precipitate [2]. Therefore, GUS activity was assayed in kanamycin-resistant callus by histochemical staining, following the methodology described by Nelson-Vasilchik and colleagues [70]. A small quantity of callus was incubated in a staining buffer (88 μM sodium phosphate buffer (pH 7.0), 4.4 μM $K_3Fe(CN)_6$, 4.4 μM $K_4Fe(CN)_6 \cdot 3H_2O$, 0.88 μM Na·EDTA·$2H_2O$ with 0.88% triton-X) containing 7 μM X-gluc in a multi-well plate. Vacuum infiltration was performed for 10 min, followed by callus incubation at 37 °C in the dark overnight. After staining, callus was rinsed in 70% ethanol three times, followed by soaking in 1:1:3 lactic acid:glycerol:PBS at room temperature in the dark for 4 h. GUS activity was recorded as indigo, blue-colored spots or sectors. Non-transformed callus was used as control.

3.9. PCR and qPCR Analysis of Transformants

Four biological replicates of putative EC transformants were used to confirm the genetic transformation with each *Agrobacterium* strain (EHA105 and LBA4404). The genomic DNA (gDNA) was extracted from 100 mg of kanamycin-resistant and non-transformed callus, previously ground in liquid nitrogen, using the NucleoSpin™ Plant II kit (Macherey-Nagel™), following the manufacturer's instructions. The concentration and quality of DNA were assessed using a NanoDrop™ One Spectrophotometer (Thermo Scientific, Waltham, MA, USA). The presence of the *gus* transgene was verified by PCR analysis using specific primers (forward 5′-CATGAAGATGCTTCG-3′ and reverse 5′-ATCCACGCCGTATTCGG-3′). Plasmid DNA (p35SGUSINT) was used as a positive control, DNA from non-transformed callus was used as a negative control and a non-template control was included. PCR reactions were carried out in a 25 μL final volume containing 12.5 μL NZYTaq II 2× Green Master Mix (NZYTech), 0.4 μM of each primer and 1 μL of genomic DNA. Amplification of a 636 bp fragment of the *gus* gene was performed by incubation at 95 °C for 3 min, followed by 35 cycles at 94 °C for 30 s, 65 °C for 30 s, 72 °C for 25 s and a final step at 72 °C for 7 min. Amplified products were separated by electrophoresis in a 1.5% (*w*/*v*) agarose gel stained with GreenSafe Premium (NZYTech) and visualized using the Gel Doc XR Imaging Systems (BioRad, Hercules, CA, USA).

Quantitative real-time PCR was carried out to quantify the *gus* insertion in the gDNA of the samples previously confirmed for *gus* transgene presence. *Gus* quantification was normalized using *ACTIN* (*ACT*) as the reference gene since, from the ones tested by Cordeiro and colleagues [17], it was the gene with the most suitable primers pair for amplification in *S. betaceum* gDNA. The primers used for gus amplification were the ones described above, and for *ACT*, the forward 5′- CCA TGT TCC CGG GTA TTG CT-3′ and reverse 5′- GTG CTG AGG GAA GCC AAG AT-3′ primers were used, as described in Cordeiro et al., 2020. Each amplification was performed in a 10 μL final volume containing 5 μL of NZYSpeedy qPCR Green Master Mix (2x; NZYTech), 0.4 μM of each specific primer and 5 ng of gDNA. The reactions were conducted under an initial activation at 95 °C for 3 min, followed by 40 cycles of 5 s at 95 °C, 20 s at 57 °C and 15 s at 68 °C. The melting curves, with a temperature gradient from 65 to 95 °C with fluorescence readings acquired at 0.5 °C increments, were recorded to further confirm the specificity of the primers. The assay included non-template controls, and all reactions were run in two technical replicates. Reactions were performed in 96-well plates and run in a CFX96 Real-Time System (Bio-Rad). The relative expression was calculated according to the Pffafl method, using non-transformed samples as the control [71]. Transformations with EHA105 and LBA4404 were compared and statistically analyzed by a *t*-test at $p < 0.05$.

4. Conclusions

An efficient protocol for *Agrobacterium*-mediated genetic transformation of tamarillo EC was established. A suitable *Agrobacterium* strain and OD, cold shock pre-treatment, correct inoculation and co-cultivation period, and the use of acetosyringone, Gln, PVP and CW were efficiently applied. Hence, all samples tested showed blue coloration in the GUS assay, the presence of the transgene in PCR and transgene integration in the qPCR, meaning that all masses tested were effectively transformed. Therefore, this protocol presents a 100% transformation efficiency rate. This transformation system constitutes a great biotechnological tool for plant gene functional analysis, such as the study of the role of genes involved in SE, revealing the molecular networks underlying plant regeneration capacity. Moreover, it can also be applied to the genetic improvement of this woody species and other important crops. For instance, through this technique and using new gene editing tools, healthier and more productive plants can be produced. In addition, metabolic pathways can be manipulated to produce secondary metabolites with applications in the food and feed industries, as well as in cosmetics and perfumery.

Supplementary Materials: The following supporting information can be downloaded at: https://www.mdpi.com/article/10.3390/plants12051202/s1, Figure S1: Tamarillo embryogenic callus after 30 days of incubation in proliferation medium and in proliferation medium supplemented with cefotaxime; Figure S2: General appearance of tamarillo embryogenic callus after 3 days co-cultured with EHA105 Agrobacterium strain; Figure S3: Tamarillo embryogenic callus during co-culture with EHA105 Agrobacterium strain, after vacuum infiltration for 10 min followed by incubation at 80 rpm and 28 °C.

Author Contributions: Conceptualization, D.C. and S.C.; methodology, D.C., R.F., B.C. and S.C.; validation, D.C. and S.C.; formal analysis, D.C.; investigation, D.C. and A.A.; resources, J.C. and S.C.; writing—original draft preparation, D.C.; writing—review and editing, D.C., J.C. and S.C.; visualization, D.C.; supervision, S.C.; funding acquisition, J.C. and S.C. All authors have read and agreed to the published version of the manuscript.

Funding: This research was funded by Foundation for Science and Technology (Portugal) through PhD fellowship SFRH/BD/136925/2018 awarded to Daniela Cordeiro, 2021.04958.BD awarded to Ricardo Ferraz and SFRH/BD/146485/2019 awarded to Bruno Casimiro. The work was carried out at the R&D Unit Center for Functional Ecology—Science for People and the Planet (UIDB/04004/2020) financed by FCT/MCTES through national funds (PIDDAC).

Institutional Review Board Statement: Not applicable.

Data Availability Statement: The data presented are contained within the article or Supplementary Materials. The datasets generated during and/or analyzed during the current study are available from the corresponding author on reasonable request.

Conflicts of Interest: The authors declare no conflict of interest.

References

1. Song, G.Q.; Prieto, H.; Orbovic, V. *Agrobacterium*-mediated transformation of tree fruit crops: Methods, progress, and challenges. *Front. Plant Sci.* **2019**, *10*, 226. [CrossRef]
2. Basso, M.F.; Arraes, F.B.M.; Grossi-de-Sa, M.; Moreira, V.J.V.; Alves-Ferreira, M.; Grossi-de-Sa, M.F. Insights into genetic and molecular elements for transgenic crop development. *Front. Plant Sci.* **2020**, *11*, 509. [CrossRef] [PubMed]
3. Altpeter, F.; Springer, N.M.; Bartley, L.E.; Blechl, A.E.; Brutnell, T.P.; Citovsky, V.; Conrad, L.J.; Gelvin, S.B.; Jackson, D.P.; Kausch, A.P.; et al. Advancing crop transformation in the era of genome editing. *Plant Cell* **2016**, *28*, 1510–1520. [CrossRef]
4. Mushtaq, R.; Shahzad, K.; Shah, Z.H.; Alsamadany, H.; Alzahrani, H.A.S.; Alzahrani, Y.; Mujtaba, T.; Ahmed, Z.; Mansoor, S.; Bashir, A. Isolation of biotic stress resistance genes from cotton (*Gossypium arboreum*) and their analysis in model plant tobacco (*Nicotiana tabacum*) for resistance against cotton leaf curl disease complex. *J. Virol. Methods* **2020**, *276*, 113760. [CrossRef]
5. Zhang, Y.; Chen, M.; Siemiatkowska, B.; Toleco, M.R.; Jing, Y.; Strotmann, V.; Zhang, J.; Stahl, Y.; Fernie, A.R. A highly efficient *Agrobacterium*-mediated method for transient gene expression and functional studies in multiple plant species. *Plant Commun.* **2020**, *1*, 100028. [CrossRef] [PubMed]
6. Travella, S.; Ross, S.M.; Harden, J.; Everett, C.; Snape, J.W.; Harwood, W.A. A comparison of transgenic barley lines produced by particle bombardment and *Agrobacterium*-mediated techniques. *Plant Cell Rep.* **2005**, *23*, 780–789. [CrossRef]
7. Tzfira, T.; Hohn, B.; Gelvin, S.B. Transfer of genetic information from *Agrobacterium* to plants. In *Reference Module in Life Sciences*; Elsevier: Amsterdam, The Netherlands, 2017.
8. Ribeiro, T.P.; Basso, M.F.; de Carvalho, M.H.; de Macedo, L.L.P.; da Silva, D.M.L.; Lourenço-Tessutti, I.T.; de Oliveira-Neto, O.B.; de Campos-Pinto, E.R.; Lucena, W.A.; da Silva, M.C.M.; et al. Stability and tissue-specific Cry10Aa overexpression improves cotton resistance to the cotton boll weevil. *Biotechnol. Res. Innov.* **2019**, *3*, 27–41. [CrossRef]
9. Wang, K. *Agrobacterium Protocols*; Wang, K., Ed.; Springer: Berlin/Heidelberg, Germany, 2015; Volume 1224, ISBN 978-1-4939-1657-3.
10. Prohens, J.; Nuez, F. The tamarillo (*Cyphomandra betacea*). *Small Fruits Rev.* **2001**, *1*, 43–68. [CrossRef]

11. Acosta-Quezada, P.G.; Raigón, M.D.; Riofrío-Cuenca, T.; García-Martínez, M.D.; Plazas, M.; Burneo, J.I.; Figueroa, J.G.; Vilanova, S.; Prohens, J. Diversity for chemical composition in a collection of different varietal types of tree tomato (*Solanum betaceum* Cav.), an andean exotic fruit. *Food Chem.* **2015**, *169*, 327–335. [CrossRef] [PubMed]
12. Wang, S.; Zhu, F. Tamarillo (*Solanum betaceum*): Chemical composition, biological properties, and product innovation. *Trends Food Sci. Technol.* **2020**, *95*, 45–58. [CrossRef]
13. Correia, S.I.; Canhoto, J.M. Biotechnology of tamarillo (*Cyphomandra betacea*): From in vitro cloning to genetic transformation. *Sci. Hortic.* **2012**, *148*, 161–168. [CrossRef]
14. Correia, S.; Lopes, M.L.; Canhoto, J.M. Somatic embryogenesis induction system for cloning an adult *Cyphomandra betacea* (Cav.) Sendt. (tamarillo). *Trees-Struct. Funct.* **2011**, *25*, 1009–1020. [CrossRef]
15. Correia, S.I.; Lopes, M.L.; Canhoto, J.M. Somatic embryogenesis in tamarillo (*Cyphomandra betacea*): Recent advances. *Acta Hortic.* **2009**, *839*, 157–164. [CrossRef]
16. Correia, S.; Cunha, A.E.; Salgueiro, L.; Canhoto, J.M. Somatic embryogenesis in tamarillo (*Cyphomandra betacea*): Approaches to increase efficiency of embryo formation and plant development. *Plant Cell Tissue Organ. Cult.* **2012**, *109*, 143–152. [CrossRef]
17. Cordeiro, D.; Rito, M.; Borges, F.; Canhoto, J.; Correia, S. Selection and validation of reference genes for qPCR analysis of miRNAs and their targets during somatic embryogenesis in tamarillo (*Solanum betaceum* Cav.). *Plant Cell Tissue Organ. Cult.* **2020**, *143*, 109–120. [CrossRef]
18. Casimiro, B.; Mota, I.; Veríssimo, P.; Canhoto, J.; Correia, S. Enhancing the production of hydrolytic enzymes in elicited tamarillo (*Solanum betaceum* Cav.) cell suspension cultures. *Plants* **2023**, *12*, 190. [CrossRef]
19. Correia, S.; Canhoto, J.M. Somatic embryogenesis of tamarillo (*Solanum betaceum* Cav.). In *Step Wise Protocols for Somatic Embryogenesis of Important Woody Plants*; Jain, S.M., Gupta, P.K., Eds.; Springer: Cham, Switzerland, 2018; pp. 171–179.
20. Dhekney, S.A.; Sessions, S.K.; Brungart-Rosenberg, M.; Claflin, C.; Li, Z.T.; Gray, D.J. Genetic modification of grapevine embryogenic cultures. *Methods Mol. Biol.* **2019**, *1864*, 191–201. [CrossRef]
21. Ganapathi, T.R.; Higgs, N.S.; Balint-Kurti, P.J.; Arntzen, C.J.; May, G.; van Eck, J.M. *Agrobacterium*-mediated transformation of embryogenic cell suspensions of the banana cultivar Rasthali (AAB). *Plant Cell Rep.* **2001**, *20*, 157–162. [CrossRef] [PubMed]
22. Leelavathi, S.; Sunnichan, V.G.; Kumria, R.; Vijaykanth, G.P.; Bhatnagar, R.K.; Reddy, V.S. A simple and rapid *Agrobacterium*-mediated transformation protocol for cotton (*Gossypium hirsutum* L.): Embryogenic calli as a source to generate large numbers of transgenic plants. *Plant Cell Rep.* **2004**, *22*, 465–470. [CrossRef] [PubMed]
23. Ratjens, S.; Mortensen, S.; Kumpf, A.; Bartsch, M.; Winkelmann, T. Embryogenic callus as target for efficient transformation of *Cyclamen persicum* enabling gene function studies. *Front. Plant Sci.* **2018**, *9*, 1–15. [CrossRef] [PubMed]
24. Rajewski, A.C.; Elkins, K.B.; Henry, A.; van Eck, J.; Litt, A. In vitro plant regeneration and *Agrobacterium tumefaciens*–mediated transformation of *Datura stramonium* (solanaceae). *Appl. Plant Sci.* **2019**, *7*, e01220. [CrossRef] [PubMed]
25. Matveeva, T.V.; Lutova, L.A. Horizontal gene transfer from *Agrobacterium* to plants. *Front. Plant Sci.* **2014**, *5*, 326. [CrossRef] [PubMed]
26. Niedbała, G.; Niazian, M.; Sabbatini, P. Modeling *Agrobacterium*-mediated gene transformation of tobacco (*Nicotiana tabacum*)-a model plant for gene transformation studies. *Front. Plant Sci.* **2021**, *12*, 1454. [CrossRef]
27. Swartwood, K.; van Eck, J. Development of plant regeneration and *Agrobacterium tumefaciens*-mediated transformation Methodology for Physalis Pruinosa. *Plant Cell Tissue Organ. Cult.* **2019**, *137*, 465–472. [CrossRef]
28. Nava, E.; Dávila, Y.; Arellano, J.; Ortiz, A.; Alvárez, L.; Marquina, S.; Villarreal, M.L. Genetic transformation of *Solanum chrysotrichum* by *Agrobacterium tumefaciens* and the production of antifungal saponins. *Vitr. Cell. Dev. Biol.-Plant* **2011**, *47*, 650–657. [CrossRef]
29. Kumar, A.; Miller, M.; Whitty, P.; Lyon, J.; Davie, P. *Agrobacterium*-mediated transformation of five wild *Solanum* species using *in vitro* microtubers. *Plant Cell Rep.* **1995**, *14*, 324–328. [CrossRef]
30. Curtis, I.S.; Power, J.B.; Hedden, P.; Phillips, A.; Lowe, K.C.; Ward, D.A.; Davey, M.R. Transformation and characterization of transgenic plants of *Solanum dulcamara* L.-incidence of transgene silencing. *Ann. Bot.* **2000**, *86*, 63–71. [CrossRef]
31. Van Eck, J.; Keen, P.; Tjahjadi, M. *Agrobacterium tumefaciens*-mediated transformation of tomato. *Methods Mol. Biol.* **2019**, *1864*, 225–234. [CrossRef]
32. Khatun, M.; Borphukan, B.; Alam, I.; Keya, C.A.; Khan, H.; Reddy, M.K.; Salimullah, M. An improved *Agrobacterium* mediated transformation and regeneration protocol for successful genetic engineering and genome editing in eggplant. *Sci. Hortic.* **2022**, *293*, 110716. [CrossRef]
33. Atkinson, R.G.; Gardner, R.C. *Agrobacterium*-mediated transformation of pepino and regeneration of transgenic plants. *Plant Cell Rep.* **1991**, *10*, 208–212. [CrossRef]
34. Peng, W.; Wang, X.; Wei, H.; Zhang, Z.; Teng, C.; Li, Q.; Lyu, K.; Lyu, S.; Fan, Y. Development of *Agrobacterium tumefaciens*-mediated transformation of *Solanum nigrum* and expression of AcMYB110 in *S. Nigrum*. *Research Square* **2022**. preprint. [CrossRef]
35. Ducreux, L.J.M.; Morris, W.L.; Taylor, M.A.; Millam, S. *Agrobacterium*-mediated transformation of *Solanum phureja*. *Plant Cell Rep.* **2005**, *24*, 10–14. [CrossRef]
36. Ugandhar, T.; Odelu, G.; Parvathi, D.; Anitha, U.D.; Venkateshwarlu, M. *Agrobacterium*-mediated genetic transformation and regeneration from leaf explants of *Solanum thorvum* (Swartz) a medicinally important plant. *Int. J. Adv. Res.* **2017**, *5*, 896–904. [CrossRef]
37. Shilpha, J.; Jayashre, M.; Joe Virgin Largia, M.; Ramesh, M. Direct shoot organogenesis and *Agrobacterium tumefaciens* mediated transformation of *Solanum trilobatum* L. *Turk. J. Biol.* **2016**, *40*, 866–877. [CrossRef]

38. Bruce, M.A.; Shoup Rupp, J.L. *Agrobacterium*-mediated transformation of *Solanum tuberosum* L., potato. *Methods Mol. Biol.* **2019**, *1864*, 203–223. [CrossRef] [PubMed]
39. Atkinson, R.G.; Gardner, R.C. Regeneration of transgenic tamarillo plants. *Plant Cell Rep.* **1993**, *12*, 347–351. [CrossRef]
40. Cohen, D.; van den Brink, R.C.; MacDiarmid, R.M.; Beck, D.L.; Forster, R.L.S. Resistance to tamarillo mosaic virus in transgenic tamarillos and expression of the transgenes in F1 progeny. *Acta Hortic.* **2000**, *521*, 43–49. [CrossRef]
41. Correia, S.; Alhinho, A.T.; Casimiro, B.; Miguel, C.M.; Oliveira, M.; Veríssimo, P.; Canhoto, J. NEP-TC a rRNA methyltransferase involved on somatic embryogenesis of tamarillo (*Solanum betaceum* Cav.). *Front. Plant Sci.* **2019**, *10*, 438. [CrossRef] [PubMed]
42. Aida, R.; Hirose, Y.; Kishimoto, S.; Shibata, M. *Agrobacterium tumefaciens*–mediated transformation of *Cyclamen persicum* Mill. *Plant Sci.* **1999**, *148*, 1–7. [CrossRef]
43. Hood, E.E.; Gelvin, S.B.; Melchers, L.S.; Hoekema, A. New *Agrobacterium* helper plasmids for gene transfer to plants. *Transgenic. Res.* **1993**, *2*, 208–218. [CrossRef]
44. Hoekema, A.; Hirsch, P.R.; Hooykaas, P.J.J.; Schilperoort, R.A. A binary plant vector strategy based on separation of *Vir*- and *T*-region of the *Agrobacterium tumefaciens* Ti-plasmid. *Nature* **1983**, *303*, 179–180. [CrossRef]
45. Vancanneyt, G.; Schmidt, R.; O'Connor-Sanchez, A.; Willmitzer, L.; Rocha-Sosa, M. Construction of an intron-containing marker gene: Splicing of the intron in transgenic plants and its use in monitoring early events in *Agrobacterium*-mediated plant transformation. *Molec. Gen. Genet.* **1990**, *220*, 245–250. [CrossRef]
46. Kole, C.; Hall, T.C. *Transgenic Temperate Fruits and Nuts*; Wiley-Blackwell: Chichester, UK, 2008; Volume 4.
47. López, E.; Proaño, K.; Jadán, M.; Mihai, R. Callus tissue induction and analysis of GUS reporter gene expression in tomato (*Solanum lycopersicum* L.) *Transformed* with *Agrobacterium tumefaciens*. *Rom. Biotechnol. Lett.* **2015**, *20*, 10205–10211.
48. Kumar, N.; Gulati, A.; Bhattacharya, A. L-glutamine and L-glutamic acid facilitate successful *Agrobacterium* infection of recalcitrant tea cultivars. *Appl. Biochem. Biotechnol.* **2013**, *170*, 1649–1664. [CrossRef] [PubMed]
49. Priya, A.M.; Pandian, S.K.; Manikandan, R. The effect of different antibiotics on the elimination of *Agrobacterium* and high frequency *Agrobacterium*-mediated transformation of indica rice (*Oryza sativa* L.). *Czech J. Genet. Plant Breed.* **2012**, *48*, 120–130. [CrossRef]
50. Chu, M.; Quiñonero, C.; Akdemir, H.; Alburquerque, N.; Pedreño, M.Á.; Burgos, L. *Agrobacterium*-mediated transformation of *Vitis* Cv. Monastrell suspension-cultured cells: Determination of critical parameters. *Biotechnol. Prog.* **2016**, *32*, 725–734. [CrossRef]
51. Sun, Z.L.; Li, X.; Zhou, W.; Yan, J.D.; Gao, Y.R.; Li, X.W.; Sun, J.C.; Fang, K.F.; Zhang, Q.; Xing, Y.; et al. *Agrobacterium*-mediated genetic transformation of chinese chestnut (*Castanea mollissima* Blume). *Plant Cell Tissue Organ. Cult.* **2020**, *140*, 95–103. [CrossRef]
52. Ma, R.; Yu, Z.; Cai, Q.; Li, H.; Dong, Y.; Oksman-Caldentey, K.M.; Rischer, H. *Agrobacterium*-mediated genetic transformation of the medicinal plant *Veratrum dahuricum*. *Plants* **2020**, *9*, 191. [CrossRef]
53. Zhang, W.J.; Dewey, R.E.; Boss, W.; Phillippy, B.Q.; Qu, R. Enhanced *Agrobacterium*-mediated transformation efficiencies in monocot cells is associated with attenuated defense responses. *Plant Mol. Biol.* **2013**, *81*, 273–286. [CrossRef]
54. Carvalho, C.H.; Zehr, U.B.; Gunaratna, N.; Anderson, J.; Kononowicz, H.H.; Hodges, T.K.; Axtell, J.D. *Agrobacterium*-mediated transformation of *Sorghum*: Factors that affect transformation efficiency. *Genet Mol. Biol.* **2004**, *27*, 259–269. [CrossRef]
55. Sheikholeslam, S.N.; Weeks, D.P. Acetosyringone promotes high efficiency transformation of *Arabidopsis thaliana* explants by *Agrobacterium tumefaciens*. *Plant Mol. Biol.* **1987**, *8*, 291–298. [CrossRef] [PubMed]
56. Uranbey, S.; Sevimay, C.S.; Kaya, M.D.; İpek, A.; Sancak, C.; Başalma, D.; Er, C.; Özcan, S. Influence of different co-cultivation temperatures, periods and media on *Agrobacterium tumefaciens*-mediated gene transfer. *Biol. Plant* **2005**, *49*, 53–57. [CrossRef]
57. Stefano, B.; Patrizia, B.; Matteo, C.; Massimo, G. Inverse PCR and quantitative PCR as alternative methods to southern blotting analysis to assess transgene copy number and characterize the integration site in transgenic woody plants. *Biochem. Genet.* **2016**, *54*, 291–305. [CrossRef] [PubMed]
58. Mason, G.; Provero, P.; Vaira, A.M.; Accotto, G.P. Estimating the number of integrations in transformed plants by quantitative real-time PCR. *BMC Biotechnol.* **2002**, *2*, 20. [CrossRef]
59. Faize, M.; Faize, L.; Burgos, L. Using quantitative real-time PCR to detect chimeras in transgenic tobacco and apricot and to monitor their dissociation. *BMC Biotechnol.* **2010**, *10*, 53. [CrossRef]
60. Yang, L.; Ding, J.; Zhang, C.; Jia, J.; Weng, H.; Liu, W.; Zhang, D. Estimating the copy number of transgenes in transformed rice by real-time quantitative PCR. *Plant Cell Rep.* **2005**, *23*, 759–763. [CrossRef]
61. You-wen, Q.; Xue-jun, G.; Bang-ruo, Q.; Lu, L.; Zhen, Z. Establishment of TaqMan real-time quantitative PCR assay for foreign gene copy numbers in transgenic soybean. *J. Northeast. Agric. Univ.* **2012**, *19*, 48–52. [CrossRef]
62. Subramanyam, K.; Subramanyam, K.; Sailaja, K.V.; Srinivasulu, M.; Lakshmidevi, K. Highly efficient *Agrobacterium*-mediated transformation of banana Cv. Rasthali (AAB) via sonication and vacuum infiltration. *Plant Cell Rep.* **2011**, *30*, 425–436. [CrossRef]
63. Murashige, T.; Skoog, F. A revised medium for rapid growth and bio assays with tobacco tissue cultures. *Physiol. Plant* **1962**, *15*, 473–497. [CrossRef]
64. Miller, J.H. *Experiments in Molecular Genetics*; Cold Spring Harbor Laboratory: Cold Spring Harbor, NY, USA, 1972; ISBN 0879691069.
65. Prakash, D.P.; Deepali, B.S.; Asokan, R.; Ramachandra, Y.L.; Anand, L.; Hanur, V.S. Effect of antibiotics and gelling agents in transformation of brinjal (*Solanum melongena* L.) Cv. Manjarigota. *J. Hortic. Sci.* **2007**, *2*, 19–25.
66. Li, Z.T.; Dhekney, S.; Dutt, M.; Aman, M.; Tattersall, J.; Kelley, K.T.; Gray, D.J. Optimizing *Agrobacterium*-mediated transformation of grapevine. *In Vitro Cell. Dev. Biol.-Plant* **2006**, *42*, 220–227. [CrossRef]

67. Cano, V.; Martínez, M.T.; San José, M.C.; Couselo, J.L.; Varas, E.; Bouza-Morcillo, L.; Toribio, M.; Corredoira, E. Regeneration of transgenic plants by *Agrobacterium*-mediated transformation of *Quercus ilex* L. somatic embryos with the gene CsTL1. *New For.* **2020**, *51*, 1003–1021. [CrossRef]
68. Ribas, A.F.; Dechamp, E.; Champion, A.; Bertrand, B.; Combes, M.; Verdeil, J.; Lapeyre, F.; Lashermes, P.; Etienne, H. *Agrobacterium*-mediated genetic transformation of *Coffea arabica* (L.) is greatly enhanced by using established embryogenic callus cultures. *BMC Plant Biol.* **2011**, *11*, 92. [CrossRef] [PubMed]
69. Bull, S.; Owiti, J.; Niklaus, M.; Beeching, J.R.; Gruissem, W.; Vanderschuren, H. *Agrobacterium*-mediated transformation of friable embryogenic calli and regeneration of transgenic cassava. *Nat. Protoc.* **2009**, *4*, 1845–1854. [CrossRef] [PubMed]
70. Plant Transformation and Genome Editing-GUS Assay Protocol. Available online: https://web.uri.edu/ctwg/ (accessed on 3 November 2022).
71. Pfaffl, M.W. A new mathematical model for relative quantification in real-time RT-PCR. *Nucleic Acids Res.* **2001**, *29*, E45. [CrossRef] [PubMed]

plants

MDPI

Article

In Vitro Mass Propagation of Coffee Plants (*Coffea arabica* L. var. Colombia) through Indirect Somatic Embryogenesis

Consuelo Margarita Avila-Victor [1], Víctor Manuel Ordaz-Chaparro [1], Enrique de Jesús Arjona-Suárez [1], Leobardo Iracheta-Donjuan [2], Fernando Carlos Gómez-Merino [1] and Alejandrina Robledo-Paz [1,*]

[1] Colegio de Postgraduados, Campus Montecillo, Carretera México-Texcoco Km. 36.5, Montecillo, Texcoco C.P. 56264, Estado de México, Mexico
[2] Instituto Nacional de Investigaciones Forestales, Agrícolas y Pecuarias, Campo Experimental Rosario Izapa, Carretera Tapachula-Cacahoatán Km. 18, Tuxtla Chico C.P. 30870, Chiapas, Mexico; iracheta.leobardo@inifap.gob.mx
* Correspondence: arobledo@colpos.mx

Abstract: *Coffea arabica* is one of the two most consumed coffee species in the world. Micropropagation through somatic embryogenesis has allowed the large-scale propagation of different coffee varieties. However, the regeneration of plants using this technique depends on the genotype. This study aimed to develop a protocol for the regeneration of *C. arabica* L. var. Colombia by somatic embryogenesis for its mass propagation. Foliar explants were cultured on Murashige and Skoog (MS) supplemented with different concentrations of 2,4-dichlorophenoxyacetic acid (2,4-D), 6-benzylaminopurine (BAP), and phytagel for inducing somatic embryogenesis. In total, 90% of the explants formed embryogenic calli with a culture medium containing 2 mg L^{-1} of 2,4-D, 0.2 mg L^{-1} BAP, and 2.3 g L^{-1} phytagel. The highest number of embryos per gram of callus (118.74) was obtained in a culture medium containing 0.5 mg L^{-1} 2,4-D, 1.1 mg L^{-1} BAP, and 5.0 g L^{-1} phytagel. In total, 51% of the globular embryos reached the cotyledonary stage when they were cultured on the growth medium. This medium contained 0.25 mg L^{-1} BAP, 0.25 mg L^{-1} indoleacetic acid (IAA), and 5.0 g L^{-1} of phytagel. The mixture of vermiculite:perlite (3:1) allowed 21% of embryos to become plants.

Keywords: coffee; somatic embryos; *in vitro* propagation; tissue culture

Citation: Avila-Victor, C.M.; Ordaz-Chaparro, V.M.; Arjona-Suárez, E.d.J.; Iracheta-Donjuan, L.; Gómez-Merino, F.C.; Robledo-Paz, A. *In Vitro* Mass Propagation of Coffee Plants (*Coffea arabica* L. var. Colombia) through Indirect Somatic Embryogenesis. *Plants* **2023**, *12*, 1237. https://doi.org/10.3390/plants12061237

Academic Editors: Zanmin Hu, Han Xiao, Yi Ren and Chengming Fan

Received: 7 February 2023
Revised: 2 March 2023
Accepted: 4 March 2023
Published: 8 March 2023

1. Introduction

Coffee is one of the most important crops worldwide. Coffee species grow positively between 800 to 2000 m above sea level in equatorial and tropical climates [1]. Coffee is native to Africa, and belongs to the Rubeaceae family [2]. Global production of coffee achieved 10.5 million t, while in Mexico it was 240,000 t [3]. The coffee species arabica (*Coffea arabica*) and robusta (*Coffea canephora* var. Robusta) account for 70% and 30% of global production, respectively [4–6]. Mexico has sustained a leading position in the list of the most important countries for the production and exportation of coffee. However, Mexico has decreased in the ranking of coffee production and exportation (ranking thirteenth and twelfth, respectively) to date [7]. Several factors have impacted coffee plantations, such as low prices in the world market, old plantations, and lack of financing and investment for culture renovation. Climate change (increasing temperature), poor soil nutrition, and diseases (highlighting coffee rust caused by the fungus *Hemileia vastatrix* Berk. and Br.) have also negatively affected coffee production in Mexico [8–10]. Coffee rust is a foliar disease, considered the most devastating for the crop, which in Mexico caused a decrease from 270,000 t to 132,000 t in 2016 [11].

A Comprehensive Coffee Care Plan (CCCP) has been implemented to increase coffee production in Mexico [12]. To achieve this increase, 300,000 ha of plantations must be renewed, demanding approximately 459 million coffee plants within the next few years [11].

As an alternative to increasing coffee production, sustainable, profitable, and environmentally friendly agriculture has been improved, with culture and renewal of coffee plantations with resistant varieties to pests or diseases instead [11]. Colombia variety has shown rust resistance and high tastiness and grain quality standards, low bearing, high production rates, phenotypic uniformity, adaptability, and good yield. Although this variety is not the most cultivated in Mexico, due to its characteristics, it may potentially improve coffee production [9].

Coffee can be propagated sexually (seed) and asexually (cuttings, grafts). The sexual method has a low capacity for multiplication and requires large cultivation areas and long periods for mass propagation. On the other hand, asexual propagation is a time-consuming, laborious, and intensive activity, making it expensive [13]. Biotechnological tools such as micropropagation allow the production of plants of selected genotypes on a large scale, at any time of the year, and in less space. Micropropagation can be carried out by axillary shoot culture, organogenesis, or somatic embryogenesis. The advantage of somatic embryogenesis over organogenesis is that a rooting stage is not required, since the individuals obtained have stem and root meristems; particularly, indirect somatic embryogenesis (previous callus stage) can generate large numbers of plants from few cells [14,15].

Several protocols have been developed to produce somatic embryos in different coffee varieties [10,12–16]; nevertheless, methodologies to induce somatic embryogenesis depend on the plant genotype. In this regard, López-Gómez et al. [16] cultivated leaf explants of 10 genotypes of *C. canephora* and 3 of *C. arabica*, on a culture medium containing 1.1 mg L^{-1} of BAP and 0.5 mg L^{-1} of 2,4-D, to induce somatic embryogenesis. The authors found that only four *C. canephora* genotypes (00-28, 95-8, 97-18, 97-20) and two *C. arabica* genotypes (2000-1128, 2000-692) were able to differentiate embryos. Likewise, Ahmed et al. (2013) tested different combinations of 2,4-D (1.0–2.0 mg L^{-1}) and BAP (0.5−3.0 mg L^{-1}) to promote the formation of somatic embryos from the foliar explants of the hybrid "Ababuna" of *C. arabica*, finding that a higher proportion of BAP, compared to 2,4-D, promoted the best response.

On the other hand, various authors consider that both growth regulators and stress are the factors that can make differentiated cells become competent to form somatic embryos [17–19]. Cells' response to stress will depend on their physiological state and the level of stress. When stress is extreme, cells can die, but if not, metabolic reprogramming occurs that allows them to adapt, survive, and form embryos [17,20].

The present study aimed to develop a protocol for the regeneration of plants of *C. arabica* L. var. Colombia by somatic embryogenesis for large-scale multiplication, to increase the number of coffee plants, and provide the plants needed for renewing the old plantations, hence maximizing coffee production in Mexico. The working hypothesis was that a greater proportion of cytokinins with respect to auxins, in combination with water stress caused by high concentrations of phytagel, would allow inducing indirect somatic embryogenesis from the foliar explants of the Colombia variety.

2. Results

2.1. Induction of Embryogenic Callus

Except for the T4 treatment (2,4-D, 1.0 mg L^{-1}; BAP, 1.0 mg L^{-1}, phytagel, 5.0 g L^{-1}), all the treatments tested promoted callus formation in more than 75% of the explants (Figure 1a), while the highest amount of calluses was observed in the explants grown in the combination of 2.0 mg L^{-1} 2,4-D, 0.2 mg L^{-1} BAP, and 2.3 g L^{-1} phytagel (T5) (Table 1).

Figure 1. Somatic embryogenesis in coffee (*Coffea arabica* L. var. Colombia). (**a**) Embryogenic callus from leaf explants after eight weeks. (**b**) Embryogenic callus production in liquid medium after 12 weeks. (**c**) Somatic embryos in different stages of development after 150 d of culture differentiation medium. (**d**) Cotyledonary embryos after four weeks of culture on maturation medium. (**e**) *In vitro* plant grown in vermiculite:perlite mixture (3:1). (**f**) *In vitro* regenerated plant grown under greenhouse conditions. ge: globular, he: heart, te: torpedo.

Table 1. Effect of different combinations of 2,4- D and BAP at two concentrations of phytagel on induction of embryogenic callus from leaf explants of *C. arabica* var. Colombia after two months of culture. All cultures were kept under dark conditions at 26 ± 2 °C.

Treatment	Growth Regulators (mg L^{-1})		Phytagel (g L^{-1})	Explants with Callus (%)	† Callus Quantity
	2,4-D	BAP			
1	0.5	1.1	2.3	90.00 a	1.03 bc
2	0.5	1.1	5.0	75.74 ab	1.00 c
3	1.0	1.0	2.3	90.00 a	1.20 b
4	1.0	1.0	5.0	69.00 b	1.15 bc
5	2.0	0.2	2.3	90.00 a	1.38 a
6	2.0	0.2	5.0	85.05 a	1.03 bc

† Explant area covered with calluses: 1: 0–20%; 2: 21–40%; 3: 41–60%; 4: 61–80%; 5: 81–100%. Means with the same letter are not statistically different (Tukey $\alpha = 0.05$).

2.2. Embryogenic Calli Production

It was possible to observe significant differences due to the effect of the position of the flask and the subculture time of the calli of the cultivar Colombia. In this way, the calli grown in the flasks (125 mL) with inclination and without subculturing for three months increased their biomass by 18.9 times (18.9 g), the maximum fresh weight accumulated in the different treatments tested (Table 2; Figure 1b). In contrast, calli that were grown upright and not subcultured, only increased their weight by five times (Table 2).

Table 2. Effect of flask position and subculture on biomass production of embryogenic calli of *C. arabica* L. var. Colombia, after three months.

Flask Position and Subculture	Biomass (g)
Inclination—without subculture	18.94 a
Vertical—without subculture	5.17 b
Vertical—with subculture	1.27 b

Means with the same letter are not statistically different (Tukey $\alpha = 0.05$).

2.3. Effect of Embryogenic Callus Induction Media on Differentiation of Somatic Embryos of C. arabica L. var. Colombia

It was possible to observe significant differences between treatments for the number of embryos per gram of fresh callus weight. The maximum value for this variable was obtained in the calli that came from the induction medium that contained 0.5 mg L^{-1} 2,4-D, 1.1 mg L^{-1} BAP, and 5.0 g L^{-1} phytagel (T2), while the lowest number of embryos was observed in the T3 treatment (1.0 mg L^{-1} 2,4-D, 1.0 mg L^{-1} BAP, and 2.3 g L^{-1} phytagel) (Table 3; Figure 1c).

Table 3. Number of embryos formed from calli of *C. arabica* var. Colombia obtained in 6 different induction media, after 17 weeks cultured in a differentiation medium.

Treatment	Growth Regulators (mg L^{-1})		Phytagel (g L^{-1})	Embryos per Gram
	2,4-D	BAP		
1	0.5	1.1	2.3	44.03 b
2	0.5	1.1	5.0	118.74 a
3	1.0	1.0	2.3	7.81 c
4	1.0	1.0	5.0	27.94 b
5	2.0	0.2	2.3	23.75 b
6	2.0	0.2	5.0	26.19 b

Means with the same letter are not statistically different (Tukey $\alpha = 0.05$). Differentiation medium: modified Yasuda et al. [21] basal salts, 1.1 mg L^{-1} BAP, 5.0 g L^{-1} phytagel, 30 g L^{-1} sucrose.

2.4. Development and Maturation

The maturation medium containing 0.25 mg L^{-1} BAP and 0.25 mg L^{-1} IAA (MD2) allowed a greater number of globular embryos to reach the cotyledonary stage (Table 4; Figure 1d), while the percentage of embryos in the initial torpedo stage that reached the cotyledonary stage was not significantly different in any of the tested media (Table 4).

Table 4. Percentage of globular and torpedo embryos of *Coffea arabica* var. Colombia that reached maturity (cotyledonary stage) after 30 days of culture on the maturation media.

Culture Medium	Globular Embryos	Torpedo Embryos
	Maturation (%)	Maturation (%)
MD1	39.24 b	40.90 a
MD2	51.72 a	45.41 a

MD1: Kinetin, 0.5 mg L^{-1}; IAA, 0.5 mg L^{-1}; MD2: BAP, 0.25 mg L^{-1}; IAA, 0.25 mg L^{-1}. Means with the same letter are not statistically different (Tukey $\alpha = 0.05$).

2.5. Conversion

The conversion percentage of the embryos into plants was significantly higher when they were placed in the vermiculite:perlite substrate mixture (21.5%) than when they were cultivated in a semi-solid culture medium (7.2%) (Figure 1e). The plants obtained in the semi-solid medium were smaller than those regenerated in the mixture of substrates; in addition, they were more easily acclimatized to *ex vitro* conditions (Figure 1f). Likewise, unlike the roots formed *in vitro*, those that were generated in the substrates were branched and had a greater amount of root hairs.

2.6. Histological Analysis of Somatic Embryogenesis

After 30 days of culture in the callus induction medium, the explants began to form calluses in the periphery and mainly in the midrib (Figure 2a). The callus was basically made up of large parenchyma cells, with vacuoles, sparsely evident nuclei, and abundant intercellular spaces. During the following 30 d, the callus cells underwent continuous divisions that allowed their biomass to increase. Then, the primary calli were transferred to the embryo differentiation medium, and small cells began to form on their surface, with dense cytoplasm, evident nuclei, and few intercellular spaces, similar to meristematic cells, which showed a certain degree of organization (proembryogenic masses). The proembryogenic masses formed began to accumulate phenolic compounds in high concentration, after 30 d of culture on the differentiation medium; the presence of said compounds caused the darkening of said masses (Figure 2a) and only those that produced phenols managed to form embryos. At 120 d of culture in the differentiation medium, embryos in the globular stage were formed from the proembryogenic masses (Figure 2b); these embryos showed a protodermis that covered the periphery of the embryo and that was made up of a series of small cells with an evident nucleus (Figure 2c). When the globular and heart embryos (Figure 2c,d) were cultured in the maturation medium, they reached the torpedo (Figure 2e,f) and cotyledonary stage after 30 and 45 d, respectively. In the torpedo and cotyledonary stage embryos, it was possible to observe a group of elongated cells with visible nuclei that were distributed throughout the embryo and that constituted the vascular procambium zone (Figure 2f).

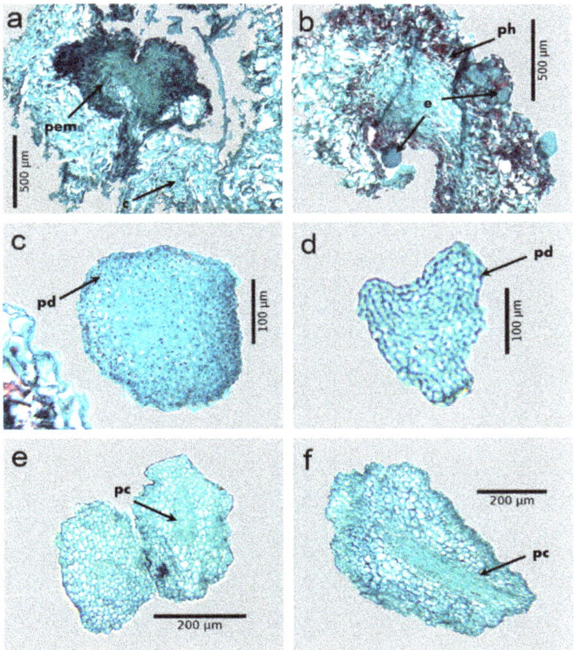

Figure 2. Histology of somatic embryogenesis of *Coffea arabica* L. var. Colombia. (**a**) Proembryogenic mass forming from a primary callus. (**b**) Somatic embryos differentiated on the proembryogenic mass that produces phenols. (**c**) Globular embryo at 120 d of culture in the differentiation medium. (**d**) Heart stage embryo after 135 d of culture in differentiation medium. (**e**) Early torpedo embryo. (**f**) Torpedo embryo after 150 d of culture on the differentiation medium, showing protodermis and procambium. c: primary callus; pem: proembryogenic mass; ph: phenols; e: embryos; pd: protodermis; pc: procambium.

3. Discussion

3.1. Induction of Embryogenic Callus

Even though all treatments tested induced callus formation in the explants of var. Colombia, after 15–30 d of culture, the medium containing 2 mg L^{-1} 2,4-D, 0.2 mg L^{-1} BAP, and 2.3 mg L^{-1} phytagel not only promoted that a high percentage of explants formed calluses but also formed calluses in greater quantities. In contrast, Gatica-Arias et al. [22] observed that 87% and 67% of the explants of the Caturra and Catuaí varieties of *C. arabica*, respectively, formed calluses in the presence of 4 mg L^{-1} kinetin, 1 mg L^{-1} 2,4-D, and 2.0 g L^{-1} Gelrite, after four months, that is, in a higher ratio of cytokinin to auxin. Likewise, Ahmed et al. [13] found that the best response in terms of callus formation (96%) was achieved when the explants of the hybrid "Ababuna" (*C. arabica*) were grown in an MS medium added with 1.0 mg L^{-1} 2, 4-D and 2.0 mg L^{-1} BAP for 90 d.

Likewise, the phytagel is a heteropolysaccharide used in the culture medium as a gelling agent. This compound is generally supplemented at a concentration ranging between 1.25–2.5 g L^{-1}. The gelling agent is the matric component of the water potential in the medium, that influences the amount of available water (the higher the concentration, the greater the retention) and the mobility of the solutes toward the cells of the explant, which will be reflected in their growth [23].

The fact that in the culture media tested containing a lower concentration of phytagel, a greater amount of calluses was formed than in those with a higher concentration (5.0 g L^{-1}), indicates that cell growth and division occurs more efficiently when there is more free water. In this regard, Meneses et al. [24] observed that the embryogenic calli of indica rice (*Oryza sativa*) grew better when the concentration of phytagel was less than 2.4 g L^{-1}.

3.2. Embryogenic Calli Production

The position and frequency of subcultures did have a significant effect, registering the highest biomass accumulation (18.9 times) when the flasks were kept tilted and no subcultures were carried out. In this regard, Gatica-Arias et al. [22] obtained 1.4 g of fresh callus weight of *C. arabica* L. cv. Catuaí from 250 mg of fresh callus weight (5.6 times) grown in 125 mL Erlenmeyer flasks with 25 mL of liquid medium, for four weeks in darkness. Likewise, de Rezende et al. [25] observed that the calli of Clone 3 [Siriema (*C. racemose* × *C. arabica*)] grown in MS salts at 50%, 1.1 mg L^{-1} 2,4-D, 1.0 mg L^{-1} indole butyric acid (IBA), and 2.0 mg L^{-1} 2-isopentenyladenine (2iP) for 63 days (with a subculture every 21 days), increased their fresh weight by 7 times, while those of Clone 28 (cross of Red Catuaí IAC 44 × Timor Hybrid CIFC 2750) increased their biomass by 9.8 times, under the same growing conditions.

In general, when cells remain in suspension, the absorption and utilization of nutrients, growth regulators, and water are greater than when they are cultured in semi-solid media [22]. However, in a liquid culture, anaerobic conditions can occur because oxygen diffuses less easily, which causes the oxygen level in the tissue to be reduced and, thus, the accumulation of biomass is less [26]. In the present investigation, when the flasks remained tilted, the cells were only temporarily submerged in the culture medium, which could have allowed the oxygen present in the flask to be more available to them, promoting their metabolism and growth to occur optimally [27].

The results obtained show that the nutrients in the culture medium can support the growth of calluses for three months without the need for it to be renewed, which allows for saving inputs and labor, reducing the risk of contamination and the probability of somaclonal variation occurring [28]. Keeping the flasks in a tilted position and without frequent subcultures can be an alternative for the successful multiplication of coffee embryogenic cultures.

3.3. Differentiation (Formation) of Somatic Embryos

The results showed that the presence of 2,4-D and BAP in a 1:2 ratio and the highest concentration of phytagel (5.0 g L^{-1}) in the culture medium had a synergistic effect to induce somatic embryogenesis in the calluses of cv. Colombia, allowing the highest number of embryos to be formed (118.7 embryos per gram of callus, before the multiplication stage). In this regard, Ahmed et al. [13] obtained the maximum number of embryos (58 embryos per magenta) from the calli of the hybrid "Ababuna" of *C. arabica*, generated in a medium that contained 1 mg L^{-1} 2,4-D and 2 mg L^{-1} BAP, which were then cultured in the presence of 4 mg L^{-1} BAP for six months.

Some authors [17–19,29] point out that the induction of somatic embryogenesis is the result of a combination between growth regulators and stress that can be caused by the removal of cells from the explant of the mother plant, as well as the conditions of *in vitro* cultures (high and low concentration of salts and gelling agent, osmotic shock, temperature, high or low light intensity, water stress, among others) [17–19]. However, it has been found that the endogenous hormone level is highly variable in the competent cells of different genotypes or not significantly different between a competent and a non-competent cell [17]. For this reason, the induction of somatic embryogenesis could not be attributed solely to growth regulators, but stress is considered a fundamental factor in this process. It has been suggested that somatic embryogenesis is a response of cells to extreme stress because in the early stages of embryo formation, the expression of different genes related to stress, oxidative stress, and cell division has been detected [17,30].

Potters et al. [31] point out that regardless of the type of stress to which cells are exposed, their division is stimulated, their growth is inhibited, and changes occur in their state of differentiation. Cells are brought from their differentiated condition to adapt to their new environment, leading to genetic and metabolic reprogramming. Under certain conditions, cells ensure their survival, leading to the formation of embryos [29].

On the other hand, 2,4-D has been used to induce somatic embryogenesis in more than 60% of the protocols [32]. It is suggested that 2,4-D, above certain concentrations, can act as a growth regulator or as an agent that generates stress. 2,4-D can modify the endogenous metabolism of IAA by affecting its binding to auxin-binding proteins [18]. Likewise, it is suggested that this regulator generates hypermethylation, which maintains cells in a state of high mitotic activity and in a proembryogenic phase [24].

Moreover, while callus induction using 2.3 g L^{-1} phytagel promoted the best response, including 5.0 g L^{-1} in the differentiation medium generated the highest number of embryos per gram of callus. This response could be because the phytagel, by acting as a matrix, decreases the availability of water and nutrients from the culture medium for the cells and at high concentrations causes water stress [33].

3.4. Development, Maturation, and Conversion

During maturation, somatic embryos undergo biochemical (accumulation of reserve substances) and morphological changes (increase in size) which are essential for their subsequent development, germination, and conversion into photoautotrophic plants [34].

The results of the present investigation show that including auxins and cytokinins in the development medium in a 1:1 ratio, allowed embryos in the globular or torpedo stage to pass to the cotyledon stage and that low concentrations (0.25 mg L^{-1}) of both BAP and IAA (MD2 medium) were sufficient to induce this response. Likewise, it was observed that embryos in the globular stage can reach the cotyledonary stage when cultured in a single medium and that it is not necessary to separate the embryos in the different stages (globular and torpedo) to promote their development since the MD2 medium works for both; this allows for reducing time, supplies, and labor.

On the other hand, different authors point out that the success in the conversion depends to a great extent on how efficient the maturation of the embryos was, as those that do not mature show low germination rates [35]. For the Colombia variety, the conversion of the embryos was higher when cultured in a mixture of perlite and vermiculite than in agar; however, the conversion rate was not greater than 25%. These results coincide with those obtained by Georget et al. [36], who found that the average conversion rate of somatic embryos to seedlings of hybrids H1 (T5296 × RS), H3 (Caturra × ET 531), H10

(T5296 × RS), H16 (T5296 × ET01A1), H17 (Catuai × ET59A2), and other *C. arabica* hybrids in a substrate (peatmoss and sand) was only 37%. The embryos of the Colombia variety grown in the mixture of substrates resulted in larger plants with a greater capacity to adapt to *ex vitro* conditions than those that remained on agar; it could be because the physical characteristics of the substrates, such as porosity, allow a better oxygen and water balance, facilitating the diffusion of nutrients and the formation of more vigorous roots [37,38]. The aforementioned indicates that germination and conversion are still a bottleneck for coffee micropropagation through somatic embryogenesis [35,36].

3.5. Histological Analysis of Somatic Embryogenesis

The histological analysis of the cultures allowed us to observe that the primary callus was made up of cells that were large, elongated, with vacuoles and large intercellular spaces as described by Bartos et al. [39] in *C. arabica* cv. Catuaí Vermelho. After the primary callus was transferred to the differentiation medium, proembryogenic masses began to form; these masses had a certain degree of organization in which the somatic proembryos arose, as was also observed in the *C. arabica* Caturra and Catuaí varieties [22,39,40].

The fact that only calli that produced phenols formed somatic embryos is consistent with the observations of Martins et al. [41], who found that the presence of phenols in the calli of *Arbutus unedo* and *A. canariensis* was a prerequisite for the formation of somatic embryos. Tonietto et al. [42] indicate that during the early stages of somatic embryogenesis, a large number of reactive oxygen substances (ROS) is generated; to defend against ROS, cells develop different enzymatic and non-enzymatic antioxidants (tocopherols, carotenoids, phenolic compounds, among others), systems that catalyze the breakdown of these oxygen molecules [43].

It was possible to observe the transition of the globular embryos to the heart, torpedo, and cotyledon phase in the following 45 days of culture in the development and maturation medium. Globular stage embryos showed an outer layer of cells covering the embryo, the protodermis, as observed by Bartos et al. [39] in the somatic embryos of *C. arabica* var. Caturra. In the torpedo stage, procambium cells connected to the stem and root meristem in embryos of the Colombia variety. In this regard, Nic-Can et al. [44] consider that the presence of the polarized fundamental meristems with differentiating procambium cells are the main anatomical characteristics of the torpedo somatic embryos of *C. canephora*. Likewise, the lack of vascular connection between the regenerated somatic embryos and the original explant was evident.

The histological analysis of the cultures obtained in the present investigation allowed us to confirm the embryogenic nature of the regenerated plants. These structures and cellular characteristics could be used as morphological markers to detect those cells that have embryogenic competence, as well as to define with more precision, the ideal time to transfer the cultures to the following stages, thereby contributing to reducing the time to obtain somatic embryos and the optimization of micropropagation protocols [39,44].

Coffee is the second-most consumed beverage in the world; *C. arabica* represents 70% of the production of this crop worldwide, which makes it the most important species of the genus. Even though there are protocols for somatic embryogenesis in different species and varieties of coffee [22,36,45,46], for *C. arabica* var. Colombia, the information is limited [47]. More research is required to establish an efficient and reproducible protocol for the propagation of plants of this variety, which in addition to being tolerant to coffee rust, has outstanding agronomic characteristics and wide adaptation to the environments where coffee is grown, as well as having a good taste quality [48]. Having an efficient micropropagation protocol for the var. Colombia will allow to have enough plants to establish and renew plantations with which economic and environmental costs are reduced, by making less use of agrochemicals (fungicides) to combat the fungus that causes rust of the coffee.

4. Materials and Methods

4.1. Disinfection

The first and second pairs of young leaves, healthy and undamaged, were obtained from one-year-old plants grown in a greenhouse at 32 ± 4 °C, 400 µmol m^{-2} s^{-1} of light intensity, with a relative humidity of 40%. The plants were watered every third day, fertilized every four months with Multicote 15-7-15® (slow-release fertilizer, Green Forest, Puebla, México), and fumigated monthly with Cupravit® (copper oxychloride, Bayer®, Querétaro, México). Then, leaves were disinfected by immersion in 0.1% fungicidal solution (Promyl®, Promotora Técnica Industrial, S.A. de C.V., Morelos, México): methyl-1-butyl carbamoyl-2-benzimidazole carbamate) for 15 min and then rinsed with sterilized distilled water. After that, leaves were placed in sodium hypochlorite (1.2% active chlorine) for 20 min and then rinsed with sterilized distilled water.

4.2. Induction of Embryogenic Callus

Under sterile conditions in the laminar air flow, segments of 1 cm^2 (explants) were obtained from disinfected leaves and placed in 90 × 15 mm Petri dishes containing 30 mL of MS (Murashige and Skoog) basal medium [49]. The media were supplemented by combining several concentrations of two growth regulators and two concentrations of phytagel (Table 5). Culture media were enriched with 200 mg L^{-1} of ascorbic acid, 100 mg L^{-1} of citric acid (antioxidants), and 30 g L^{-1} of sucrose; pH was adjusted to 5.7 and then, media were sterilized in an autoclave at 121 °C for 20 min.

Table 5. Treatments used to induce the embryogenic calli from leaf explants of *C. arabica* var. Colombia.

Treatment	Growth Regulators (mg L^{-1})		Phytagel (g L^{-1})
	2,4-D	BAP	
T1	0.5	1.1	2.3
T2	0.5	1.1	5.0
T3	1.0	1.0	2.3
T4	1.0	1.0	5.0
T5	2.0	0.20	2.3
T6	2.0	0.20	5.0

2,4-D, 2,4-dichlorophenoxyacetic acid; BAP, 6-benzylaminopurine.

The cultures were incubated in a growth chamber under dark conditions at 26 ± 2 °C. After two months, the following were evaluated: (a) the percentage of explants that generated calluses and (b) the amount of calluses that formed on each explant (explant area covered with calluses). A scale to evaluate the amount of calluses was used, considering the percentage of area covered with calluses as follows: 1: 0–20%; 2: 21–40%; 3: 41–60%; 4: 61–80%; 5: 81–100%. The experiment for callus induction had a factorial arrangement with two factors: growth regulator combinations (3), and phytagel concentration (2), resulting in six treatments. Each treatment was reproduced in ten repetitions, each (Petri dish) containing six explants.

4.3. Embryogenic Calli Production

Five hundred milligrams were placed in 125 mL flasks with 50 mL of full MS medium supplemented with 1 mg L^{-1} 2,4-D, 1 mg L^{-1} BAP, and 30 g L^{-1} sucrose; the flasks were shaken (orbital shaker, Thomas Scientific®, Swedesboro, NJ, USA) at 110 rpm in darkness.

To evaluate the increase in biomass of the calli, three conditions were tested: the first consisted of cultivating the calli in 125 mL flasks and keeping them upright in an orbital shaker, performing subcultures every four weeks for three months. In the second and third conditions, the calli were cultured in 125 mL flasks without performing subcultures; after three months the weight of the calluses was recorded. The difference between the second and third conditions consisted of placing the flasks on the shaker in a vertical or inclined position (30° approximately), respectively.

A completely randomized experimental design with three treatments and ten repetitions was used; one flask was considered as a repetition.

4.4. Differentiation (Formation) of Somatic Embryos

After 60 d, the embryogenic callus formed in each of the induction media was separated and placed in Petri dishes of 60 × 15 mm with 15 mL of differentiation medium, which contained the basal salts of Yasuda et al., modified [21], 1.1 mg L^{-1} BAP, 5.0 g L^{-1} phytagel, and 30 g L^{-1} sucrose. The cultures were incubated at 26 ± 2 °C in darkness and after four months, the number of embryos per gram of callus was quantified.

The experiment had a factorial arrangement with two factors: growth regulator combinations in the induction medium (3), and the concentration of phytagel (2), giving rise to six treatments; each treatment consisted of ten repetitions and one repetition was made up of a Petri dish with calluses (500 mg).

4.5. Development and Maturation of Embryos

The embryos in the globular state, heart, and initial torpedo were placed in flasks of 30 mL with 10 mL of two culture media made with the basal salts of Yasuda et al., modified [21], 5.0 g L^{-1} phytagel, 30 g L^{-1} sucrose plus 0.5 mg L^{-1} 6-furfurylaminopurine (kinetin), and 0.5 mg L^{-1} indoleacetic acid (IAA) (MD1 medium) or 0.25 mg L^{-1} BAP and 0.25 mg L^{-1} IAA (medium MD2). The number of cotyledonary stage embryos was evaluated after six weeks. A completely randomized experimental design with two treatments was used. Each treatment (culture medium) consisted of 15 repetitions, considering as an experimental unit a flask with 10 embryos.

4.6. Conversion

The cotyledonary stage embryos were transferred to 250 mL flasks containing 30 mL with half MS medium, 2.3 g L^{-1} phytagel, and 20 g L^{-1} sucrose or a mixture of vermiculite: perlite (3:1, particle size 0.5 mm); the mixture was moistened with 30 mL of the previously described medium without agar. The cultures were kept at 26 ± 2 °C and 16 h of fluorescent cold white light (60 µmol m^{-2} s^{-1}). After two months, the number of embryos that became plants was counted. The experimental design was a completely randomized one with two treatments and ten repetitions; a flask with ten embryos was considered as an experimental unit.

4.7. Histology

Embryogenic calli at different stages of development were fixed in a mixture of 10% formaldehyde, 5% acetic acid, 52% ethanol, and 33% water (*v*/*v*); then, they were dehydrated with a series of ethanols (30, 40, 50, 70, 85, 100%) and xylenes (ethyl alcohol 50% -xylene 50%, pure xylene) and embedded in paraffin. Thick sections of 10 µm were made with a rotary microtome (American Optical®, model Spencer 820, Vernon Hills, IL, USA). The samples were treated with xylene (100%), and a series of alcohols (ethanol 50, 70, 85, 100%) to eliminate paraffin; then, they were stained with fixed O-safranin-green and infiltrated and embedded in synthetic resin [50]. The samples were observed with the help of a Carl Zeiss® (Oberkochen, Germany) optical microscope (Tessovar model) and the images were captured with a Paxcam® (Villa Park, IL, USA) digital camera.

4.8. Statistical Analysis

The data were statistically processed using an analysis of variance (ANOVA); the means of the treatments were compared utilizing a Tukey test with a significance level of 5%. The statistical program SAS version 9.0 was used.

5. Conclusions

The culture conditions established in the present investigation allowed the development of a protocol by which it is possible to obtain *in vitro* plants of *C. arabica* var. Colombia, through indirect somatic embryogenesis. The osmotic stress generated by the high concentration of phytagel in the medium increased the ability of the explants to form embryos. The low concentration of nutrients in the callus multiplication medium and the temporary contact of the cells with the culture medium made it possible to multiply them efficiently. The regenerated embryos were able to grow into plants and establish themselves under *ex vitro* conditions.

Author Contributions: Conceptualization, C.M.A.-V., V.M.O.-C., E.d.J.A.-S., L.I.-D., F.C.G.-M. and A.R.-P.; data curation, C.M.A.-V.; formal analysis, C.M.A.-V., E.d.J.A.-S. and A.R.-P.; funding acquisition, A.R.-P.; investigation, C.M.A.-V. and A.R.-P.; methodology, C.M.A.-V. and A.R.-P.; project administration, A.R.-P.; resources, V.M.O.-C. and A.R.-P.; software, C.M.A.-V. and E.d.J.A.-S.; supervision, L.I.-D. and A.R.-P.; validation, C.M.A.-V., V.M.O.-C., L.I.-D., F.C.G.-M. and A.R.-P.; visualization, C.M.A.-V., F.C.G.-M. and A.R.-P.; writing—original draft, C.M.A.-V.; writing—review and editing, C.M.A.-V., O.C, E.d.J.A.-S., L.I.-D., F.C.G.-M. and A.R.-P. All authors have read and agreed to the published version of the manuscript.

Funding: This research was funded by the National Council of Science and Technology (CONACYT), Mexico (1146055).

Data Availability Statement: All data generated in this study are included in the tables and figures.

Acknowledgments: The authors thank the National Council of Science and Technology (CONACYT) for the scholarship granted to Consuelo Margarita Avila Victor to carry out her doctoral studies (1146055). Likewise, we would like to thank Jorge M. Valdéz Carrasco for the technical support in the image processing.

Conflicts of Interest: The authors declare no conflict of interest.

References

1. Jaramillo, R.A. La Agroclimatología Del Cafeto. In *Clima Andino y Café en Colombia*; CENICAFE: Manizales, Colombia, 2005; pp. 149–157. Available online: https://biblioteca.cenicafe.org/bitstream/10778/859/1/Portada.pdf (accessed on 7 February 2023).
2. Farah, A.; Dos Santos, T.F. *The Coffee Plant and Beans: An Introduction*; Elsevier Inc.: Amsterdam, The Netherlands, 2015; ISBN 9780124167162. Available online: https://www.researchgate.net/profile/Adriana_Farah2/publication/267967496_The_coffee_plant_and_beans_Introduction_Chapter_1/links/546541710cf2f5eb17ff3901/The-coffee-plant-and-beans-Introduction-Chapter-1.pdf (accessed on 7 February 2023).
3. International Coffee Organization. World Coffee Production. Available online: https://www.ico.org/prices/po-production.pdf (accessed on 23 February 2023).
4. Davis, A.P. Psilanthus Mannii, the Type Species of Psilanthus, Transferred to Coffea. *Nord. J. Bot.* **2011**, *29*, 471–472. [CrossRef]
5. Wintgens, J.N. *Coffee: Growing, Processing, Sustainable Production. A Guidebook for Growers, Processors, Traders, and Researchers*; WILEY-VCH Verlag GMBH & Co. KGaA: Weinheim, Germany, 2004; ISBN 3527307. [CrossRef]
6. Flores, D. Mexico Coffee Annual Production Flat with Quality Exports and Robusta Bean Imports Both Rising. Available online: https://apps.fas.usda.gov/newgainapi/api/report/downloadreportbyfilename?filename=Coffee%20Annual_Mexico%20City_Mexico_5-14-2012.pdf (accessed on 17 November 2022).
7. Servicio de Información Agroalimentaria y Pesquera (SIAP). *Panorama Agroalimentario 2021 | Servicio de Información Agroalimentaria y Pesquera | Gobierno | Gob.Mx*; SIAP: Ciudad de México, México, 2021; Available online: https://www.researchgate.net/figure/Fuente-Servicio-de-Informacion-Agroalimentaria-y-Pesquera-SIAP-Recuperado-del_fig1_348994273 (accessed on 7 February 2023).
8. Renard Hubert, M.-C.; Larroa Torres, R.M. Política Pública y Sustentabilidad de Los Territorios Cafetaleros En Tiempos de Roya: Chiapas y Veracruz. *Estud. Latinoam.* **2017**, 95–113. [CrossRef]
9. Flores, D. Coffee Plan on Track to Achieve Goals. Available online: https://apps.fas.usda.gov/newgainapi/api/report/downloadreportbyfilename?filename=Coffee%20Annual_Mexico%20City_Mexico_5-15-2018.pdf (accessed on 15 July 2022).
10. Hernandez, O. ¿Un Futuro Sin Café? El Cambio Climático Disminuirá En Un 50% La Superficie Cultivable de Café Para 2050. Mira Cuales Aquí. Available online: https://www.iadb.org/es/mejorandovidas/el-efecto-mas-impensado-del-cambio-climatico (accessed on 23 February 2023).
11. Flores, D.; Harrison, T. Mexico Launches New Policies as Rust Continues to Impact Production. Available online: https://apps.fas.usda.gov/newgainapi/api/report/downloadreportbyfilename?filename=Coffee%20Annual_Mexico%20City_Mexico_5-13-2016.pdf (accessed on 28 October 2022).
12. SAGARPA Convención Internacional Del Café México 2015 | Servicio de Información Agroalimentaria y Pesquera | Gobierno | Gob.Mx. Available online: https://www.gob.mx/siap/convencion-internacional-del-cafe-mexico-2015 (accessed on 1 February 2023).

13. Ahmed, W.; Feyissa, T.; Disasa, T. Somatic Embryogenesis of a Coffee (*Coffea Arabica* L.) Hybrid Using Leaf Explants. *J. Hortic. Sci. Biotechnol.* **2013**, *88*, 469–475. [CrossRef]
14. Su, Y.H.; Zhang, X.S. Plant Signaling & Behavior Auxin Gradients Trigger de Novo Formation of Stem Cells during Somatic Embryogenesis. *Plant Signal. Behav.* **2009**, *4*, 574–576. [CrossRef] [PubMed]
15. Khan, S.; Al-Qurainy, F.; Nadeem, M. Biotechnological Approaches for Conservation and Improvement of Rare and Endangered Plants of Saudi Arabia. *Saudi J. Biol. Sci.* **2012**, *19*, 1–11. [CrossRef] [PubMed]
16. López-Gómez, P.; Iracheta-Donjuan, L.; Castellanos-Juárez, M.; Méndez-López, I.; Sandoval-Esquivez, A.; Aguirre-Medina, J.F.; Carmen Ojeda-Zacarías, M.; Gutiérrez-Díez, A. Influencia Del Explante y Medio de Cultivo En La Embriogénesis Somática En Hojas de Café. *Rev. Fitotec. Mex* **2010**, *33*, 205–213. [CrossRef]
17. Zavattieri, M.A.; Frederico, A.M.; Lima, M.; Sabino, R.; Arnholdt-Schmitt, B. Induction of Somatic Embryogenesis as an Example of Stress-Related Plant Reactions. *Electron. J. Biotechnol.* **2010**, *13*, 1–9. [CrossRef]
18. Fehér, A. Why Somatic Plant Cells Start to Form Embryos? In *Somatic Embryogenesis*; Springer: Berlin/Heidelberg, Germany, 2005; Volume 2, pp. 85–101. [CrossRef]
19. Patnaik, D.; Mahal, A.; Khurana, P. *Effect of Water Stress and Heavy Metals on Induction of Somatic Embryogenesis in Wheat Leaf Base Cultures*; NIScPR: New Delhi, India, 2005; Volume 43. Available online: https://pubmed.ncbi.nlm.nih.gov/16121718/ (accessed on 7 February 2023).
20. Pasternak, T.P.; Prinsen, E.; Ayaydin, F.; Miskolczi, P.; Potters, G.; Asard, H.; Van Onckelen, H.A.; Dudits, D.; Fehér, A. The Role of Auxin, PH, and Stress in the Activation of Embryogenic Cell Division in Leaf Protoplast-Derived Cells of Alfalfa. *Plant Physiol.* **2002**, *129*, 1807–1819. [CrossRef]
21. Yasuda, T.; Fujii, Y.; Yamaguchi, T. Embryogenic Callus Induction from Coffea Arabica Leaf Explants by Benzyladenine. *Plant Cell Physiol.* **1985**, *26*, 595–597. [CrossRef]
22. Gatica-Arias, A.M.; Arrieta-Espinoza, G.; Espinoza Esquivel, A.M. Plant Regeneration via Indirect Somatic Embryogenesis and Optimisation of Genetic Transformation in Coffea Arabica L. Cvs. Caturra and Catuaí. *Electron. J. Biotechnol.* **2008**, *11*, 1–12. [CrossRef]
23. López-Escamilla, A.L.; López-Herrera, M.; Loaiza-Alanís, C.; López-Escamilla, A.L.; López-Herrera, M.; Loaiza-Alanís, C. Efecto de Diferentes Agentes Gelificantes En La Germinación y Desarrollo in Vitro de Plántulas de Echinocactus Platyacanthus Link et Otto (Cactaceae). *Polibotánica* **2016**, 153–166. [CrossRef]
24. Meneses, A.; Flores, D.; Muñoz, M.; Arrieta, G.; Espinoza, A.M. Effect of 2,4-D, Hydric Stress and Light on Indica Rice (Oryza Sativa) Somatic Embryogenesis. *Rev. Biol. Trop.* **2005**, *53*, 361–368. [CrossRef] [PubMed]
25. de Rezende, J.C.; de Carvalho, C.H.S.; Pasqual, M.; Santos, A.C.R.; de Carvalho, S.M. Calli Induction in Leaf Explants of Coffee Elite Genotypes. *Ciência Rural* **2011**, *41*, 384–389. [CrossRef]
26. Curtis, W.R. Application of Bioreactor Design Principles to Plant Micropropagation. In *Liquid Culture Systems for In Vitro Plant Propagation*; Springer: Berlin/Heidelberg, Germany, 2005; Volume 9781402031, pp. 21–40. ISBN 9781402032004. [CrossRef]
27. de Feria, M.; Jiménez, E.; Barbón, R.; Capote, A.; Chávez, M.; Quiala, E. Effect of Dissolved Oxygen Concentration on Differentiation of Somatic Embryos of Coffea Arabica Cv. Catimor 9722. *Plant Cell. Tissue Organ Cult.* **2003**, *72*, 1–6. [CrossRef]
28. Bobadilla Landey, R.; Cenci, A.; Guyot, R.; Bertrand, B.; Georget, F.; Dechamp, E.; Herrera, J.-C.; Aribi, J.; Lashermes, P.; Etienne, H. Assessment of Genetic and Epigenetic Changes during Cell Culture Ageing and Relations with Somaclonal Variation in Coffea Arabica. *Plant Cell. Tissue Organ Cult.* **2015**, *122*, 517–531. [CrossRef]
29. Fehér, A. Somatic Embryogenesis—Stress-Induced Remodeling of Plant Cell Fate. *Biochim. Biophys. Acta-Gene Regul. Mech.* **2015**, *1849*, 385–402. [CrossRef] [PubMed]
30. Thibaud-Nissen, F.; Shealy, R.T.; Khanna, A.; Vodkin, L.O. Clustering of Microarray Data Reveals Transcript Patterns Associated with Somatic Embryogenesis in Soybean. *Plant Physiol.* **2003**, *132*, 118–136. [CrossRef]
31. Potters, G.; Pasternak, T.P.; Guisez, Y.; Palme, K.J.; Jansen, M.A.K. Stress-Induced Morphogenic Responses: Growing out of Trouble? *Trends Plant Sci.* **2007**, *12*, 98–105. [CrossRef]
32. Karami, O.; Saidi, A. The Molecular Basis for Stress-Induced Acquisition of Somatic Embryogenesis. *Mol. Biol. Rep.* **2010**, *37*, 2493–2507. [CrossRef]
33. Huang, Y.; Begum, M.; Chapman, B.; Hocking, A.D. Effect of Reduced Water Activity and Reduced Matric Potential on the Germination of Xerophilic and Non-Xerophilic Fungi. *Int. J. Food Microbiol.* **2010**, *140*, 1–5. [CrossRef]
34. Aslam, J.; Khan, S.A.; Cheruth, A.J.; Mujib, A.; Sharma, M.P.; Srivastava, P.S. Somatic Embryogenesis, Scanning Electron Microscopy, Histology and Biochemical Analysis at Different Developing Stages of Embryogenesis in Six Date Palm (*Phoenix Dactylifera* L.) Cultivars. *Saudi J. Biol. Sci.* **2011**, *18*, 369–380. [CrossRef] [PubMed]
35. Etienne, H.; Bertrand, B.; Georget, F.; Lartaud, M.; Montes, F.; Dechamp, E.; Verdeil, J.-L.; Barry-Etienne, D. Development of Coffee Somatic and Zygotic Embryos to Plants Differs in the Morphological, Histochemical and Hydration Aspects. *Tree Physiol.* **2013**, *33*, 640–653. [CrossRef] [PubMed]
36. Georget, F.; Courtel, P.; Garcia, E.M.; Hidalgo, M.; Alpizar, E.; Breitler, J.-C.; Bertrand, B.; Etienne, H. Somatic Embryogenesis-Derived Coffee Plantlets Can Be Efficiently Propagated by Horticultural Rooted Mini-Cuttings: A Boost for Somatic Embryogenesis. *Sci. Hortic.* **2017**, *216*, 177–185. [CrossRef]

37. Barry-Etienne, D.; Bertrand, B.; Vasquez, N.; Etienne, H. Comparison of Somatic Embryogenesis-Derived Coffee (*Coffea Arabica* L.) Plantlets Regenerated in Vitro or Ex Vitro: Morphological, Mineral and Water Characteristics. *Ann. Bot.* **2002**, *90*, 77–85. [CrossRef] [PubMed]
38. Jimarez-Montiel, M.J.; Robledo-Paz, A.; Ordaz-Chaparro, V.M.; Trejo-Tellez, L.I.; Molina-Moreno, J.C. Protocol for the Reduction of Costs in Habanero Chili (Capsicum Chinense Jacq.) Micropropagation. *Phyton* **2018**, *87*, 94–104. [CrossRef]
39. Bartos, P.M.C.; Gomes, H.T.; Gomes, S.M.; Vasconcelos Filho, S.C.; Teixeira, J.B.; Scherwinski-Pereira, J.E. Histology of Somatic Embryogenesis in *Coffea Arabica* L. *Biologia* **2018**, *73*, 1255–1265. [CrossRef]
40. Ribas, A.F.; Dechamp, E.; Champion, A.; Bertrand, B.; Combes, M.-C.; Verdeil, J.-L.; Lapeyre, F.; Lashermes, P.; Etienne, H. Agrobacterium-Mediated Genetic Transformation of *Coffea Arabica* (L.) Is Greatly Enhanced by Using Established Embryogenic Callus Cultures. *BMC Plant Biol.* **2011**, *11*, 92. [CrossRef]
41. Martins, J.F.; Correia, S.I.; Canhoto, J.M. Somatic Embryogenesis Induction and Plant Regeneration in Strawberry Tree (*Arbutus Unedo* L.). *Methods Mol. Biol.* **2016**, *1359*, 329–339. [CrossRef]
42. Tonietto, Â.; Sato, J.H.; Teixeira, J.B.; de Souza, E.M.; Pedrosa, F.O.; Franco, O.L.; Mehta, A. Proteomic Analysis of Developing Somatic Embryos of Coffea Arabica. *Plant Mol. Biol. Report.* **2012**, *30*, 1393–1399. [CrossRef]
43. Kulbat, K. The Role of Phenolic Compounds in Plant Resistance. *Biotechnol. Food Sci.* **2016**, *80*, 97–108. [CrossRef]
44. Nic-Can, G.I.; López-Torres, A.; Barredo-Pool, F.; Wrobel, K.; Loyola-Vargas, V.M.; Rojas-Herrera, R.; De-la-Peña, C. New Insights into Somatic Embryogenesis: LEAFY COTYLEDON1, BABY BOOM1 and WUSCHEL-RELATED HOMEOBOX4 Are Epigenetically Regulated in Coffea Canephora. *PLoS ONE* **2013**, *8*, e72160. [CrossRef] [PubMed]
45. Menéndez-Yuffá, A.; Barry-Etienne, D.; Bertrand, B.; Georget, F.; Etienne, H. A Comparative Analysis of the Development and Quality of Nursery Plants Derived from Somatic Embryogenesis and from Seedlings for Large-Scale Propagation of Coffee (*Coffea Arabica* L.). *Plant Cell. Tissue Organ Cult.* **2010**, *102*, 297–307. [CrossRef]
46. Molina, D.M.; Aponte, M.E.; Cortina, H.; Moreno, G. The Effect of Genotype and Explant Age on Somatic Embryogenesis of Coffee. *Plant Cell. Tissue Organ Cult.* **2002**, *71*, 117–123. [CrossRef]
47. Avila-Victor, C.M.; Martinez-Infante, A.; Ordaz-Chaparro, V.M.; Arjona-Suarez, E.J. Embriogénesis Somática Directa e Indirecta En Coffea Arabica Var. Colombia. *Agroproductividad* **2018**, *11*, 30–35. Available online: https://core.ac.uk/download/pdf/249320109.pdf (accessed on 7 February 2023).
48. Escamilla, P.E.; Ruiz, R.O.; Díaz, P.G.; Landeros, S.C.; Platas, R.D.E.; Zamarripa, C.A.; González, H.V.A. El Agroecosistema Café Orgánico En México. *Manejo Integr. Plagas Agroecol. (Costa Rica)* **2005**, *76*, 5–16. Available online: https://repositorio.catie.ac.cr/handle/11554/6135 (accessed on 7 February 2023).
49. Murashige, T.; Skoog, F. A Revised Medium for Rapid Growth and Bio Assays with Tobacco Tissue Cultures. *Physiol. Plant.* **1962**, *15*, 473–497. [CrossRef]
50. Márquez Guzmán, J.; Wong, R.; Pérez Pacheco, M.; Lopez Curto, L.; Murguía Sánchez, G. *Tecnicas Para El Estudio Del Desarrollo En Angiospermas*; Libro de Laboratorio: Ciudad de México, México, 2016; ISBN 9789703227488. Available online: https://chapingo.orex.es/bib/95483 (accessed on 7 February 2023).

plants

MDPI

Article

Characterization of the NAC Transcription Factor in Passion Fruit (*Passiflora edulis*) and Functional Identification of *PeNAC-19* in Cold Stress

Yi Xu [1,2,3], Pengfei Li [4], Funing Ma [1,2,3], Dongmei Huang [1], Wenting Xing [1], Bin Wu [1], Peiguang Sun [1], Binqiang Xu [1] and Shun Song [1,2,3,*]

1 State Key Laboratory of Biological Breeding for Tropical Crops, Tropical Crops Genetic Resources Institute, Chinese Academy of Tropical Agricultural Sciences, Germplasm Repository of Passiflora, Hainan Province, Hainan 571101, China
2 Sanya Research Institute, Chinese Academy of Tropical Agricultural Sciences, Sanya 571101, China
3 Hainan Yazhou Bay Seed Laboratory, Sanya 571101, China
4 College of Tropical Crops, Yunnan Agricultural University, Kunming 650201, China; lpengfei24@163.com
* Correspondence: songs@catas.cn

Abstract: The NAC (NAM, ATAF and CUC) gene family plays an important role in plant development and abiotic stress response. However, up to now, the identification and research of the NAC (*PeNAC*) family members of passion fruit are still lacking. In this study, 25 *PeNACs* were identified from the passion fruit genome, and their functions under abiotic stress and at different fruit-ripening stages were analyzed. Furthermore, we analyzed the transcriptome sequencing results of *PeNACs* under four various abiotic stresses (drought, salt, cold and high temperature) and three different fruit-ripening stages, and verified the expression results of some genes by qRT-PCR. Additionally, tissue-specific analysis showed that most *PeNACs* were mainly expressed in flowers. In particular, *PeNAC-19* was induced by four various abiotic stresses. At present, low temperatures have seriously endangered the development of passion fruit cultivation. Therefore, *PeNAC-19* was transformed into tobacco, yeast and Arabidopsis to study their function of resisting low temperature. The results show that *PeNAC-19* responded to cold stress significantly in tobacco and Arabidopsis, and could improve the low temperature tolerance of yeast. This study not only improved the understanding of the *PeNAC* gene family characteristics and evolution, but also provided new insights into the regulation of the *PeNAC* gene at different stages of fruit maturation and abiotic stresses.

Keywords: NAC transcription factor; passion fruit; abiotic stress; fruit maturity stage; gene expression

Citation: Xu, Y.; Li, P.; Ma, F.; Huang, D.; Xing, W.; Wu, B.; Sun, P.; Xu, B.; Song, S. Characterization of the NAC Transcription Factor in Passion Fruit (*Passiflora edulis*) and Functional Identification of *PeNAC-19* in Cold Stress. *Plants* **2023**, *12*, 1393. https://doi.org/10.3390/plants12061393

Academic Editors: Zanmin Hu, Han Xiao, Yi Ren and Chengming Fan

Received: 20 February 2023
Revised: 7 March 2023
Accepted: 13 March 2023
Published: 21 March 2023

1. Introduction

Transcription factors (TFs) play an important role in regulating cell signaling, cell morphogenesis and plant resistance to external environmental stresses [1,2]. TFs regulate gene expression by binding to specific promoter cis-acting elements to activate or repress the transcriptional level of target genes [3,4]. In plants, the NAC (NAM, ATAF1/2 and CUC2) transcription factor family is named after three proteins: petunia apical meristem (NAM), *Arabidopsis thaliana* ATAF1/2 and cup cotyledon (CUC) [5,6], which is one of the largest and most plant-specific TF families. Typical NAC proteins include a highly conserved N-terminal region (NAC domain), while the C-terminal region contains a relatively distinct transcriptional activation/repression region (TAR or TRR) [5,7,8], which is a highly diverse transcriptional regulatory region [9], may be involved in protein–protein interactions and contribute to its regulatory specificity [10]. The N-terminus of the NAC protein contains a conserved domain of 150–160 amino acids involved in DNA binding, dimerization and localization [11], which is further divided into five subdomains (A-E), of which the A, C and D subdomains are highly conserved [12].

Since the first report of the NAC protein in 1996 [5], NAC protein families have been identified in several plant species, such as *Arabidopsis thaliana* [13], *Oryza sativa* [14], *Musa acuminata* [15], *Medicago sativa* [16], *Dimocarpus longan* [17], *Actinidia chinensis* [18], *Populus trichocarpa* [19], *Vitis vinifera* [20] and *Pyrus pyrifolia* [21]. Due to their powerful functions in plants, the NAC family has been extensively studied in recent years. The related reports indicate that NAC family members play an essential role in response to plant abiotic stress.

The overexpression of *OsNAC10* in rice can improve the drought tolerance and grain yield of plants [22]. The overexpression of *OsNAC6/SNAC2* can improve the drought tolerance, salt tolerance and cold tolerance of rice seedlings [23,24]. The transcription factor *MbNAC25* of Siberian crab apple (*Malus baccata* Borkh) can improve cold tolerance in transgenic Arabidopsis [25]. Both *SNAC2* of rice and *PbeNAC1* of pear (*Pyrus betulifolia* Bunge) confer cold tolerance in rice and pears [26]. In tomatoes, *SlNAM1* is induced by chilling stress and can improve the chilling resistance in transgenic plants [27].

In addition, there are also related studies showing that the NAC transcription factor is related to fruit development and ripening. Fruit development and ripening are complex processes regulated by various factors such as gene regulation, hormones, light and temperature The processes in the development and ripening of fruit are regulated by many factors such as light, temperature, hormone and gene regulation. In particular, they are regulated by a variety of transcription factors that affect the expression levels of downstream target genes [28]. Some studies have shown that NAC transcription factors can regulate the expression of the genes involved in hormone biosynthesis and signal transduction during fruit development and maturation [29]. In spruce, the overexpression of *PaNAC03* affects plant embryonic development [30]. Furthermore, in *Vitis vinifera*, the development of seeds and fruits is affected by the interaction between *VvNAC26* and *VvMADS9* [31]. In tomatoes, *SlNAC1* plays an important role in the softening process [32]. In strawberries (*Fragaria chiloensis*), the FcNAC1 protein is involved in pectin metabolism to soften fruit [33].

Passion fruit (*Passiflora edulis* Sim) is a perennial evergreen vine of the *Passiflora* genus of Passifloraceae, a tropical and rare fruit tree, which contains more than 100 kinds of aroma in its fruit pulp. It is native to central and northern South America, and is widely distributed throughout America, Australia and Africa [34]. According to the reports, *Passiflora* has about 520 varieties, most of which are used for ornamental purposes, and only a small number of 60 species can be eaten [35]. At present, the main countries for passion fruit are Brazil, Colombia, Ecuador, Australia, Vietnam, China, etc. Due to its unique flavor and short growth period (4–6 months), the planting area under cultivation gradually increased. Abiotic stresses seriously affect the normal development of the passion fruit industry. Therefore, it is important to excavate the function genes of stress resistance in passion fruit and analyze their mechanisms of action [36]. In this study, using the high-quality genomic data of passion fruit [37], 25 members of the passion fruit NAC (*PeNAC*) family were identified. In contrast to another previously published result [38], they used another genome [39]. The result of genome assemblies vary widely in the HR genome [37] and the MER genome [39]: contig N50 (3.1 Mb and 70 kb), complete BUSCOs (91.56% and 88.1%) and scaffold N50 (148,138.5 Mb and 126.4 Mb). More importantly, we also identified the expression patterns of this gene in different fruit-ripening stages and abiotic stresses, and validated them by qPCR. Additionally, one of the *PeNAC* genes exhibited resistance to cold stress. These results provide useful information for the genetic improvement of fruit quality and the improvement of the abiotic stress resistance of passion fruit, and lay a good foundation for the study of the regulation mechanism of fruit quality.

2. Results

2.1. Identification of the Passion Fruit NAC Family

In this research, 25 *PeNACs* have been identified. In addition, we analyzed the characteristics of the *PeNACs* (Table 1). The length of the *PeNAC* CDS sequence ranged from 174 bp (*PeNAC-21*) to 6126 bp (*PeNAC-12*). The identified *PeNACs* encoded proteins ranging from 57 amino acids of *PeNAC-21* to 2041 amino acids of *PeNAC-12*. The MW ranged from 6.74Da (*PeNAC-21*) to 229.5Da (*PeNAC-12*). The isoelectric point ranged from 3.84 (*PeNAC-9*) to 10.44 (*PeNAC-7*). Subcellular localization predicted that all genes were located in the nucleus.

Table 1. Basic information of NAC genes identified in passion fruit. MW: molecular weight. PI: isoelectric point.

Gene	Gene ID	CDS Length (bp)	Protein Length (aa)	MW (Da)	PI	Subcellular Localization
PeNAC-1	P_edulia010001655.g	1377	458	50.19	4.78	Nucleus
PeNAC-2	P_edulia010001845.g	786	261	29.7	8.76	Nucleus
PeNAC-3	P_edulia010003708.g	1524	507	56.98	6.28	Nucleus
PeNAC-4	P_edulia010004544.g	1938	645	72.96	4.61	Nucleus
PeNAC-5	P_edulia020006444.g	891	296	33.61	7.99	Nucleus
PeNAC-6	P_edulia030008739.g	927	308	35.6	8.44	Nucleus
PeNAC-7	P_edulia030009215.g	639	212	24.27	10.44	Nucleus
PeNAC-8	P_edulia030009267.g	1182	393	43.74	5.1	Nucleus
PeNAC-9	P_edulia030009488.g	975	324	35.12	3.84	Nucleus
PeNAC-10	P_edulia040010645.g	939	312	36.11	8.1	Nucleus
PeNAC-11	P_edulia050011226.g	468	155	17.98	4.41	Nucleus
PeNAC-12	P_edulia060013132.g	6126	2042	229.5	7.11	Nucleus
PeNAC-13	P_edulia060013466.g	594	197	22.43	10.39	Nucleus
PeNAC-14	P_edulia060013670.g	891	296	33.2	9	Nucleus
PeNAC-15	P_edulia060013771.g	576	191	22.03	4.92	Nucleus
PeNAC-16	P_edulia060014061.g	579	192	22.67	7.83	Nucleus
PeNAC-17	P_edulia060014082.g	246	81	9.47	4.57	Nucleus
PeNAC-18	P_edulia060015325.g	801	266	30.06	6.71	Nucleus
PeNAC-19	P_edulia060015528.g	840	279	30.93	4.39	Nucleus
PeNAC-20	P_edulia060015714.g	1284	427	48.17	4.89	Nucleus
PeNAC-21	P_edulia070017373.g	174	57	6.74	4.48	Nucleus
PeNAC-22	P_edulia080019184.g	927	308	34.59	7.69	Nucleus
PeNAC-23	P_eduliaContig140022926.g	1524	507	56.99	6.21	Nucleus
PeNAC-24	P_eduliaContig140022928.g	636	211	24.02	4.25	Nucleus
PeNAC-25	P_eduliaContig140022930.g	1551	516	57.88	4.49	Nucleus

2.2. Phylogenetic Analysis of PeNACs Protein

To study the classification and evolutionary relationships of NAC proteins in passion fruit, a phylogenetic tree was constructed by the protein sequences of 25 *PeNACs* (passion fruit), 29 *AtNACs* (Arabidopsis) and 23 *OsNACs* (rice) (Figure 1). According to the kinship of the members, *PeNACs* were divided into three subfamilies: group 1 (*PeNAC-2/4/5/8/9/14/16/18/20/22*), group 2 (*PeNAC-1/10/12/17/24/25*) and group 3 (*PeNAC-3/6/7/11/13/15/19/21/23*).

The homologous genes of NAC in passion fruit and Arabidopsis can be inferred due to the fact that passion fruit and Arabidopsis are both dicotyledones. For group 1, *PeNAC-4* was the best orthology matches of At1G56010.2. *PeNAC-8* was the most homogeneous genes of At3G01600.1. *PeNAC-9* were phylogenetically closest to At3G10490.3 and At3G10490.4. *PeNAC-14* exhibited the closest relationship with At1G25580.1. For group 2, *PeNAC-1* was the best orthology match of At3G46565.1. *PeNAC-12* was the most homogeneous gene of At1G32770.1 and At2G46770.1. For group 3, *PeNAC-21* was the best orthology match of At1G12260.2. *PeNAC-11* was phylogenetically closest to At3G18400.1. *PeNAC-15* exhibited the closest relationship with At5G41090.1 and At3G56520.1.

2.3. Expression Pattern of NACs in Passion Fruit

The result of the RNA-seq data showed that the NACs have different response degrees to various abiotic stresses (Figure 2). Nineteen genes were induced by drought stress. The expression of *PeNAC-2/5/10/14/15/17/18/25* reached the highest levels when the soil water content was 10%. The expression level of some genes is induced to increase with the degree of salt stress, such as *PeNAC-2/3/10/11/16/23/25*. The transcript levels of *PeNAC-1/6/7/8/13* did not change much with the increase in salt stress. Under cold stress, most genes were upregulated. Among them, the expression of *PeNAC-1/3/4/6/7/13/20* displayed significant changes. Under high temperature stress, the expression of the *PeNAC-1/4/5/8/9/11/16/17/18/19/24* increased with the degree of stress. The expression of some genes (*PeNAC-6/7/10/13/14/15/20*) displayed no visible change. The results show that most of the *PeNACs* can respond to various abiotic stresses (Figure 3).

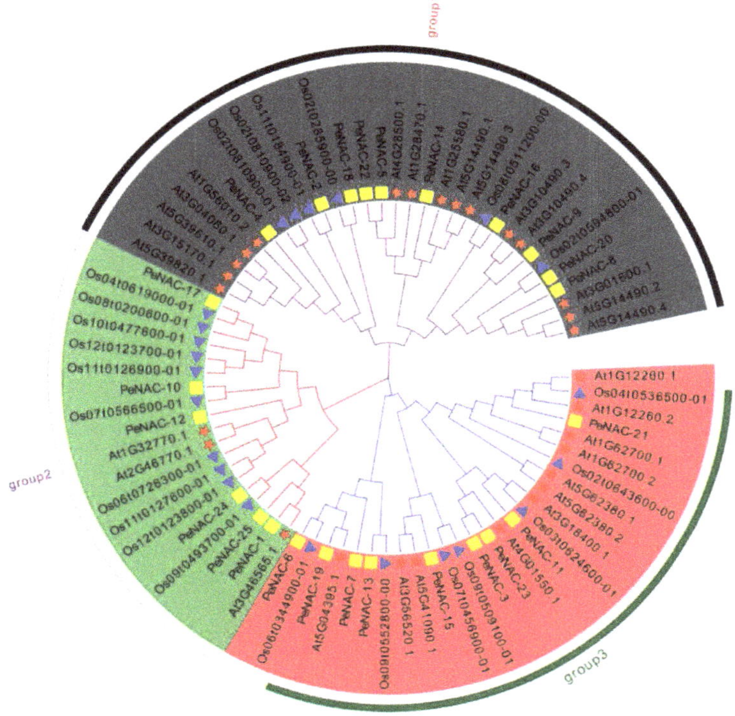

Figure 1. Phylogenetic evolutionary tree of NACs among the passion fruit, Arabidopsis and rice was constructed by ClustalX 2.0 and MEGA 7.0. The square represents the NACs in passion fruit; the five-pointed star represents the NACs in Arabidopsis; the triangle represents the NACs in rice.

We performed transcript sequencing during three periods: the seventh day before fruit harvest (Time1), the harvest period (Time2) and the seventh day after fruit harvest (Time3) in passion fruit [37], wherein we analyzed the expression levels of the *PeNACs* (Figure 4). The results show that most of the genes demonstrated the highest expression levels at Time1, and their expression decreased as the fruit matured, with the lowest expression in the third period, such as *PeNAC3/8/17/20/23/24/25*. Among them, *PeNAC22* is most expressed in the third period of fruit maturity. Some genes were verified by qRT-PCR and the results were consistent with the transcript sequencing (Figure 5). This indicated that this gene may be associated with fruit maturity.

The expression of some *PeNACs* in different tissues of passion fruit has been analyzed (Figure 6). Among them, four were mainly expressed in the flower (*PeNAC-6*, *PeNAC-17*, *PeNAC-19*, *PeNAC-23*). *PeNAC-1* was mainly expressed in the fruit. Additionally, *PeNAC-4/13/20* were mainly expressed in the root. The results show that the gene was expressed in different parts.

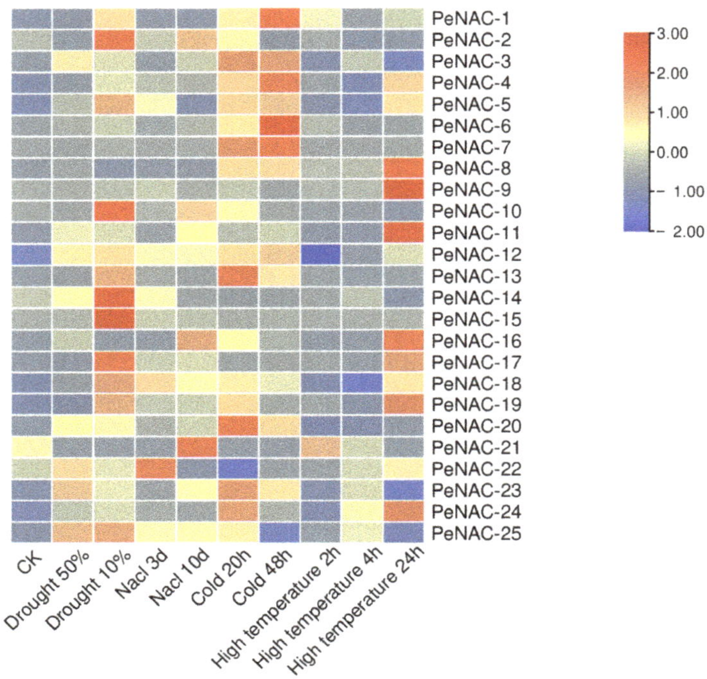

Figure 2. Expression profiles of *PeNACs* genes responding to drought (soil moisture of 50% and 10%), salt (300 mM NaCl for 3 d and 10 d), cold (0 °C for 20 h and 48 h) and high temperature (45 °C for 2 h, 4 h and 24 h). The plants of about 1 m and with 8–10 functional leaves were used for abiotic stress treatments. The details are shown in Table S1. A low expression level is shown in blue and a high expression level is shown in red. CK means control check.

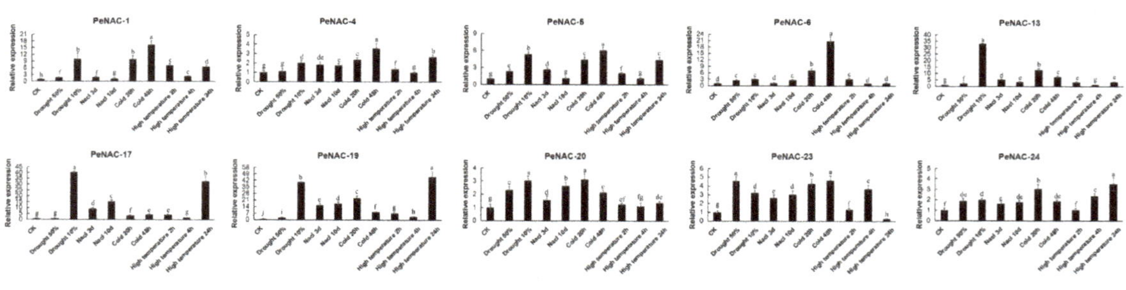

Figure 3. Expression analysis of 10 *PeNACs* under different abiotic stresses in the passion fruit. The experimental materials are identical to those used for RNA-seq under the abiotic stresses. The details are shown in Table S2. Data are means ± SD of n = 3 biological replicates. Means denoted by the same letter are not significantly different at $p < 0.05$ as determined by Duncan's multiple range test.

2.4. Cold Stress Analysis in Transgenic Tobacco

We further transiently transformed tobacco with the promoter of *PeNAC-19* (Figure 7). The *PeNAC-19p*-transformed tobacco and control were subjected to low-temperature stress treatment at 4 °C. The results show that the gene was highly induced during 2 h and 24 h treatments.

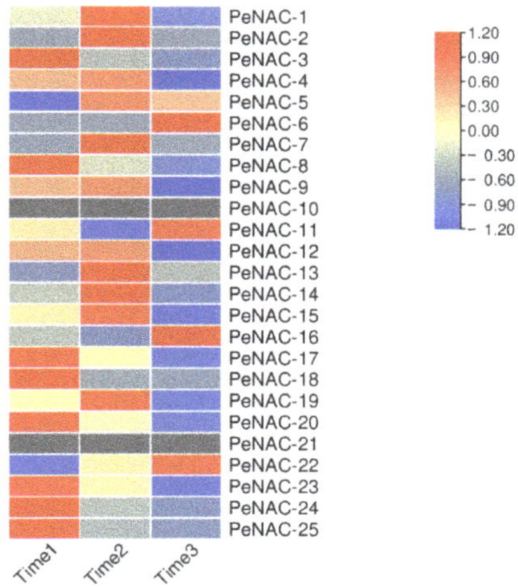

Figure 4. Expression profiles of *PeNACs* genes during three fruit-ripening periods (Time1: the 7th day before fruit harvest, Time2: the harvest period, Time3: the 7th day after fruit harvest). The details are shown in Table S3. A low expression level is shown in blue and a high expression level is shown in red. The heat map was generated using TBtools.

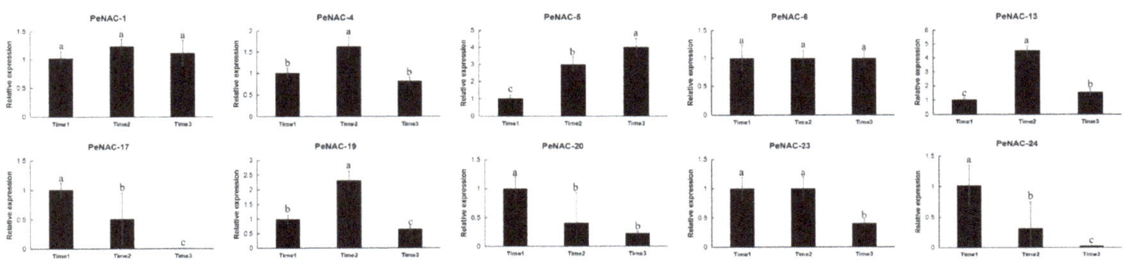

Figure 5. Expression analysis of 10 *PeNACs* during three fruit-ripening periods in the passion fruit. The experimental materials are identical to those used for RNA-seq during three fruit-ripening periods. The details are shown in Table S4. Data are means ± SD of n = 3 biological replicates. Means denoted by the same letter are not significantly different at $p < 0.05$ as determined by Duncan's multiple range test.

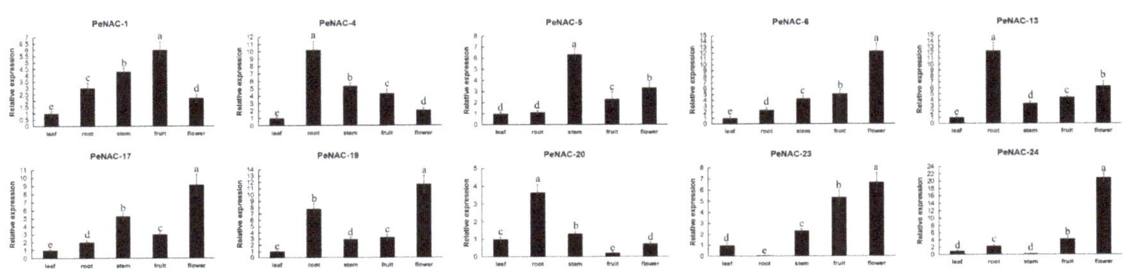

Figure 6. Expression analysis of 10 *PeNACs* in the leaf, root, stem, fruit and flower of the passion fruit. The leaf, root, stem, fruit, flower and fruits were obtained from the healthy passion fruit (*Passiflora edulis*). The details are shown in Table S5. Data are means ± SD of n = 3 biological replicates. Means denoted by the same letter are not significantly different at $p < 0.05$ as determined by Duncan's multiple range test.

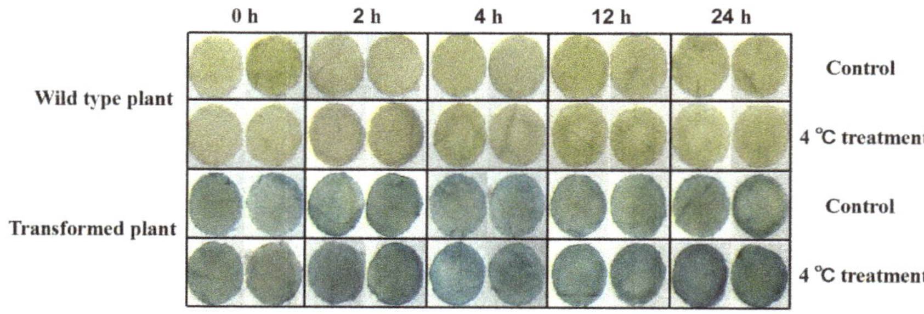

Figure 7. Gus staining was performed on transgenic tobacco under normal and cold treatment conditions. Tobacco was treated for 0, 2, 4, 12 and 24 h at low temperature (3 replicates per treatment) and then processed into discs (diameter 0.5 cm). The leaf discs that floated under normal growth temperature (25 °C) were used as control.

2.5. Functional Complementation Validation of PeNAC-19

In Figure 8, the pYES2-*PeNAC-19* and the pYES2 empty vector (control) were transformed into INVSC1 (*Saccharomyces cerevisiae*) for the cold-stress experiment (−20 °C for 0 h, 24 h, 48 h and 72 h). The results indicate that the growth of both the empty vector and the transgenic yeast is more and more restricted with increasing treatment time. When treated at −20 °C for 72 h, the *PeNAC-19*-transformed yeast grew better than the control. This suggests that *PeNAC-19* plays a certain role under the cold stress.

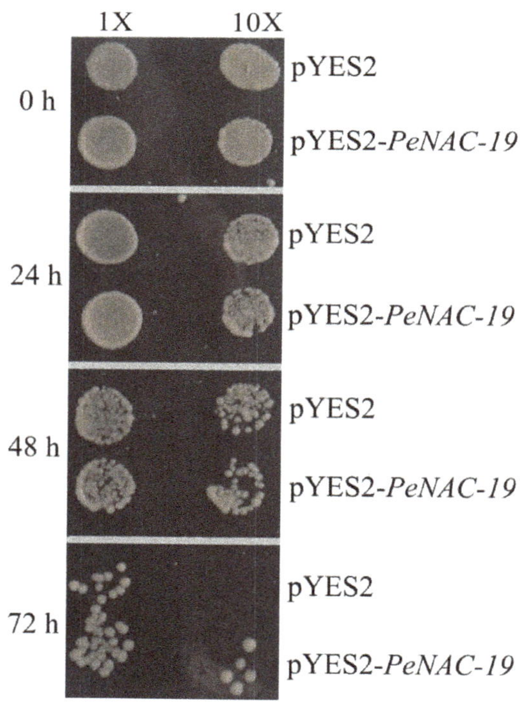

Figure 8. Growth status of the *Saccharomyces cerevisiae* INVSc1 strain expressing pYES2–*PeNAC-19* and pYES2 (control) under cold stress (−20 °C). 1× means the original bacterial fluid, 10× is the fluid diluted 10 times.

2.6. Response of Transgenic Arabidopsis to Low-Temperature Stress

As shown in Figure 9A, under normal growth conditions, GUS staining was mainly concentrated in stems, and under low-temperature stress, GUS staining was enhanced and mainly concentrated in the leaves and roots of the seedling. The GUS enzyme activity test was carried out, and we found that GUS activities are 3.4-fold higher than the control. The results show that *PeNAC-19* was induced by low-temperature stress.

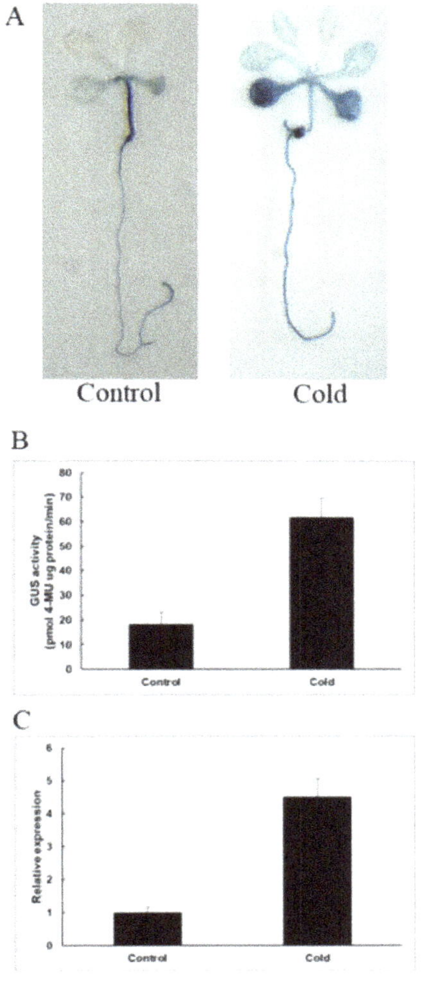

Figure 9. Induction and expression pattern of *PeNAC-19* under low temperature (4 °C for 36 h). (**A**) GUS staining of transgenic Arabidopsis. (**B**) GUS enzyme activity analysis of transgenic Arabidopsis thaliana. (**C**) Expression of *PeNAC-19* under the cold stress. Means denoted by the same letter are not significantly different at $p < 0.05$ as determined by Duncan's multiple range test.

3. Materials and Methods

3.1. Identification of NAC Genes in Passion Fruit

The passion fruit genome data were downloaded from the NGDC database (https://ngdc.cncb.ac.cn/gwh/Genome/557/show, accessed on 2 April 2021). Additionally, the PeNACs were initially screened and identified using hmmsearch and local blast (Output E value: 1×10^{-5}). In addition, analysis was performed using the NAC with the highest comparison value to further identify possible NACs in the passion fruit database. The protein sequences identified by the two methods described above were integrated and resolved to remove redundancy. NAC protein sequences in Arabidopsis and rice were downloaded from databases (http://plants.ensembl.org/Arabidopsis_thaliana/Info/Index, accessed on 2 June 2022, http://plants.ensembl.org/Oryza_sativa/Info/Index, accessed on 3 June 2022). The phylogenetic trees were constructed using MEGA 7.0 software (https://www.megasoftware.net/, accessed on 12 June 2022) [40]. The members of *PeNACs* were finally obtained through the screening of the above methods.

3.2. Gene Identification

The physical and chemical properties of protein were calculated using the ProtParam website (http://www.expasy.org/tools/protparam.html, accessed on 15 June 2022) (https://meme-suite.org/meme/tools/meme, accessed on 20 June 2022), NetPhos 3.1 Server (http://www.cbs.dtu.dk/services/NetPhos/, accessed on 22 June 2022) and WoLF PSORT 0.2 software (https://www.genscript.com/wolf-psort.html, accessed on 29 June 2022).

3.3. Plant Materials and Growth Conditions

For the abiotic stress treatment, the two-months-old healthy passion fruit seedlings (*Passiflora edulis*) were chosen. They were grown in soil under a growth chamber (30 °C; 200 μ mol·m^{-2}·s^{-1} light intensity; 12 h light/12 h dark cycle; 70% relative humidity) to a height of about 1 m and with 8–10 functional leaves, which were subjected to various abiotic stress treatments [36]. For the sampling of fruit during ripening, the fruits were chosen in the 7th day before fruit harvest (Time1), the harvest period (Time2), and the 7th day after fruit harvest (Time3) in passion fruit [37]. Each treatment was repeated three times with 10 plants/fruits at a time, and the material was RNA extracted and RNAseq.

3.4. Heat Map Construction

Transcriptome data used for heat map construction are shown in Tables S1 and S3. The analysis was conducted using TBtools software (https://bio.tools/tbtools, accessed on 10 July 2022) [40].

3.5. Cloning and Vector Construction of PeNAC-19 and the Promoter

The full-length cDNA of *PeNAC-19* and a 2000 bp DNA fragment before the start codon of the *PeNAC-19* was amplified from the passion fruit (*Passiflora edulis*). Healthy passion fruit seedlings were used to construct a single-stranded cDNA template.

To examine whether *PeNAC-19* responds to low-temperature stress, the expression vectors have been constructed. The PCR products of the *PeNAC-19* ORF were inserted into the pCAMBIA1304 (Cambia, Australia) expression vector, which is called pCAMBIA1304-*PeNAC-19*. The PCR products of the *PeNAC-19* promoter were cloned into the pCAMBIA1304 vector called pCAMBIA1304-*PeNAC-19p*. The vectors were transferred into the *Agrobacterium* strain EHA105.

3.6. Cold Stress Treatment in Wild-Type and Transgenic Plants

The two-month-old tobacco leaves were used for transient expression experiments. The *Agrobacterium* transformed with pCAMBIA1304-*PeNAC-19*p was shaken at 28 °C to OD 600 = 0.8–1.0, and the bacterial solution was infected with an equal volume of MES buffer (10 mM MES; 10 mM $MgCl_2$; 200 mM MAS) and then kept in the dark at room temperature for 2–3 h. After that, the solution had been injected into the back of the tobacco which was left to culture for 2–3 d. The plants were treated at 4 °C and 25 °C for 0, 2, 4, 12 and 24 h. The leaf discs with a diameter of 0.5 cm were cut out for further staining testing.

The *Agrobacterium* transformed with pCAMBIA1304-*PeNAC-19* and pCAMBIA1304-*PeNAC19*p were shaken at 28 °C in YEB medium, and then added to 1/2 MS solution until OD 600 = 0.8–1.0. The transformation of *Arabidopsis thaliana* was conducted using the inflorescence dip method. The seeds of *Arabidopsis thaliana* were disinfected with 75%

ethanol and spreading on the medium (1/2MS, 25 μg/mL hygromycin B). After culturing in the dark at 4 °C for 3 days, the seeds were transferred to a culture at 23 °C for growth. After ten days, the normal growth seedlings were the T1 generation. Additionally, the T2 generation grown in the selection medium were for further experimental analysis. The 14-day-old transgenic Arabidopsis with about 6–8 leaves was treated in an incubator at 4 °C and 23 °C for 36 h.

3.7. GUS Activity Detection

The fresh transgenic tobacco and Arabidopsis samples under the normal and cold conditions were placed in X-Gluc solution (GBT, St. Louis, MO, USA) [41] for histochemical analysis. Gus enzyme activity was determined by 4-methyl umbelliferate glucuronide fluorescence method [42].

3.8. Functional Complementation of *PeNAC-19* in Yeast

The pYES2–*PeNAC-19* vector was constructed using the full-length cDNA of *PeNAC-19*. Then, the pYES2–*PeNAC-19* and pYES2 vector (control) were transfected into the INVSc1 Strain (*Saccharomyces cerevisiae*). To perform the yeast complementation assays, the yeast liquid was cultured in SD-Ura liquid medium at 30 °C until OD600 = 1.0 and treated under the cold stress [43]. The experiment was repeated three times.

3.9. RNA Extraction, Transcriptome Sequencing and qRT-PCR

Total RNA was extracted from frozen samples with plant RNA isolation kit (Fuji, China, Chengdu), and three biological replicates were performed. The cDNA was used for transcriptome sequencing analysis and quantitative real-time polymerase chain reaction (qRT-PCR). The SYBR® Premix Ex Taq™ kit (TaKaRa, Tokyo, Japan) was used to detect in qRT-PCR. Relative expression levels were calculated using the $2^{-\Delta\Delta CT}$ method and normalized to the *PeNACs*.

4. Discussion

Passion fruit (*Passiflora edulis* Sim), a perennial vine, is rich in nutrients and contains a variety of amino acids and vitamins. Due to its short growth cycle and good economic value, it is very popular [36,37]. However, only a few studies have investigated its function [44]. Previous studies have shown that transcription factors can regulate plant growth and development and enhance plant responses to abiotic stresses [45]. The NAC family, which is one of the largest transcription factor families, plays an important role in plant growth, development and abiotic stress responses [46].

In this study, we have reported 25 *PeNACs* in passion fruit, and performed the expression patterns analysis under different abiotic stresses and during different fruit-ripening stages. At the same time, a low-temperature stress-inducible gene, *PeNAC-19*, was screened for transient expression in tobacco, transgenic Arabidopsis and in vivo in yeast. These results provide evidence for the response of *PeNAC* members under cold stress. In physical and chemical property terms, the isoelectric point of the PeNAC protein ranges from 3.84 to 10.44, with an average of 6.38, which is consistent with *BjuNAC* [47]. All *PeNAC* family members are subcellularly localized in the nucleus by prediction. This was consistent with the localization of most transcription factors.

In general, the expression level of a gene determines its function [48]. Transcription factors usually play a key role in regulating the expression of tissue-specific genes [49]. In this study, we have selected some *PeNAC* genes for qRT-PCR analysis of each tissue, and the results show that some genes exhibited higher expression in the roots, such as *PeNAC-4/13/20*. Similar results were also shown in other plants, such as orchardgrass (*Dactylis glomerata* L.) [45], *Fagopyrum tataricum* [50], *Panicum miliaceum* [51] and *Triticum aestivum* [52]. In orchards, *DgNAC046/087/103* had the highest expression in the stems, which may play an important role in stem development. In this research, the expression of *PeNAC-6/19/17/23* is higher in flowers, which may be related to the development of floral organs. In addition, the overexpression of the tissue's specifically expressed NAC gene, poplar *NAC15*, was able to promote wood formation [53]. In orchardgrass, *DgNAC* genes are extensively involved in tissue development [45]. The NAC domain transcription factor *PdWND3A* affects lignin biosynthesis [54].

The relationship between NAC family members and plant abiotic stress has also been reported in many species. In *B. juncea* var. *tumida*, some NAC genes such as *BjuNAC112*, *BjuNAC178*, *BjuNAC184* and *BjuNAC240* can respond to high-temperature stress [55]. In addition, in tobacco, the overexpression of *LpNAC13* in *Lilium pumilum* reduced the tolerance to drought stress, but positively regulated the response to salt stress [47]. In *Cerasus humilis*, *ChNAC1* positively regulates the expression of abscisic acid (ABA)-responsive genes, and its overexpressed plants exhibited higher drought tolerance [25,56]. Arabidopsis AtNTL6 expressed the highest expression at low temperature for 18 h [57]. *OsNAC6/SNAC2* overexpression could improve the drought, salt and low-temperature tolerance of rice [23,24]. SlNAC1 from *Suaeda liaotungensis* and *VvNAC1* from *Vitis vinifera* L. can positively regulate the cold resistance of transgenic plants [58,59]. In wheat (*Triticum aestivum* L.), *TaNAC2/47/67* can improve cold tolerance in transgenic plants [60–62]. *GmNAC20* transgenic rice has stronger salt tolerance and cold tolerance by regulating the expression of abiotic stress-response genes [63].

In this research, one member of the *PeNACs*, *PeNAC-19*, was induced in the different abiotic stresses. *PeNAC-19* was induced by four various abiotic stresses. It may be a candidate gene for stress-resistant breeding. At present, low temperatures have seriously endangered the development of passion fruit cultivation. Therefore, we first focus on the relationship between *PeNAC-19* and cold stress. The results show that *PeNAC-19* responded to cold stress significantly in tobacco and Arabidopsis, and could improve the low-temperature tolerance of yeast. This study lays a foundation for the functional study of *PeNAC* gene family members under fruit ripening and abiotic stress.

5. Conclusions

In this study, the NAC family members in passion fruit were identified and analyzed, and the transcriptome results at different fruit-ripening stages and abiotic stresses were verified by qRT-PCR. The expression of one of the NAC genes (*PeNAC-19*) was induced by cold stress. Further verification of this gene showed that it could enhance the ability of transgenic tobacco, Arabidopsis and yeast to resist cold stress. The results lay a good foundation for further studies on the ability of the passion fruit to resist to abiotic stresses.

Supplementary Materials: The following supporting information can be downloaded at https://www.mdpi.com/article/10.3390/plants12061393/s1, Table S1. The transcriptome data of passion fruit under the drought, salt, cold and high-temperature treatment. Table S2. qPCR data of PeNACs under the drought, salt, cold and high-temperature treatment. Table S3. The transcriptome data of passion fruit in the three stages of fruit ripening. Table S4. qPCR data of PeNACs in the three stages of fruit ripening. Table S5. qPCR data of PeNACs in different tissues. Table S6. A list of the oligo primers of *PeNACs* used for qRT-PCR.

Author Contributions: Experiments were performed by P.L., F.M., W.X., D.H., B.W. and P.S. who analyzed the data; Y.X. and S.S. drafted the manuscript; F.M. and B.X. supervised the experiments and finalized the manuscript. All authors have read and agreed to the published version of the manuscript.

Funding: This work was sponsored by National Natural Science Foundation of China (32260737), Project of Sanya Yazhou Bay Science and Technology City (SCKJ-JYRC-2022-93, SCKJ-JYRC-2022-84) and Hainan Provincial Natural Science Foundation of China (321RC1088, 321MS091).

Data Availability Statement: The raw sequence data have been deposited in the National Genomics Data Center (NGDC), and the link is https://ngdc.cncb.ac.cn/gwh/Genome/557/show accessed on 2 April 2021. All datasets are available from the corresponding author upon reasonable request.

Conflicts of Interest: The authors declare no conflict of interest.

References

1. Riaño-Pachón, D.M.; Ruzicic, S.; Dreyer, I.; Mueller-Roeber, B. PlnTFDB: An integrative plant transcription factor database. *BMC Bioinform.* **2007**, *8*, 42. [CrossRef] [PubMed]
2. Zhang, H.; Jin, J.; Tang, L.; Zhao, Y.; Gu, X.; Gao, G.; Luo, J. PlantTFDB 2.0: Updateand improvement of the comprehensive plant transcription factor database. *Nucleic Acids Res.* **2011**, *39* (Suppl. S1), D1114–D1117. [CrossRef] [PubMed]
3. Riechmann, J.L.; Heard, J.; Martin, G.; Reuber, L.; Jiang, C.-Z.; Keddie, J.; Adam, L.; Pineda, O.; Ratcliffe, O.; Samaha, R. Arabidopsis transcription factors: Genome-wide comparative analysis among eukaryotes. *Science* **2000**, *290*, 2105–2110. [CrossRef] [PubMed]
4. Wray, G.A.; Hahn, M.W.; Abouheif, E.; Balhoff, J.P.; Pizer, M.; Rockman, M.V.; Romano, L.A. The evolution of transcriptional regulation in eukaryotes. *Mol. Biol. Evol.* **2003**, *20*, 1377–1419. [CrossRef]
5. Aida, M.; Ishida, T.; Fukaki, H.; Fujisawa, H.; Tasaka, M. Genes involved in organ separation in Arabidopsis: An analysis of the cup-shaped cotyledon mutant. *Plant Cell.* **1997**, *9*, 841–857. [CrossRef]
6. Souer, E.; van Houwelingen, A.; Kloos, D.; Mol, J.; Koes, R. The no apical meristemgene of Petunia is required for pattern formation in embryos and flowersand is expressed at meristem and primordia boundaries. *Cell* **1996**, *85*, 159–170. [CrossRef]
7. Kikuchi, K.; Ueguchi-Tanaka, M.; Yoshida, K.; Nagato, Y.; Matsusoka, M.; Hirano, H.-Y. Molecular analysis of the NAC gene family in rice. *Mol. Gen. Genet.* **2000**, *262*, 1047–1051. [CrossRef]
8. Puranik, S.; Sahu, P.P.; Srivastava, P.S.; Prasad, M. NAC proteins: Regulation and role in stress tolerance. *Trends Plant Sci.* **2012**, *17*, 369–381. [CrossRef]
9. Welner, D.; Deeba, F.; Lo Leggio, L.; Skriver, K. NAC transcription factors: From structure to function in stress-associated networks. In *Plant Transcription Factors*; Gonzalez, D.H., Ed.; Academic Press: Boston, MA, USA, 2015; pp. 199–212.
10. Jensen, M.K.; Skriver, K. NAC transcription factor gene regulatory and protein–protein interaction networks in plant stress responses and senescence. *IUBMB Life* **2014**, *66*, 156–166. [CrossRef]
11. Mathew, I.E.; Agarwal, P. May the fittest protein evolve: Favoring the plant-specific origin and expansion of NAC transcription factors. *BioEssays* **2018**, *40*, 1800018. [CrossRef]
12. Zhu, G.; Chen, G.; Zhu, J.; Zhu, Y.; Lu, X.; Li, X.; Hu, Y.; Yan, Y. Molecular characterization and expression profiling of NAC transcription factors in *Brachypodium distachyon* L. *PLoS ONE* **2015**, *10*, e0139794. [CrossRef]
13. Ooka, H.; Satoh, K.; Doi, K.; Nagata, T.; Otomo, Y.; Murakami, K.; Matsubara, K.; Osato, N.; Kawai, J.; Carninci, P.; et al. Comprehensive analysis of NAC family genes in *Oryza sativa* and *Arabidopsis thaliana*. *DNA Res.* **2003**, *10*, 239–247. [CrossRef] [PubMed]
14. Fang, Y.; You, J.; Xie, K.; Xie, W.; Xiong, L. Systematic sequence analysis and identification of tissue-specific or stress-responsive genes of NAC transcription factor family in rice. *Mol. Genet. Genom.* **2008**, *280*, 547–563. [CrossRef] [PubMed]
15. Li, B.; Fan, R.Y.; Yang, Q.S.; Hu, C.H.; Sheng, O.; Deng, G.M.; Dong, T.; Li, C.Y.; Peng, X.X.; Bi, F.C.; et al. Genome-Wide Identification and Characterization of the NAC Transcription Factor Family in Musa Acuminata and Expression Analysis during Fruit Ripening. *Int. J. Mol. Sci.* **2020**, *21*, 634. [CrossRef] [PubMed]
16. Min, X.Y.; Jin, X.Y.; Zhang, Z.S.; Wei, X.Y.; Ndayambaza, B.; Wang, Y.R.; Liu, W.X. Genome-Wide Identification of NAC Transcription Factor Family and Functional Analysis of the Abiotic Stress-Responsive Genes in *Medicago sativa* L. *J. Plant Growth Regul.* **2020**, *39*, 324–337. [CrossRef]
17. Nigarish, M.; Chen, Y.K.; Chen, X.H.; Azher, N.M.; Junaid, I.; Muhammad, R.H.; Shen, X.; Lin, Y.L.; Xu, X.H.; Lai, Z.X. Genome-wide identification and comprehensive analyses of NAC transcription factor gene family and expression patterns during somatic embryogenesis in *Dimocarpus longan* Lour. *Plant Physiol. Biochem.* **2020**, *157*, 169–184.
18. Feng, J.D.; Qiang, J.Z.; Hui, F.H.; Chen, L.; Liao, G.L.; He, Y.Q.; Huang, C.H.; Xu, X.B. Genome-wide identification and comprehensive analysis of NAC family genes involved in fruit development in kiwifruit (Actinidia). *BMC Plant Biol.* **2021**, *21*, 44.
19. Hu, R.; Qi, G.; Kong, Y.; Kong, D.; Gao, Q.; Zhou, G. Comprehensive analysis of NAC domain transcription factor gene family in *Populus trichocarpa*. *BMC Plant Biol.* **2010**, *10*, 145. [CrossRef] [PubMed]
20. Wang, N.; Zheng, Y.; Xin, H.; Fang, L.; Li, S. Comprehensive analysis of NAC domain transcription factor gene family in *Vitis vinifera*. *Plant Cell Rep.* **2013**, *32*, 61–75. [CrossRef] [PubMed]
21. Ahmad, M.; Yan, X.; Li, J.; Yang, Q.; Jamil, W.; Teng, Y.; Bai, S. Genome wide identification and predicted functional analyses of NAC transcription factors in Asian pears. *BMC Plant Biol.* **2018**, *18*, 214. [CrossRef]
22. Jeong, J.S.; Kim, Y.S.; Baek, K.H.; Jung, H.; Ha, S.H.; Choi, Y.D.; Kim, M.; Kim, J.K. Root-specific expression of OsNAC10 improves drought tolerance and grain yield in rice under field drought conditions. *Plant Physiol.* **2010**, *153*, 185–197. [CrossRef] [PubMed]
23. Nakashima, K.; Tran, L.S.P.; Van Nguyen, D.; Fujita, M.; Maruyama, K.; Todaka, D.; Ito, Y.; Hayashi, N.; Shinozaki, K.; Yamaguchi-Shinozaki, K. Functional analysis of a NAC-type transcription factor OsNAC6 involved in abiotic and biotic stress-responsive gene expression in rice. *Plant J.* **2007**, *51*, 617–630. [CrossRef] [PubMed]
24. Hu, H.; You, J.; Fang, Y.; Zhu, X.; Qi, Z.; Xiong, L. Characterization of transcription factor gene SNAC2 conferring cold and salt tolerance in rice. *Plant Mol. Biol.* **2008**, *67*, 169–181. [CrossRef] [PubMed]
25. Han, D.G.; Du, M.; Zhou, Z.Y.; Wang, S.; Li, T.M.; Han, J.X.; Xu, T.L.; Yang, G.H. Overexpression of a *Malus baccata* NAC Transcription Factor Gene MbNAC25 Increases Cold and Salinity Tolerance in Arabidopsis. *Int. J. Mol. Sci.* **2020**, *21*, 1198. [CrossRef]

26. Jin, C.; Li, K.Q.; Xu, X.Y.; Zhang, H.P.; Chen, H.X.; Chen, Y.H.; Hao, J.; Wang, Y.; Huang, X.S.; Zhang, S.L. A novel NAC transcription factor, PbeNAC1, of *Pyrus betulifolia* confers cold and drought tolerance via interacting with PbeDREBs and activating the expression of stress-responsive genes. *Front. Plant Sci.* **2017**, *8*, 1049. [CrossRef]
27. Li, X.D.; Zhuang, K.Y.; Liu, Z.M.; Yang, D.Y.; Ma, N.N.; Meng, Q.W. Overexpression of a novel NAC-type tomato transcription factor, SlNAM1, enhances the chilling stress tolerance of transgenic tobacco. *J. Plant Physiol.* **2016**, *204*, 54–65. [CrossRef]
28. Giovannoni, J. Molecular Biology of Fruit Maturation and Ripening. *Annu. Rev. Plant Biol.* **2001**, *52*, 725–749. [CrossRef]
29. Wang, J.F.; Wang, Y.P.; Zhang, J.; Ren, Y.; Li, M.Y.; Tian, S.W.; Yu, Y.T.; Zuo, Y.; Gong, G.Y.; Zhang, H.Y.; et al. The NAC transcription factor ClNAC68 positively regulates sugar content and seed development in watermelon by repressing ClINV and ClGH3.6. *Hortic. Res.* **2021**, *8*, 214. [CrossRef]
30. Kerstin, D.; Johanna, W.J.; Miguel, N.G.; Almuth, H.; Karl, L.; Ines, E.; Malin, E. Overexpression of PaNAC03; a stress induced NAC gene family transcription factor in Norway spruce leads to reduced flavonol biosynthesis and aberrant embryo development. *BMC Plant Biol.* **2017**, *17*, 6.
31. Zhang, S.L.; Dong, R.Z.; Wang, Y.W.; Li, X.M.; Ji, M.M.; Wang, X.P. NAC domain gene VvNAC26 interacts with VvMADS9 and influences seed and fruit development. *Plant Physiol. Biochem.* **2021**, *164*, 63–72. [CrossRef]
32. Ma, N.; Feng, H.; Meng, X.; Li, D.; Yang, D.; Wu, C.; Meng, Q. Overexpression of tomato SlNAC1 transcription factor alters fruit pigmentation and softening. *BMC Plant Biol.* **2014**, *14*, 351. [CrossRef]
33. Carrasco-Orellana, C.; Stappung, Y.; Mendez-Yañez, A.; Allan, A.C.; Espley, R.V.; Plunkett, B.J.; Moya-Leon, M.A.; Herrera, R. Characterization of a ripeningrelated transcription factor FcNAC1 from *Fragaria chiloensis* fruit. *Sci. Rep.* **2018**, *8*, 10524. [CrossRef]
34. Huang, D.; Xu, Y.; Wu, B.; Ma, F.N.; Song, S. Comparative analysis of basic quality of passion fruits (*Passiflora edulis* sims) in Guangxi, Guizhou and Fujian, China. *Banglandesh J. Bot.* **2019**, *48*, 901–906.
35. Costa, J.L.; Jesus, O.N.D.; Oliverira, G.A.F.; Oliverira, E.J.D. Effect of selection on genetic variability in yellow passion fruit. *Crop Breed. Appl. Biotechnol.* **2012**, *12*, 253–260. [CrossRef]
36. Song, S.; Zhang, D.; Ma, F.; Xing, W.; Huang, D.; Wu, B.; Chen, J.; Chen, D.; Xu, B.; Xu, Y. Genome-Wide Identification and Expression Analyses of the Aquaporin Gene Family in Passion Fruit (*Passiflora edulis*), Revealing PeTIP3-2 to Be Involved in Drought Stress. *Int. J. Mol. Sci.* **2022**, *23*, 5720. [CrossRef]
37. Xia, Z.; Huang, D.; Zhang, S.; Wang, W.; Ma, F.; Wu, B.; Xu, Y.; Xu, B.; Chen, D.; Zou, M.; et al. Chromosome-scale genome assembly provides insights into the evolution and flflavor synthesis of passion fruit (*Passiflflora edulis* Sims). *Hortic. Res.* **2021**, *8*, 14. [CrossRef]
38. Yang, Q.; Li, B.; Rizwan, H.M.; Sun, K.; Zeng, J.; Shi, M.; Guo, T.; Chen, F. Genome-wide identification and comprehensive analyses of NAC transcription factor gene family and expression analysis under *Fusarium kyushuense* and drought stress conditions in *Passiflora edulis*. *Front. Plant Sci.* **2022**, *13*, 972734. [CrossRef]
39. Ma, D.; Dong, S.; Zhang, S.; Wei, X.; Xie, Q.; Ding, Q. Chromosome-level reference genome assembly provides insights into aroma biosynthesis in passion fruit (*Passiflora edulis*). *Mol. Ecol. Resour.* **2021**, *21*, 955–968. [CrossRef]
40. Tamura, K.; Dudley, J.; Nei, M.; Kumar, S. MEGA4, molecular evolutionary genetics analysis MEGA software version 4.0. *Mol. Biol. Evol.* **2007**, *24*, 1596–1599. [CrossRef]
41. Jefferson, R.A.; Kavanagh, T.A.; Bevan, M.W. GUS fusions: Beta-glucuronidase as a sensitive and versatilegene fusion marker in higher plants. *EMBO J.* **1987**, *6*, 3901–3907. [CrossRef]
42. Xu, Y.; Jin, Z.Q.; Xu, B.Y.; Li, J.Y.; Li, Y.J.; Wang, X.Y.; Wang, A.B.; Hu, W.; Huang, D.M.; Wei, Q.; et al. Identification of transcription factors interacting with a 1274bp promoter of MaPIP1;1 which confers high-level gene expression and drought stress Inducibility in transgenic *Arabidopsis thaliana*. *BMC Plant Biol.* **2020**, *20*, 278. [CrossRef]
43. Wang, B.F.; Wang, Y.C.; Zhang, D.W.; Li, H.Y.; Yang, C.P. Verification of the resistance of a LEA gene from Tamarix expression in *Saccharomyces cerevisiae* to abiotic stresses. *J. For. Res.* **2008**, *19*, 58–62. [CrossRef]
44. Li, X.; Cai, K.; Pei, X.; Li, Y.; Hu, Y.; Meng, F.; Song, X.; Tigabu, M.; Ding, C.; Zhao, X. Genome-Wide Identification of NAC Transcription Factor Family in *Juglans mandshurica* and Their Expression Analysis during the Fruit Development and Ripening. *Int. J. Mol. Sci.* **2021**, *22*, 12414. [CrossRef]
45. Yang, Z.F.; Nie, G.; Feng, G.Y.; Han, J.T.; Huang, L.K.; Zhang, X.Q. Genome-wide identification, characterization and expression analysis of the NAC transcription factor family in orchardgrass (*Dactylis glomerata* L.). *BMC Genom.* **2021**, *22*, 178. [CrossRef]
46. Nakashima, K.; Takasaki, H.; Mizoi, J.; Shinozaki, K.; Yamaguchi-Shinozaki, K. NAC transcription factors in plant abiotic stress responses. *Biochim. Biophys. Acta* **2012**, *1819*, 97–103. [CrossRef]
47. Wang, Y.; Cao, S.J.; Guan, C.J.; Kong, X.; Wang, Y.P.; Cui, Y.; Liu, B.; Zhou, Y.W.; Zhang, Y.N. Overexpressing the NAC transcription factor LpNAC13 from *Lilium pumilum* in tobacco negatively regulates the drought response and positively regulates the salt response. *Plant Physiol. Biochem.* **2020**, *149*, 96–110. [CrossRef]
48. Zhou, Q.; Zhang, S.; Chen, F.; Liu, B.; Wu, L.; Li, F.; Zhang, J.; Bao, M.; Liu, G. Genome-wide identification and characterization of the SBP-box gene family in Petunia. *BMC Genom.* **2018**, *19*, 193. [CrossRef]
49. Naya, F.J.; Stellrecht, C.; Tsai, M.J. Tissue-specific regulation of the insulin gene by a novel basic helix-loop-helix transcription factor. *Genes Dev.* **1995**, *9*, 1009–1019. [CrossRef]
50. Liu, M.; Ma, Z.; Sun, W.; Huang, L.; Wu, Q.; Tang, Z.; Bu, T.; Li, C.; Chen, H. Genome-wide analysis of the NAC transcription factor family in Tartary buckwheat (*Fagopyrum tataricum*). *BMC Genom.* **2019**, *20*, 113. [CrossRef]

51. Shan, Z.; Jiang, Y.; Li, H.; Guo, J.; Dong, M.; Zhang, J.; Liu, G. Genome-wide analysis of the NAC transcription factor family in broomcorn millet (*Panicum miliaceum* L.) and expression analysis under drought stress. *BMC Genom.* **2020**, *21*, 96. [CrossRef]
52. Guerin, C.; Roche, J.; Allard, V.; Ravel, C.; Mouzeyar, S.; Bouzidi, M.F. Genome-wide analysis, expansion and expression of the NAC family under drought and heat stresses in bread wheat (*T. aestivum* L.). *PLoS ONE* **2019**, *14*, e0213390. [CrossRef] [PubMed]
53. Yao, W.; Zhang, D.; Zhou, B.; Wang, J.; Li, R.; Jiang, T. Over-expression of poplar NAC15 gene enhances wood formation in transgenic tobacco. *BMC Plant Biol.* **2020**, *20*, 12. [CrossRef]
54. Yang, Y.; Yoo, C.G.; Rottmann, W.; Winkeler, K.A.; Collins, C.M.; Gunter, L.E.; Jawdy, S.S.; Yang, X.; Pu, Y.; Ragauskas, A.J. PdWND3A, a wood-associated NAC domain-containing protein, affects lignin biosynthesis and composition in Populus. *BMC Plant Biol.* **2019**, *19*, 486. [CrossRef]
55. Jiang, L.; Sun, Q.; Wang, Y.; Chang, P.; Kong, H.; Luo, C.; He, X. Genome-wide identification and characterization of NAC genes in *Brassica juncea* var *tumida*. *PeerJ* **2021**, *9*, e11212. [CrossRef] [PubMed]
56. Wang, F.; Wang, J.W.; Sun, L.J.; Song, X.S. The molecular cloning and functional characterization of ChNAC1; a NAC transcription factor in Cerasus humilis. *Plant Growth Regul.* **2019**, *89*, 331–343. [CrossRef]
57. Seo, P.J.; Park, C.M. A membrane-bound NAC transcription factor as an integrator of biotic and abiotic stress signals. *Plant Signal. Behav.* **2010**, *5*, 481–483. [CrossRef]
58. Li, X.L.; Yang, X.; Hu, Y.X.; Yu, X.D.; Li, Q.L. A novel NAC transcription factor from *Suaeda liaotungensis* K. enhanced transgenic Arabidopsis drought, salt, and cold stress tolerance. *Plant Cell Rep.* **2014**, *33*, 767–778. [CrossRef]
59. Le, H.G.; Profizi, C.; Courteaux, B.; Rabenoelina, F.; Gérard, C.; Clément, C.; Baillieul, F.; Cordelier, S.; Dhondtcordelier, S. Grapevine NAC1 transcription factor as a convergent node in developmental processes, abiotic stresses, and necrotrophic/biotrophic pathogen tolerance. *J. Exp. Bot.* **2013**, *64*, 4877–4893.
60. Mao, X.; Zhang, H.; Qian, X.; Li, A.; Zhao, G.; Jing, R. TaNAC2, a NAC-type wheat transcription factor conferring enhanced multiple abiotic stress tolerances in Arabidopsis. *J. Exp. Bot.* **2012**, *63*, 2933–2946. [CrossRef]
61. Zhang, L.; Zhang, L.; Xia, C.; Zhao, G.; Jia, J.; Kong, X. The novel wheat transcription factor TaNAC47 enhances multiple abiotic stress tolerances in transgenic plants. *Front. Plant Sci.* **2015**, *6*, 1174. [CrossRef]
62. Mao, X.; Chen, S.; Li, A.; Zhai, C.; Jing, R. Novel NAC transcription factor TaNAC67 confers enhanced multi-abiotic stress tolerances in Arabidopsis. *PLoS ONE* **2014**, *9*, e84359. [CrossRef] [PubMed]
63. Yarra, R.; Wei, W. The NAC-type transcription factor GmNAC20 improves cold, salinity toleranced lateral root formation in transgenic rice plants. *Funct. Integr. Genom.* **2021**, *21*, 473–487. [CrossRef] [PubMed]

plants

Article

A Set of Molecular Markers to Accelerate Breeding and Determine Seed Purity of CMS Three-Line Hybrids in *Brassica napus*

Yanfeng Zhang [1], Ran An [1], Min Song [1,2], Changgen Xie [2], Shihao Wei [1], Daojie Wang [3], Yuhong Dong [1], Qingli Jia [1], Shuhua Huang [1,*] and Jianxin Mu [1,*]

1 Hybrid Rapeseed Research Center of Shaanxi Province, Yangling 712100, China; zhangyfcl@126.com (Y.Z.); anrants@126.com (R.A.); songmin914@nwafu.edu.cn (M.S.); shihao100@126.com (S.W.); dyh9919@163.com (Y.D.); yljiaqingli@163.com (Q.J.)
2 State Key Laboratory of Crop Stress Biology in Arid Areas, Northwest A&F University, Yangling 712100, China; changen.xie@nwsuaf.edu.cn
3 Key Laboratory of Plant Stress Biology, State Key Laboratory of Cotton Biology, School of Life Sciences, Henan University, Kaifeng 475004, China; wangdj@henu.edu.cn
* Correspondence: hsh813@126.com (S.H.); jxmsxyc@163.com (J.M.)

Abstract: Cytoplasmic male sterility (CMS) is the main mechanism employed to utilize the heterosis of *Brassica napus*. CMS three-line rapeseed hybrids have dramatically enhanced yield and brought about the global revolution of hybrid varieties, replacing conventional crop varieties. Over the last half century, China has led the development of hybrid *Brassica napus* varieties. Two sterile lines, *polima* (*pol*) and *shaan* 2A, were of particular importance for the establishment of three-line hybrid systems in rapeseed, which has opened up a new era of heterosis utilization. However, in current breeding practices, it takes up to three years to identify the restorer or maintainer relationship and the cytoplasmic type of any inbred material. This greatly affects the breeding speed of new varieties and inhibits the rapid development of the rapeseed industry. To address this problem, we developed a set of molecular markers for the identification of fertile cytoplasmic gene N and sterile cytoplasmic gene S, as well as for the fertile nucleus gene R and sterile nucleus gene r, based on differences in the gene sequences between the CMS line, maintainer line and restorer line of *Brassica napus*. Combining these markers can accurately identify the CMS line, maintainer and restorer of both the *pol* and *shaan* systems, as well as their hybrids. These markers can not only be used to identify of the maintainer and restorer relationship of inbred materials; they can also be used as general molecular markers to identify the CMS-type hybrid purity of *pol* and *shaan* systems.

Keywords: *Brassica napus*; cytoplasmic male sterile (CMS); molecular markers; seed purity identification

Citation: Zhang, Y.; An, R.; Song, M.; Xie, C.; Wei, S.; Wang, D.; Dong, Y.; Jia, Q.; Huang, S.; Mu, J. A Set of Molecular Markers to Accelerate Breeding and Determine Seed Purity of CMS Three-Line Hybrids in *Brassica napus*. *Plants* **2023**, *12*, 1514. https://doi.org/10.3390/plants12071514

Academic Editors: Zanmin Hu, Han Xiao, Yi Ren and Chengming Fan

Received: 6 March 2023
Revised: 28 March 2023
Accepted: 29 March 2023
Published: 30 March 2023

1. Introduction

The discovery of CMS was a milestone advancement in the development of global rapeseed cultivation, introducing a new era of heterosis utilization in *Brassica napus* (*B. napus*). More than 10 CMS types have been discovered in *B. napus*, among them *pol* [1], *shaan* [2,3], *ogu* [4,5], *nap* [6], *tour* [7], *nca* and *hau* [8]. The *pol* and *shaan* types belong to the same type of CMS system, sharing common fertility restorers [9]. In 1972, Fu et al. discovered the first male sterile line of rapeseed in the world, named *pol* CMS [10]. In 1985, Li et al. used the CMS line *shaan* 2A to breed the three-line hybrid variety Qinyou No. 2, which was the first widely promoted hybrid variety of rapeseed in the world [3,11]. Through over 40 years of continuous development, China has realized the hybridization of *B. napus* varieties, in which the *pol* and *shaan* CMS systems play a crucial role as the backbone of the sterile lines for the heterosis utilization of rapeseed.

CMS is a type of male sterility with a sterile phenotype controlled by the interaction between a cytoplasmic sterility gene S and a pair of recessive nuclear genes r; the genotype is named S (r r) [12]. Its maintainer with a fertile phenotype contains a cytoplasmic fertility gene N and a pair of recessive nuclear genes r, named N (r r) [12]. When the CMS plants are pollinated with pollen from the maintainer plants, the harvested seeds still show cytoplasmic male sterility. The restorer with a fertile phenotype contains a pair of dominant fertile nuclear genes R and has the two genotypes, S (R R) and N (R R). When the CMS plants are pollinated by the restorer, the harvested seeds are hybrids with an S (R r) genotype [12]. In breeding practice, the CMS line can be easily identified in the flowering phase by its extremely degenerate stamens. However, the identification of the maintainer and the restorer lines is difficult since both phenotypes generate normal stamens [12,13]. In order to identify the lines, they both need to pollinate the respective CMS line [13]. If all offspring are normal plants, the pollinated plant is a restorer, and if all offspring are sterile plants, the pollinated plant is a maintainer. If the pollinated plant produces both normal fertile plants and male sterile plants in equal amounts, it is identified as a heterozygous plant with an Rr genotype [13]. Plants with the Rr genotype cannot be used for breeding and generally account for about half of the total number, making this breeding system inefficient [12]. To identify the cytoplasm type of the restorer, a female restorer needs to be pollinated by the maintainer to produce F_2-generation offspring [12]. If the F_2-generation plants are all normal fertile plants, the restorer has N cytoplasm [12]. If the F_2-generation results in about 1/4 sterile plants, the restorer has S cytoplasm. In general, it takes up to three years to establish the restorer and maintainer relationships of rapeseed plants [8]. This restricts the cultivation of new varieties. Therefore, developing a set of CMS specific molecular markers is of great importance to accelerate the breeding of new rapeseed varieties.

Through the application of molecular biology techniques, some AFLP (amplified fragment length polymorphism), RAPD (random amplified polymorphic DNA) and SSR (simple sequence repeat) molecular markers of *pol* CMS have been developed, and its restorer gene has been explored [14–16]. Furthermore, the molecular mechanism of *pol* CMS has been uncovered [1,17–20]. The mitochondrial genome of the *pol* CMS plant has an approximately 4.5 kb fragment more than that of its maintainer plant, which contains a specific open-reading frame, ORF224, upstream of the *ATP6* gene [18]. When co-transcribed with *ATP6*, it generates a new gene, *orf224-atp6*, which is largely responsible for the *pol* CMS formation [20]. Liu et al. located a PPR (pentatricopeptide repeat) gene in the restorer plant and proved that the PPR gene alone is the restorer gene of the *pol* system [1]. It is assumed that the post-transcriptional regulation of this gene cleaves the *ORF224-ATP6* gene, thereby degrading the ORF224 fragment and recovering male fertility [1,21,22]. Additionally, a functional molecular marker for the restorer of the *pol* system has been developed based on the restorer gene PPR. This marker achieves 100% reliability and is widely used for the identification of restorer plants and for the assessment of seed purity in *pol* CMS-type hybrids [23]. However, using only the restorer gene marker is not sufficient to accurately identify the restorer lines. The method is unable to distinguish between homozygous and heterozygous restorers, or between hybrids and their restorers. Therefore, it is necessary to develop a set of accurate and reliable molecular markers to identify the components of the *pol* and *shaan* CMS three-line hybrid systems.

In this study, we developed a set of molecular markers based on the gene sequence differences between the three lines of the *pol* CMS and *shaan* CMS *B. napus* systems. The markers are able to distinguish between sterile cytoplasm S and fertile cytoplasm N, and sterile nuclear gene r and fertile nuclear gene R. Using this set of molecular markers, the CMS lines of *pol* and *shaan* and their respective maintainers, restorers and hybrids can be identified accurately and quickly. The markers can be used not only to identify the restorers or maintainers of rapeseed materials, but also to assess the seed purity of *pol* and *shaan* CMS type hybrids.

2. Results

2.1. Developing Cytoplasmic Molecular Markers for N and S Cytoplasm Identification of the Pol and Shaan CMS Three-Line Hybrids

Mitochondrial resequencing of the CMS lines *pol* A and *shaan* 2A and their maintainers *pol* B and *shaan* 2B in *B. napus* revealed that the mitochondrial genomes of *pol* A and *shaan* 2A, unlike those of *pol* B and *shaan* 2B, both had the same exogenous fragment insertions, which were caused by a 4443 bp exogenous fragment replacing an 813 bp fragment on the mitochondrion of fertile cytoplasm N. Since there was no homology between the 4445 bp sequence and the 813 bp sequence, the specific reverse primers S-R and N-R were developed from the 4445 bp fragment and the 813 bp fragment, respectively. In addition, a common forward primer, SN-F, was designed for upstream of the fragment-replacement region (Figure 1A).

Using N- and S-type cytoplasm DNA as templates, PCR amplification showed that mixtures of the three primers, SN-F, S-R and N-R, could clearly distinguish between S cytoplasm and N cytoplasm; N-type cytoplasm amplified a clear 655 bp band, while S-type cytoplasm amplified a clear 1287 bp band (Figure 1B). In addition, S cytoplasm had two types of bands: the first one, which accounted for a very small proportion of the identified samples, had only one clear 1287 bp band, while the second one, which was the prominent type, had a clear 1287 bp band and a weak 655 bp band.

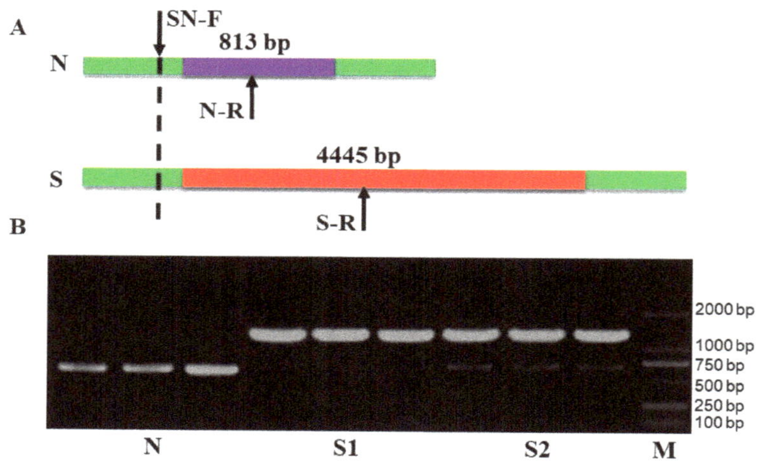

Figure 1. Identification of sterile cytoplasm S and fertile cytoplasm N of *pol*- and *shaan*-type CMS three-line hybrids by molecular markers. (**A**) The design schematic of S and N primers. (**B**) The identification results of S and N. The identical sequences of N and S genes are boxed in green. The inconsistent sequences of N and S genes are boxed in purple and red, respectively. N, fertile cytoplasm gene N; S, sterile cytoplasm gene S; M, DNA marker.

2.2. Developing Cytoplasmic Molecular Markers for N and S Cytoplasm Identification of the Pol and Shaan CMS Three-Line Hybrids

Based on specific single nucleotide polymorphisms (SNPs) of the sterile nuclear gene r and fertile nuclear gene R of CMS systems, the R-specific forward primer R-F and r-specific forward primer r-F were designed. At the same time, their common reverse primer Rr-R was designed. The R-F and Rr-R primers were used to identify the R gene, while r-F and Rr-R primers were used to identify the r gene. Because the amplification bands of R and r were the same size, PCR amplification and agarose electrophoresis for R and r identification were performed separately.

Figure 2 shows that all three lines (sterile line, maintainer line and restorer line) and their hybrids could be identified using PCR and a combination of the S/N, R and r molecular markers. The sterile line had S cytoplasm and only contained an r band; thus, it was identified as S (r r) genotype line. The maintainer exhibited N cytoplasm and also only contained an r band; thus, its genotype was identified as N (r r). Restorer 1 had S cytoplasm and only contained an R band and hence has an S (R R) genotype, while restorer 2 had N cytoplasm with only an R band and was identified as the N (R R) genotype. The hybrid, which had S cytoplasm and contained both an R band and an r band at same time, had an S (R r) genotype. These results are consistent with the genetic pattern of the three lines of the CMS system and reveal that the combination of S/N, R and r molecular markers can accurately identify the *B. napus* CMS sterile lines, maintainers and restorers, as well as their hybrids.

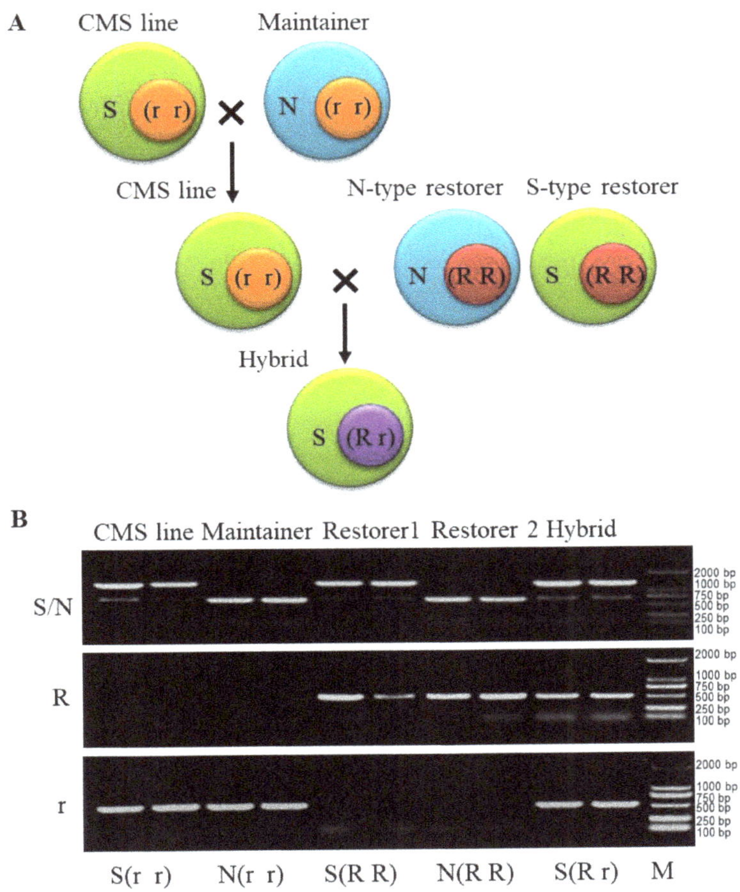

Figure 2. Identification of CMS three lines and its hybrids in *B. napus* by the molecular markers. (**A**) Schematic diagram of genetic pattern of CMS three lines and their hybrids in *B. napus*. (**B**) Identification of CMS three lines and their hybrids in *B. napus* by the molecular markers. N, fertile cytoplasm gene N; S, sterile cytoplasm gene S; R, fertility nuclear restorer gene; r, recessive nuclear sterile gene; M, DNA marker.

2.3. The Molecular Markers Are Suitable for the Identification of the Three Lines in the Pol- and Shaan-Type CMS Systems of B. Napus

The *pol* CMS and *shaan* CMS are two important systems for the heterosis utilization widely applied in rapeseed production [24]. In order to verify that our set of molecular markers can be applied in the identification of both CMS systems, we extracted the DNA of the pol (*pol* A and its maintainer *pol* B) and *shaan* (*shaan* 2A and its maintainer *shaan* 2B) systems as templates for PCR amplification.

The results show that the bands of *pol* A and *shaan* 2A both exhibited S (r r) genotypes. Similarly, the bands of *pol* B and *shaan* 2B both exhibited N (r r) genotypes. Thus, this set of molecular markers can be used as common markers to identify both the *pol* and *shaan* CMS systems simultaneously (Figure 3).

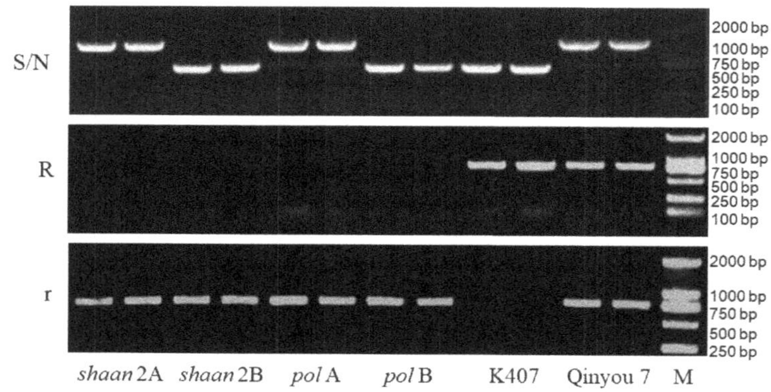

Figure 3. Identification of *pol* A and *shaan* 2A and their maintainers by the molecular markers. N, fertile cytoplasm gene N; S, sterile cytoplasm gene S; R, fertility nuclear restorer gene; r, recessive nuclear sterile gene; M, DNA marker.

2.4. Identification of Rapeseed Germplasm with the CMS Three-Line Molecular Markers

In order to verify the accuracy of this set of molecular markers, the CMS sterile lines (samples 4 and 6), maintainers (sample 2) and restorers (samples 1, 3 and 5) were taken and examined in the flowering stage. PCR amplification confirmed that samples 4 and 6 were of the CMS genotype S (r r), sample 2 was of the maintainer genotype N (r r), sample 1 was of the restorer genotype N (R R) and samples 3 and 5 were of genotype the S (R R) (Figure 4). These results are consistent with the known phenotypes of the test samples. They also clarified the S or N cytoplasmic type of the restorers. Thus, these markers can be used for the fast and accurate identification of rapeseed germplasms.

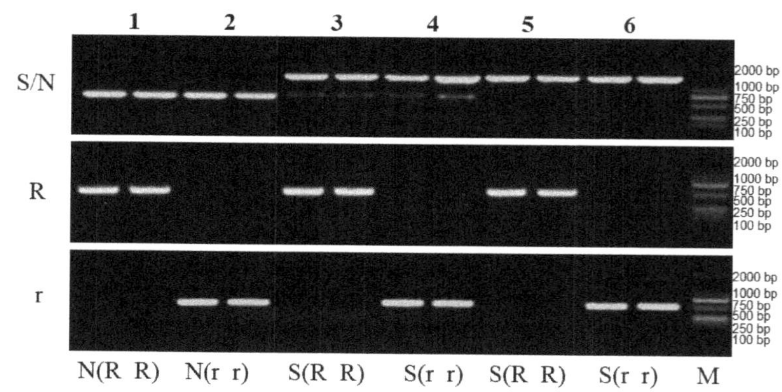

Figure 4. Identification of rapeseed germplasm with the CMS three-line molecular markers. N, fertile cytoplasm gene N; S, sterile cytoplasm gene S; R, fertility nuclear restorer gene; r, recessive nuclear sterile gene; M, DNA marker. The Arabic numbers (1 to 6) indicates different types of the three CMS lines: CMS sterile lines (samples 4 and 6), maintainers (sample 2) and restorers (samples 1, 3 and 5).

2.5. Identification of Rapeseed Germplasm with the CMS Three-Line Molecular Markers

The rapid and accurate identification of seed purity is of great importance to reduce the risk of low-quality seeds, to ensure seed safety and to safeguard farmers' rights and interests. In the past, it usually took one year to identify seed purity through field planting. In order to shorten the identification time, molecular markers such as RAPD and SSR have been applied to the identification of hybrid purity [25,26]. However, these markers usually only correspond to a specific variety, and the change in variety requires the re-development of molecular markers. At present, the *pol* and *shaan* CMS-type hybrids still represent the majority of rapeseed in China [12]. This set of CMS molecular markers can be used as universal markers to identify hybrids purity.

DNA samples extracted from hybrid seedlings that were germinated for at least four days were used as templates for PCR amplification. The analysis of the results shows that two samples (3 and 14) were restorer seeds, sample 6 was a sterile seed, sample 11 was a maintainer seed and the remaining samples were hybrid seeds (Figure 5). The proportion of hybrids, sterile lines, maintainers and restorers in the test samples was 73.3%, 6.6%, 6.6% and 13.3%, respectively.

In conclusion, these molecular markers can be used to identify the germplasms and the seed purity of *pol* or *shaan* CMS-type hybrids. For seed purity, the entire identification can be performed in less than one week, which greatly shortens the process time and ensures the quality and safety of rapeseed hybrids.

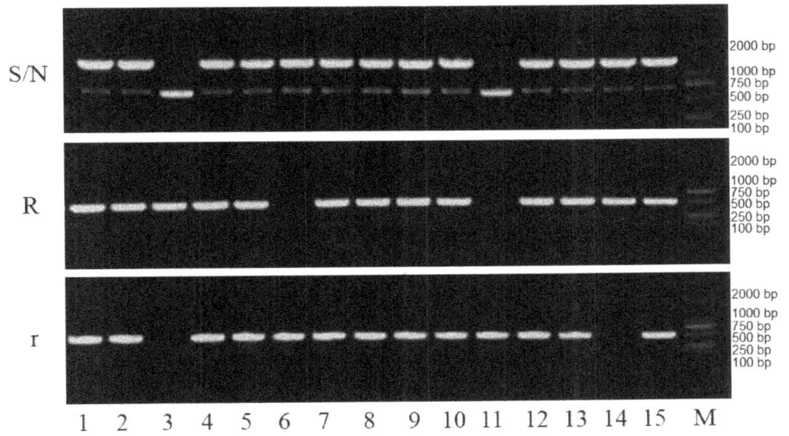

Figure 5. Identification of seed purity for CMS three-line hybrids with the CMS three-line molecular markers. N, fertile cytoplasm gene N; S, sterile cytoplasm gene S; R, fertility nuclear restorer gene; r, recessive nuclear sterile gene; M, DNA marker. The Arabic numbers (1 to 15) indicates different hybrid seedlings.

3. Discussion

The identification of the relationship between the maintainer and restorer in rapeseed is crucial for cultivating new varieties, which is directly related to transferring a sterile line or breeding a restorer [1,27–30]. In this study, the S and N molecular markers developed are derived from the indel differences in mitochondrial DNA between fertile cytoplasmic N and sterile cytoplasmic S, while the R and r molecular markers developed are derived from core SNP differences between the restorer gene R and the sterility gene r in the nucleus [12,31]. The combination of these molecular markers can rapidly and accurately identify the CMS lines, maintainer lines, restorer lines and hybrids of the *pol* and *shaan* CMS systems [2,31]. This can be applied to the identification of both the maintainer or restorer relationship in inbred lines, and the rapeseed purity of the cytoplasmic three-line rapeseed hybrids. This is of great significance for accelerating the cultivation of rapeseed varieties and ensuring the seed quality and safety. For seed purity, the entire identification can be performed in less than one week.

Although this set of CMS three-line molecular markers can accurately identify the purity of rapeseed inbred lines and its hybrids, attention should be paid to obtaining accurate and reliable identification results by strictly following the described experimental protocol. Inconsistent DNA concentrations will lead to unstable PCR bands, resulting in inaccurate identification results. The Taq enzyme is also an important factor affecting the identification results. There is only a slight difference in the three bases between the specific molecular markers of the nuclear sterile gene r and the nuclear restorer gene R. When amplified by the common Taq enzyme, some restorer samples (which should not have any bands) will show a light band, which is significantly different from the clearly visible R band. From experience, we know that this light band is an artefact. If amplified by some high-fidelity Taq enzymes, the brightness of this non-specific band will increase, which can cause misinterpretation of the result. Therefore, we recommend that the MIX products with common Taq enzymes be used as much as possible during the identification process.

CMS-type hybrid systems usually consist of three types of seeds: the hybrid seed; the sterile seed, which is a serious negative factor affecting the purity of hybrids and must be removed [12,27]; and the restorer seed. The maintainer seed is rarely mixed into hybrids and is generally ignored. Hence, in order to improve identification speed and save costs, the identification of S- and N-type markers can be omitted, and the R and r molecular markers are sufficient to determine seed purity. However, if seed purity is too low and needs to be re-established, we suggest using S/N molecular markers to determine whether the reason for the impurity is present because the sterile line is mixed with maintainer seed or the restorer line is mixed with (R r) heterozygous plants. If the low purity is caused by the impurity of the parents, the parents can be purified according to the identification results to improve the quality of subsequent seed production. Overall, this set of markers has strong applicability and important application value and can greatly shorten the time required to identify the restorer and maintainer relationships of rapeseed plants. Furthermore, it can also be used as a universal marker to identify the purity of *pol* and *shaan* CMS three-line hybrid, which is important in ensuring the seed quality and safety of hybrids.

4. Materials and Methods

4.1. Materials

The rapeseed CMS lines *shaan* 2A [2,3] and *pol* A [1] of rapeseed and their maintainers *shaan* 2B and *pol* B [3] were sourced from Dianrong Li. The rapeseed CMS hybrid Qinyou 7 of rapeseed and its sterile line *shaan* 3A, its maintainer *shaan* 3B and its restorer K407 were sourced from Dianrong Li. Some other rapeseed inbred lines and three-line hybrids, also sourced from Dianrong Li, were used for identification.

4.2. Protocol for High Throughput Screening and Rapid DNA Extraction for Rapeseed

We added one stainless steel ball (5.0 mm diameter) into each hole of a 96-well medium-hole plate (30 mm height). For each plant sample, we took one piece of round leaf using a hole puncher with a 3 mm aperture. We loaded each sample into a matching hole using tweezers and filled in the sample register (Figure S1). We covered the plate with a silicone cover and transferred it into an ice box with a temperature of 4 °C. Samples were either directly taken to the lab for DNA extraction or stored at −20 °C. The seed DNA was extracted after germination for at least 4 days. For each sample, 200 μL 1 × DNA extraction solution (10 mmol·L^{-1} Tris (pH 9.5), 0.5 mmol·L^{-1} EDTA and 0.1 mol·L^{-1} KCl) was added into the well, and we covered the plate with a silicone cover. The plate was then put into a grinder (Tissuelyser-192, Shanghai Jingxin), where the samples were ground for 2 min at 30 Hz, repeated once. The plate was then placed into a centrifuge (Eppendorf 5810R with four plate buckets) and the samples were centrifuged for 5 min at 3000 rpm. The plate was stored at 25 °C for 1 h or overnight at 4 °C. The supernatant was then used as a template for PCR amplification.

4.3. Primer Design

Sequence analysis of cytoplasmic genes of the CMS lines and their maintainers was performed using the Vector NTI software. According to indel (insertion-deletion) differences in mitochondrial DNA between S cytoplasm and N cytoplasm, three primers, SN-F, N-R and S-R, were designed using the Primer Premier 5 software (Table 1, Figure S2). The cytoplasmic fertile gene N was identified using the SN-F and N-R primer pair with a 655 bp amplification band, while the cytoplasmic sterile gene S was identified using the SN-F and S-R primer pair with a 1287 bp amplification band. Due to the sharp difference in size between the amplified N and S fragments, S and N genotypes can be clearly separated when mixing all three primers into the same reaction system for PCR amplification.

To identify the sterile nuclear gene r and fertile nuclear gene R among the CMS three-line hybrids, we designed the R-specific primer R-F, the r-specific primer r-F and a common reverse primer Rr-R based on specific single-nucleotide polymorphisms (SNPs) (Table 1, Figure S3).

Table 1. Primers of molecular markers for identification of *pol* and *shaan* CMS three-lines as well as their hybrids in *B. napus*.

Primer Name	Sequence (5′→3′)	Base Number
SN-F	TTCATACGGCGAGAGTCATTG	22 bp
N-R	CAAGACCATAGAATAGGAGAACCAC	22 bp
S-R	GCTCGTTCGGCTACTTATCTTG	22 bp
R-F	GGGATGCGATCCTGATATTTG	21 bp
r-F	GGGATGCGATCCTGATATCG	20 bp
Rr-R	CTCCAAAAGGACCAGAAAGCA	21 bp

4.4. PCR Reaction System and Procedure

PCR reaction system (16 μL): 1 μL DNA template, 8 μL 2 × Es Taq Mix (CW0690, CW Bio), 6.0 μL ddH2O, forward and reverse primers, each 0.5 μL. For S/N identification, add 5.5 μL ddH2O, SN-F, N-R and S-R 0.5 μL, respectively.

PCR procedure: Pre-denaturation at 95 °C for 5 min, denaturation at 94 °C for 40 s, annealing at 60 °C for 40 s, extension at 72 °C for 60 s, cycle 30 times, final extension at 72 °C for 10 min and preservation at 16 °C. In the present study, the Tm value of N/S and R was at 60 °C, while that of r was at 61 °C.

Electrophoresis: PCR products were detected using 1% agarose gel electrophoresis, 140 V, 15 min. DM2000 DNA ladder (CW0632, CW Bio) was used in the present study to determine the size of the DNA fragment. The gel was stained using ethidium bromide solution (1.0 mg/100 mL). Photographs were taken with the gel imaging system (Bio-Rad).

5. Conclusions

In this study, a set of molecular markers for both *pol* and *shaan* CMS three lines of *B. napus* were developed. Combining these markers can accurately and rapidly identify the CMS line, maintainer line and restorer line of both the *pol* and *shaan* CMS systems, as well as their hybrids. These markers can not only be used to identify the maintainer and restorer relationship of inbred materials; they can also be used as general molecular markers to identify the purity of *pol* and *shaan* CMS-type hybrid systems, which is important in ensuring the seed quality and safety of hybrids.

Supplementary Materials: The following supporting information can be downloaded at: https://www.mdpi.com/article/10.3390/plants12071514/s1, Figure S1: Schematic diagram of the sample loading; Figure S2: The nucleotide acid sequence of N and S regions; Figure S3: The alignment of R and r nucleotide acid sequences.

Author Contributions: Conceptualization, Y.Z., J.M. and S.H.; Data curation, Y.Z. and Q.J.; Funding acquisition, Y.Z. and J.M.; Investigation, C.X., S.W., D.W. and Y.D.; Methodology, Y.Z.; Supervision, Y.Z. and J.M.; Validation, Y.Z., R.A. and M.S.; Writing—original draft, Y.Z. and S.H.; Writing—review and editing, Y.Z. and S.H. All authors have read and agreed to the published version of the manuscript.

Funding: This research was funded by the Scientific Research and Sharing Platform Construction project of Shaanxi Province, grant number 2021PT-036, the Crop Breeding Key Project of "Two Chains" Fusion of Shaanxi Province, grant number 2021LLRH-07, the Key Research and Development Program of Henan Province, grant number 221111110200, the Key Research and Development Program of Yangling Seed Industry Innovation Center, grant number Ylzy-yc-2021-02, and the Shaanxi Creative Talents Promotion Plan-Technological Innovation Team, grant number 2020TD-051.

Institutional Review Board Statement: Not applicable.

Informed Consent Statement: Not applicable.

Data Availability Statement: Not applicable.

Acknowledgments: We are grateful to Dianrong Li at the Hybrid Rapeseed Research Center of Shaanxi Province for providing the *pol* and *shaan CMS* lines.

Conflicts of Interest: The authors declare no conflict of interest.

References

1. Liu, Z.; Yang, Z.; Wang, X.; Li, K.; An, H.; Liu, J.; Yang, G.S.; Fu, T.D.; Yi, B.; Hong, D.F. A Mitochondria-Targeted PPR Protein Restores *pol* Cytoplasmic Male Sterility by Reducing orf224 Transcript Levels in Oilseed Rape. *Mol. Plant* **2016**, *9*, 1082–1084. [CrossRef] [PubMed]
2. Havlíčková, L.; Čurn, V.; Jozová, E.; Kučera, V.; Klíma, M. Sequence analysis of the mtDNA region correlated with Shaan 2A cytoplasmic male sterility in rapeseed (*Brassica napus* L.). *Czech J. Genet. Plant Breed.* **2012**, *48*, 139–142. [CrossRef]
3. Wang, Y.F.; Ma, S.M.; Wang, M.; Zheng, X.Q.; Gu, M.; Hu, S.W. Sequence analysis of the gene correlated with cytoplasmic male sterility (CMS) in rape-seed (*Brassica napus*) Polima and Shaan 2A. *Chin. Sci. Bull.* **2002**, *47*, 122–126.
4. Yu, H.L.; Li, Z.Y.; Yang, L.M.; Liu, Y.M.; Zhuang, M.; Zhang, L.G.; Lv, H.H.; Li, Z.S.; Han, F.Q.; Liu, X.P.; et al. Morphological and molecular characterization of the second backcross progenies of Ogu-CMS Chinese kale and rapeseed. *Euphytica* **2017**, *213*, 55. [CrossRef]
5. Wang, C.D.; Lezhneva, L.; Arnal, N.; Quadrado, M.; Mireau, H. The radish Ogura fertility restorer impedes translation elongation along its cognate CMS-causing mRNA. *Proc. Natl. Acad. Sci. USA* **2021**, *118*, e2105274118. [CrossRef] [PubMed]
6. Engelke, T.; Hirsche, J.; Roitsch, T. Metabolically engineered male sterility in rapeseed (*Brassica napus* L.). *Theor. Appl. Genet.* **2011**, *122*, 163–174. [CrossRef]
7. Bhatia, R.; Dey, S.S.; Sharma, K.; Singh, S.; Kumar, S.; Pramanik, A.; Parkash, C.; Kumar, R. Back-cross introgression of 'Tour' cytoplasm from Brassica napus through in vitro embryo rescue reveals partial restoration of sterility in *B. oleracea*. *Sci. Hortic.* **2021**, *282*, 110014. [CrossRef]
8. Wan, Z.J.; Bing, J.; Tu, J.X.; Ma, C.Z.; Shen, J.X.; Yi, B.; Wen, J.; Huang, T.; Wang, X.J.; Fu, T.D. Genetic characterization of a new cytoplasmic male sterility system (*hau*) in *Brassica juncea* and its transfer to *B. napus*. *Theor. Appl. Genet.* **2008**, *116*, 355–362. [CrossRef]
9. Liu, J.M.; Li, M.T.; Wang, H.; Yu, L.J.; Li, D.R. Sequence analysis and expression of orf224 gene associated with two types of cytoplasmic male sterility in *Brassica napus* L. *Z. Nat. C J. Biosci.* **2010**, *65*, 395–402. [CrossRef]
10. Fu, T.; Yang, G.S.; Yang, X.N. Studies on "Three Line" Polima Cytoplasmic Male Sterility Developed in *Brassica napus* L. *Plant Breed.* **2010**, *104*, 115–120.
11. Ning, L.Y.; Wang, H.; Li, D.R.; Lin, Z.W.; Li, Y.H.; Zhao, W.G.; Chao, H.B.; Miao, L.Y.; Li, M.T. Transcriptomic and Proteomic Analysis of Shaan2A Cytoplasmic Male Sterility and Its Maintainer Line in *Brassica napus*. *Front. Plant Sci.* **2019**, *10*, 252. [CrossRef] [PubMed]
12. Chen, L.; Liu, Y.G. Male sterility and fertility restoration in crops. *Annu. Rev. Plant Biol.* **2014**, *65*, 579–606. [CrossRef]
13. Bohra, A.; Jha, U.C.; Adhimoolam, P.; Bisht, D.; Singh, N.P. Cytoplasmic male sterility (CMS) in hybrid breeding in field crops. *Plant Cell Rep.* **2016**, *35*, 967–993. [CrossRef]
14. Liu, Z.W.; Fu, T.D.; Tu, J.X.; Chen, B.Y. Inheritance of seed colour and identification of RAPD and AFLP markers linked to the seed colour gene in rapeseed (*Brassica napus* L.). *Theor. Appl. Genet.* **2005**, *110*, 303–310.

15. Li, Y.; Liu, Z.; Cai, Q.; Yang, G.S.; He, Q.B.; Liu, P.W. Identification of a SSR marker linked to the fertility-restoring gene for polima cytoplasmic male sterile line in *Brassica napus*. *Afr. J. Biotechnol.* **2011**, *10*, 9563–9569.
16. Zeng, F.Q.; Yi, B.; Tu, J.X.; Fu, T.D. Identification of AFLP and SCAR markers linked to the male fertility restorer gene of *pol* CMS (*Brassica napus* L.). *Euphytica* **2009**, *165*, 363–369. [CrossRef]
17. Wang, B.Q.; Farooq, Z.; Chu, L.; Liu, J.; Wang, H.D.; Guo, J.; Tu, J.X.; Ma, C.Z.; Dai, C.; Wen, J.; et al. High-generation near-isogenic lines combined with multi-omics to study the mechanism of polima cytoplasmic male sterility. *BMC Plant Biol.* **2021**, *21*, 130. [CrossRef]
18. L'Homme, Y.; Stahl, R.J.; Li, X.Q.; Hameed, A.; Brown, G.G. Brassica nap cytoplasmic male sterility is associated with expression of a mtDNA region containing a chimeric gene similar to the *pol* CMS-associated orf224 gene. *Curr. Genet.* **1997**, *31*, 325–335. [CrossRef] [PubMed]
19. Singh, M.; Brown, G.G. Suppression of Cytoplasmic Male Sterility by Nuclear Genes Alters Expression of a Novel Mitochondrial Gene Region. *Plant Cell* **1991**, *3*, 1349–1362.
20. An, H.; Yang, Z.H.; Yi, B.; Wen, J.; Shen, J.X.; Tu, J.X.; Ma, C.Z.; Fu, T.D. Comparative transcript profiling of the fertile and sterile flower buds of *pol* CMS in *B. napus*. *BMC Genom.* **2014**, *15*, 258. [CrossRef] [PubMed]
21. Yuan, M.; Yang, G.S.; Fu, T.D.; Li, Y. Transcriptional control of orf224/atp6 by the *pol* CMS restorer Rfp gene in *Brassica napus* L. *Yi Chuan Xue Bao* **2003**, *30*, 469–473.
22. Barkan, A.; Small, I. Pentatricopeptide Repeat Proteins in Plants. *Annu. Rev. Plant Biol.* **2014**, *65*, 415–442. [CrossRef] [PubMed]
23. Yang, Z.H. *Map-Based Cloning and Function Analysis of the Restorer Gene of POLIMA Cytoplasmic Male Sterility in Brassica napus*; Huazhong Agricultural University: Wuhan, China, 2016.
24. Liao, Z.Q.; Kuang, C.G.; Xu, L.F.; Zhou, C.Y. The Main Styles of Cytoplasmic Male Sterility in *Brassica napus* L. and Their Applications in China. *Chin. Agric Sci. Bull.* **2010**, *26*, 105–110.
25. Havlickova, L.; Jozová, E.; Rychlá, A.; Klíma, M.; Kučera, V.; Čurn, V. Genetic Diversity Assessment in Winter Oilseed Rape (*Brassica napus* L.) Collection Using AFLP, ISSR and SSR Markers. *Czech J. Genet. Plant Breed.* **2014**, *20*, 216–225. [CrossRef]
26. Koutu, G.; Shrivastava, A.; Samaiya, R.; Mishra, D. Morphological characterization and assessment of genetic purity of rice hybrids using SSR markers. *Oryza* **2016**, *53*, 21–26.
27. Zhang, W.; Shi, H.R.; Zhou, Y.; Liang, X.Y.; Luo, X.; Xiao, C.; Li, Y.; Xu, P.; Wang, J.S.; Gong, W.Z.; et al. Rapid and Synchronous Breeding of Cytoplasmic Male Sterile and Maintainer Line Through Mitochondrial DNA Rearrangement Using Doubled Haploid Inducer in *Brassica napus*. *Front. Plant Sci.* **2022**, *13*, 871006. [CrossRef] [PubMed]
28. Xing, M.; Guan, C.Y.; Guan, M. Comparative Cytological and Transcriptome Analyses of Anther Development in Nsa Cytoplasmic Male Sterile (1258A) and Maintainer Lines in *Brassica napus* Produced by Distant Hybridization. *Int. J. Mol. Sci.* **2022**, *23*, 2004. [CrossRef] [PubMed]
29. Hu, Q.; Andersen, S.B.; Dixelius, C.; Hansen, L.N. Production of fertile intergeneric somatic hybrids between *B. napus* and *S. arvensis* for the enrichment of rapeseed gene pool. *Plant Cell Rep.* **2002**, *21*, 147–152.
30. Yamagishi, H.; Bhat, S.R. Cytoplasmic male sterility in *Brassicaceae* crops. *Breed. Sci.* **2014**, *64*, 38–47. [CrossRef]
31. Ning, L.Y.; Wang, H.; Li, D.R.; Li, Y.H.; Chen, K.; Chao, H.B.; Li, H.X.; He, J.J.; Li, M.T. Genome-wide identification of the restorer-of-fertility-like (RFL) gene family in *Brassica napus* and expression analysis in *shaan*2A cytoplasmic male sterility. *BMC Genom.* **2020**, *21*, 765. [CrossRef] [PubMed]

plants

MDPI

Communication

A Method for Electroporation of Cre Recombinase Protein into Intact *Nicotiana tabacum* Cells

Yuichi Furuhata, Emiko Egi, Tomi Murakami and Yoshio Kato *

Biomedical Research Institute, National Institute of Advanced Industrial Science and Technology (AIST), 1-1-1 Higashi, Tsukuba 305-8566, Japan; y-furuhata@aist.go.jp (Y.F.)
* Correspondence: y-kato@aist.go.jp; Tel.: +81-29-861-3014

Abstract: The Cre/*lox* recombination system has become a powerful technology for gene function analysis in a broad spectrum of cell types and organisms. In our previous report, Cre protein had been successfully delivered into intact *Arabidopsis* thaliana cells using electroporation. To expand the feasibility of the method of protein electroporation to other plant cells, here we attempt the protein electroporation into tobacco-derived BY-2 cells, one of the most frequently used plant cell lines for industrial production. In this study, we successfully deliver Cre protein into BY-2 cells with intact cell walls by electroporation with low toxicity. Targeted *lox*P sequences in the BY-2 genome are recombined significantly. These results provide useful information for genome engineering in diverse plant cells possessing various types of cell walls.

Keywords: genome engineering; protein delivery; electroporation; Cre recombinase

1. Introduction

Site-specific DNA recombinases have empowered researchers to manipulate various genes across a broad spectrum of cell types and organisms. In particular, the Cre/*lox* recombination system has become a powerful technology for gene function analysis. In many studies, Cre-mediated recombination has been used to conditionally mutate genes in diverse species [1,2]. Cre recombinase specifically recognizes a 34-bp sequence designated as *lox* (the 'locus of crossover') and catalyzes DNA strand exchange between these two sites. Depending on the relative orientation of the *lox* sites to each other, recombination reactions can result in deletions, inversions, insertions, or translocations of chromosomal DNA with high fidelity [2]. Moreover, molecular engineering of Cre-*lox* system allows us to use multiple or orthogonal combinations of recombinase and the recognition sequence different from the original pair to facilitate highly advanced genome engineering such as Tre-*lox*LTR for removal of HIV [3] or Brainbow imaging for neurons [4].

Despite the usefulness, the effectiveness of the Cre/*lox* system depends largely on the availability of a method for delivery of Cre recombinase into cells. In plants, agrobacterium-based gene transfer (AMGT) has been widely used to introduce foreign DNA into plant cells and is the system of choice for genetic transformation of model plants including *Arabidopsis* and *Nicotiana*. However, owing to the biological nature of agrobacterium, transgenes derived from the AMGT vector are randomly integrated into the host genome. This might result in the disruption of host gene expression, since an AMGT-based Cre/*lox* system requires the introduction of a Cre gene in addition to the target *lox* gene. Multiple gene introduction also requires more drug-resistant genes hat their repertoire is limited.

To overcome the limitation of these DNA-based deliveries, protein-based delivery has been attracting attention for biotechnology applications. Previous work in our laboratory involved the introduction of Cre protein into *A. thaliana* cells using electroporation, where we demonstrated the removal of a specific gene fragment in the plant genome [5,6]. Using a suitable buffer, we succeeded in delivering Cre proteins into *Arabidopsis* cells with low cytotoxicity and a high efficiency of over 80%. To evaluate further the reported methodology

Citation: Furuhata, Y.; Egi, E.; Murakami, T.; Kato, Y. A Method for Electroporation of Cre Recombinase Protein into Intact *Nicotiana tabacum* Cells. *Plants* **2023**, *12*, 1631. https://doi.org/10.3390/plants12081631

Academic Editors: Zanmin Hu, Han Xiao, Yi Ren and Chengming Fan

Received: 10 March 2023
Revised: 6 April 2023
Accepted: 10 April 2023
Published: 12 April 2023

applicability to other plant cells, here we attempted the protein electroporation into tobacco-derived BY-2 cells. BY-2 cells have become one of the most frequently used plant cell lines for protein production, including for industrial and drug applications, because of their high proliferative nature and scalability [7]. In this study, we successfully delivered Cre protein into BY-2 cells with the intact cell wall by electroporation. The optimal voltage field strength and duration at poring pulse for BY-2 are lower than that for *A. thaliana* cells, reflecting that these cell types have different physical properties on their cell membranes and cell walls.

2. Results

We first established a reporter cell line (BY-2-xGxFL) to evaluate the delivery efficiency of Cre protein into cells to induce genomic DNA recombination. The reporter cell harbors the gene cassette encoding green fluorescent protein (GFP), followed by a transcription termination signal and a sequence encoding Firefly luciferase (FLuc). We integrated a binary vector T-DNA from pCambiaN-xGxFL into BY-2 with AMGT and selected a hygromycin-resistant clone. The reporter cells originally express GFP under the control of 35S promoter and do not express FLuc. As the result of Cre-induced recombination of two *lox*P sites, which locates up and down stream of GFP, BY-2-xGxFL cells come to express FLuc and, thus, quantified as luminescence (Figure 1). This method is much more reliable and quantitative as the proof method for intracellular delivery compared with the microscopic observation of fluorescently labeled protein, which may experience background fluorescence [8]. A luminescence signal is acquired only when Cre proteins get internalized into cells and mediate the recombination reaction at the host cell genome.

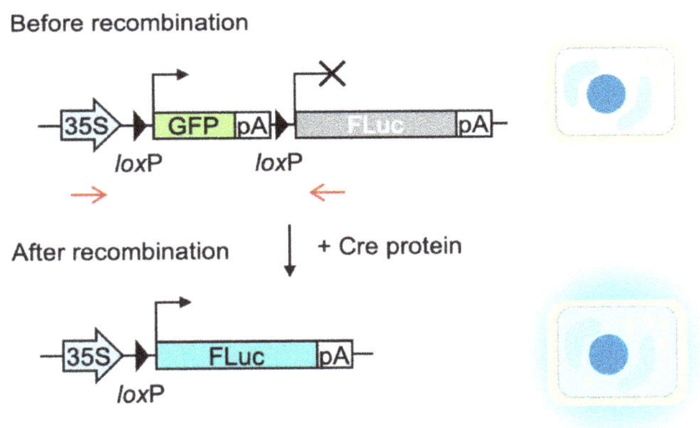

Figure 1. Schematic of Cre reporter design. GFP, pA, and FLuc indicate green fluorescent protein, terminator polyadenylation signal, and firefly luciferase, respectively. Before recombination, the reporter gene is transcribed to synthesize GFP (black arrow) but not FLuc (X). Cre protein catalyzes the recombination between the two *lox*P sites flanking the GFP coding sequence, resulting in FLuc expression that exhibits luminescence upon addition of D-luciferin. The red arrows are the primer-binding sites for genomic PCR.

To deliver Cre protein into BY-2 cells with an intact cell wall, the purified Cre protein was subjected to the reporter cell line BY-2-xGxFL with electroporation. Cre proteins were genetically fused with an N-terminal nuclear localization signal (NLS) and hexa-histidine tag. In our previous reports, we observed efficient delivery for *A. thaliana* T87 cells at the condition with 375 V/cm of field strength, 10 msec of poring pulse, and 5 times of multiple pulses in the presence of an Opti-MEMI buffer using an NEPA21 TYPE II apparatus [5]. We initially attempted the same condition for tobacco BY-2 cells as T87 cells; however, the BY-2 cells were largely damaged. Then we re-examined the electroporation conditions for BY-2 cells with a milder field strength of the poring pulse. The cytotoxicity caused by the poring pulse increased in response to increases in field strength when BY-2 cells were subjected to electroporation and was severe even at 200 V/cm of the field strength (Figure 2a). We obtained the optimal condition at 50 V/cm of the field strength and 20 ms duration of the poring pulse (Figure 2b). At this condition, reporter cells after treatment of protein electroporation exhibited less cytotoxicity and the highest luminescence among tested.

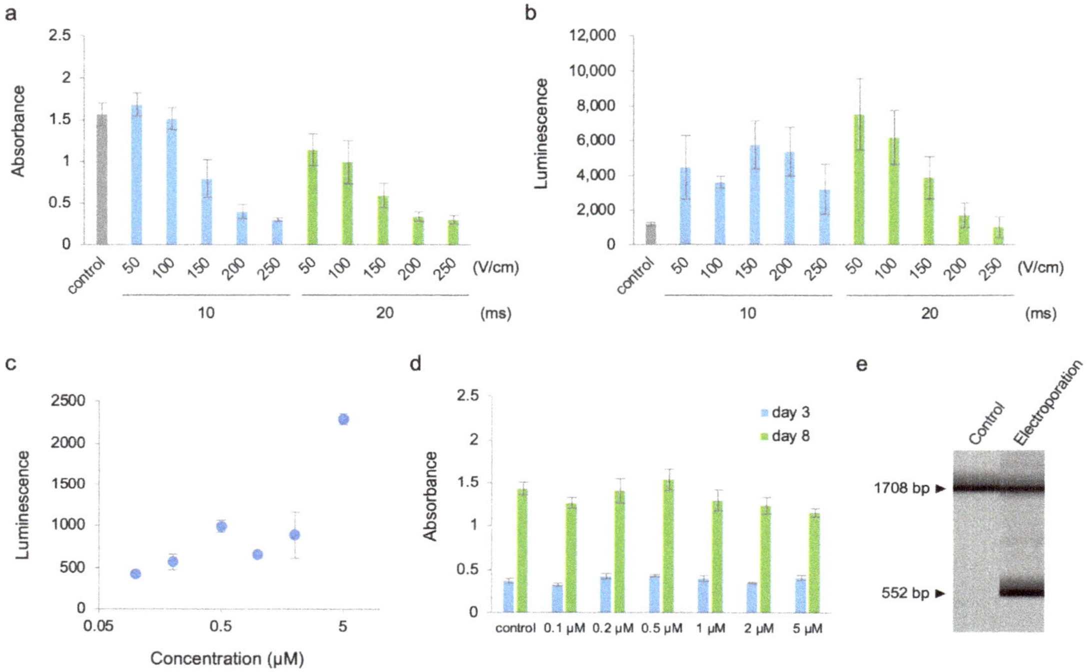

Figure 2. Electroporation-mediated Cre protein delivery into BY-2 cells. (**a**) Effect of poring pulse conditions on the viability of BY-2-xGxFL cells. CCK8 assay was performed 2 d after electroporation in electroporated BY-2-xGxFL cells using different poring pulse conditions: field strength (50, 100, 150, 200, or 250 V/cm), duration (10 or 20 ms) by measuring the absorbance at 450 nm. Control indicates untreated BY-2-xGxFL cells. Values shown are the mean ± SE of $n = 3$. (**b**) Effect of poring pulse conditions on electroporation efficiency. Firefly luciferase activity was determined by the luminescence measurement of catalyzed D-luciferin in electroporated BY-2-xGxFL cells using different poring pulse conditions: field strength (50, 100, 150, 200, or 250 V/cm), duration (10 or 20 ms). In all cases, 5 μM Cre protein dissolved in Opti-MEMI was used for electroporation. Control indicates untreated BY-2-xGxFL cells. Values shown are the mean ± SE of $n = 3$. (**c**) Effect of Cre protein concentration on on electroporation efficiency. Firefly luciferase activity as determined by the luminescence measurement of catalyzed D-luciferin in electroporated BY-2-xGxFL cells with different concentrations of Cre protein dissolved in Opti-MEMI; 0.1, 0.2, 0.5, 1.0, 2.0, or 5.0 μM. Values shown are the mean ± SE of $n = 3$. (**d**) Viability of electroporated BY-2-xGxFL cells with different concentrations of Cre protein. CCK8 assay was performed 2 d after electroporation by measuring the absorbance at 450 nm. Control indicates untreated BY-2-xGxFL cells. Values shown are the mean ± SE of $n = 3$. (**e**) Microchip electrophoresis analysis of genomic DNA PCR products from BY-2-xGxFL cells. The representative image in three independent experiments is shown. The 1708 bp and 552 bp fragments represent the reporter gene cassette before and after Cre-mediated recombination, respectively.

Next, we assessed the effect of Cre protein concentration on the electroporation efficiency. Cells were electroporated with 0.01–5 μM of Cre protein using the optimal electroporation program. As a result, Cre protein delivery increased luminescence concomitantly with the concentration of Cre protein (Figure 2c). Higher concentration of Cre protein administration did not largely affect the cytotoxicity over 3–8 days after electroporation (Figure 2d). These results reveal that protein concentration was an important factor in the efficacy of protein electroporation into cells.

To further validate the Cre protein delivery, we analyzed genomic DNA from the reporter cells with or without protein electroporation. A pair of PCR primers was designed to bind the regions of both the 35S promoter and the gene for FLuc (Figure 1; red arrows). The resultant PCR products, estimated as 1708 bp for the non-treatment control and 552 bp for recombined sequence, were quantified as electrogram by capillary electrophoresis. As shown in Figure 2e, electroporated cells exhibited a significant ratio on the PCR product, indicating that the electroporation at the optimal condition delivered the Cre protein into cells and induced the recombination at the targeted DNA sequence.

3. Discussion

Using the electroporation method, we introduced the Cre protein into tobacco cells, and observed the activity of the introduced protein through the genome recombination event. The characteristics of the protein of interest, concentration, or electroporation conditions can greatly affect the efficiency of delivery depending on the types of cargo proteins and target cells. In a previous report, Cedeño et al. attempted the delivery of ERD proteins into tobacco cells to analyze the in-cell NMR through the cell wall using electroporation [9]. Fluorescently labeled ERD proteins (250 uM, 20–30 kDa, and 5.1–5.4 of pI) were electroporated with a 750 V/cm and 20 ms poring pulse condition. Under our experimental condition, BY-2 cells were severely damaged with the poring pulse over 250 V/cm and 20 ms, although the conditions are not identical because of the different electroporation devices. Researchers may need to consider the copy number of the target sequences and to screen optimal parameters according to the specific cell types.

Unlike the introduction of DNA, the introduction of proteins into plant cells with intact cell walls brings various advantages. In particular, DNA-free genome modification or editing technology is expected to play an important role in plant engineering. For example, long-term expression of artificial nucleases such as CRISPR-Cas9 or TALENs induces harmful actions for the host genome by off-target effects. The Cre-*lox*P system as a method of removing unnecessary genes is an appropriate option for BY-2 cells, which are promising in the production of drug substances or biopharmaceuticals, since null segregations by crossing are not applicable. Thus, it may be possible to prevent such side effects simply by introducing the artificial nuclease proteins or by removing the gene cassette with the Cre protein. Based on our results, we anticipate that BY-2 cells will be more industrially valuable through the gene network engineering by direct protein delivery.

4. Methods

4.1. Plasmid Construct Preparation

To construct pCambiaN-xGxFL, used for generation of BY-2 reporter cells, a DNA fragment encoding Firefly luciferase was amplified with the PCR primer set (5′- TACGAAGTTATC TAGACCATGGAAGATGCCAAAAACATT-3′ and 5′- GTCACCAATTCACACGTGTTACAC GGCGATCTTGCCGCC-3′) and integrated into pCambiaN-xGxGUS[5] at XbaI and PmlI sites to replace the GUS gene. The complete sequence of pCambiaN-xGxFL is shown in Figure S1.

4.2. Preparation of Cre Protein

HNCre protein was expressed using *Escherichia coli* strain BL21(DE3; Nippon Gene, Tokyo, Japan) carrying pET-HNCre, encoding a hexa-histidine tag and nuclear localization signal at the N-terminus followed by Cre recombinase. This Cre gene carries the A207T mutation, while its recombination activity is comparable to that of the wild type [5].

The overnight starter culture of *E. coli* BL21(DE3) cells were diluted 100-fold into LB broth (Lennox; Merck, Darmstadt, Germany) medium supplemented with 10 μg/mL kanamycin and incubated at 37 °C until A_{600} reached 0.6. The cells were then incubated with 0.1 mM isopropyl β-D-1-thiogalactopyranoside for another 2.5 h at 37 °C to induce protein expression and then harvested. Cells were then lysed with Digital Sonifier 250D Advanced (Branson, CT, USA) and applied to an Ni-NTA column with elution buffer (50 mM Tris-HCl, 500 mM NaCl, 10% glycerol, and 500 mM imidazole, pH 8.0). HNCre protein was further purified using a gel filtration column (HiPrep 16/60 Sephacryl S-200 HR; GE healthcare, Chicago, IL, USA) with Buffer A (20 mM HEPES, 500 mM NaCl, 10% glycerol, and 1 mM dithiothreitol, pH 7.4). The purified HNCre protein was flash-frozen in liquid N_2 and stored at −80 °C until further use.

4.3. BY-2 Cell Line and Agrobacterium Tumefaciens (Rhizobium radiobacter) Culture

The *Nicotiana tabacum* BY-2 cell line was obtained from the RIKEN Bio Resource Center (Ibaraki, Japan). BY-2 cells were cultured in a liquid LS culture medium at 22 °C while shaking (120 rpm) in the dark. The LS culture medium consists of the following ingredients: 30 g/L sucrose, 200 mg/L KH_2PO_4, 1× Murashige and Skoog plant salt mixture (FUJIFILM Wako Pure Chemical, Osaka, Japan), 40 g/L myo-inositol and 400 mg/L thiamine hydrochloride, and 904 nM 2,4-dichlorophenoxyacetic acid, with pH 5.7 adjusted with KOH solution. Cells were maintained using weekly 30- to 60-fold dilutions.

4.4. Agrobacterium-Mediated Transformation of BY-2 Cells

BY-2-xGxFL cells, the BY-2 cell line carrying Cre-responsive *Firefly* luciferase reporter cassette, were generated using AMGT, as reported previously, with minor modifications (An, Plant. Physiol, 1985). In brief, BY-2 cells were mixed with *Agrobacterium* cells harboring pCambiaN-xGxFL in the presence of 100 μM acetosyringone. Two days post-infection, the cells were washed using culture medium supplemented with 200 μg/mL cefotaxime to eliminate EHA105 and were transferred onto LS agar (LS medium with 0.7% agar) with 200 μg/mL cefotaxime and 40 μg/mL hygromycin and cultured at 28 °C in the dark for 2 weeks. A single clone was then picked and cultured in liquid LS medium containing 20 μg/mL hygromycin.

4.5. Electroporation

BY-2-xGxFL cells were collected in a tube and washed once with Opti-MEM I (Thermo Fisher Scientific, Waltham, MA, USA). Cells with 20 μL of PCV were suspended in 200 μL of Opti-MEM I with HNCre protein on ice. HNCre protein concentrations were 5 μM to evaluate the poring pulse voltage and length effect, were 0.1, 0.2, 0.5, 1, 2, or 5 μM to evaluate the protein concentration effect, and were 2 μM to evaluate the recombination efficiency with genomic PCR. Cells were then transferred into a 4 mm wide cuvette (Nepa Gene, Chiba, Japan). To evaluate the effect of voltage and pulse length of the poring pulse, the poring pulse program was 50, 100, 150, 200, or 250 V/cm (20, 40, 60, 80, or 100 V setting/0.4 cm width), 10 or 20 ms, 5 times, 10% of decay rate, and a 50 ms interval. In the other experiments, the poring pulse program was 50 V/cm (20 V setting/0.4 cm width), 20 ms, 5 times, 10% of decay rate, and a 50 ms interval. The transfer pulse program was 20 V, 20 ms, 20 times, 10% of decay rate, and a 50 ms interval for all the experiments. Electroporation was performed using the NEPA21 Type II (Nepa Gene). Immediately after electroporation, the cells were washed once with LS medium and cultured in 2 mL of LS medium at 22 °C with shaking (120 rpm) in the dark until they were used for further experiments.

4.6. Luminescence Quantification in Cells

Two days after electroporation, the volume of each cell culture was reduced to 800 μL by discarding the supernatant, and 80 μL of each cell suspension was transferred to a white 96-well plate (Thermo Fisher Scientific). Subsequently, cells were added to 40 μL of

Dual-Glo Luciferase Reagent (Promega, Madison, WI, USA) and shaken with a microplate mixer (NS-4P, ASONE, Osaka, Japan) at 25 °C for 5 min. The plate was then turned around and shaken for another 5 min. After incubation, luminescence was measured using a microplate reader (Infinite 200 Pro; Tecan, Zurich, Switzerland).

4.7. Cell Counting Kit-8 (CCK8) Assay

Cell number was analyzed using the Cell Counting Kit-8 (Dojindo Laboratories, Kumamoto, Japan). Two and seven days after electroporation, 100 μL each of the cell suspensions was transferred to a 96-well plate to which 10 μL of CCK8 solution was added. The cells were incubated at 22 °C with shaking (120 rpm) in the dark for 3 h. Absorbance at 450 nm was measured using a microplate reader (Infinite 200 Pro; Tecan) to evaluate the number of cells.

4.8. Genomic PCR of Cre-Responsive Reporter Cassette

BY-2 cells 7 d after electroporation were incubated with 40 μL of 250 mM NaOH and 0.1% Tween 20 at 98 °C for 10 min to extract genomic DNA [5]. The samples were then treated with 20 μL of 1 M Tris-HCl (pH 6.5) and centrifuged. Using the extracted genomic DNA, the integrated sequence of Cre-responsive reporter cassette in BY-2 genome was amplified via PCR using a KOD One PCR Master Mix (TOYOBO; Osaka, Japan, 35 cycles each of 98 °C for 10 s, 62 °C for 5 s, and 68 °C for 1 min 30 s) with the primers 5′-TGGAGCACGACACACTTGTCTAC-3′ and 5′-AGCTGCTCGCCGGCGGTCCC-3′. The PCR products were analyzed with a 1 kb Plus DNA Ladder (New England Biolabs, Ipswich, MA, USA) using MultiNA (MCE202, SHIMADZU, Kyoto, Japan).

4.9. Quantification and Statistical Analyses

All measurements are presented as mean ± standard error (SE). p values in figures were obtained by a Student's t-test. Sample sizes are indicated in the figure legends.

Supplementary Materials: The following supporting information can be downloaded at: https://www.mdpi.com/article/10.3390/plants12081631/s1, Figure S1: DNA sequence of the pCambiaN-xGxFL plasmid.

Author Contributions: Y.F. and Y.K. designed the research; E.E. and T.M. purified the proteins; Y.F., E.E. and T.M. performed the experiments; Y.F., E.E. and Y.K. analyzed the data; Y.F. and Y.K. drafted the manuscript. All authors reviewed and edited the manuscript. Y.K. is the contact for all correspondence and material requests. All authors have read and agreed to the published version of the manuscript.

Funding: New Energy and Industrial Technology Development Organization (NEDO, Kanagawa, Japan).

Institutional Review Board Statement: Not applicable.

Informed Consent Statement: Not applicable.

Data Availability Statement: The data of this study are available from Yoshio Kato upon request.

Acknowledgments: This work was supported by the New Energy and Industrial Technology Development Organization (NEDO, Kanagawa, Japan).

Conflicts of Interest: The authors declare no conflict of interest.

References

1. Branda, C.S.; Dymecki, S.M. Talking about a revolution: The impact of site-specific recombinases on genetic analyses in mice. *Dev. Cell* **2004**, *6*, 7–28. [CrossRef] [PubMed]
2. Nagy, A. Cre recombinase: The universal reagent for genome tailoring. *Genesis* **2000**, *26*, 99–109. [CrossRef]
3. Ankrum, J.A.; Dastidar, R.G.; Ong, J.F.; Levy, O.; Karp, J.M. Performance-enhanced mesenchymal stem cells via intracellular delivery of steroids. *Sci. Rep.* **2014**, *4*, 203–209. [CrossRef] [PubMed]
4. Livet, J.; Weissman, T.A.; Kang, H.; Draft, R.W.; Lu, J.; Bennis, R.A.; Sanes, J.R.; Lichtman, J.W. Transgenic strategies for combinatorial expression of fluorescent proteins in the nervous system. *Nature* **2007**, *450*, 56–62. [CrossRef] [PubMed]

5. Furuhata, Y.; Sakai, A.; Murakami, T.; Morikawa, M.; Nakamura, C.; Yoshizumi, T.; Fujikura, U.; Nishida, K.; Kato, Y. A method using electroporation for the protein delivery of Cre recombinase into cultured Arabidopsis cells with an intact cell wall. *Sci. Rep.* **2019**, *9*, 2163. [CrossRef] [PubMed]
6. Furuhata, Y.; Sakai, A.; Kato, Y. Protein electroporation of Cre recombinase into cultured Arabidopsis cells with an intact cell wall. *Protoc. Exch.* **2019**, 1–9. [CrossRef]
7. Doran, P.M. Therapeutically important proteins from in vitro plant tissue culture systems. *Curr. Med. Chem.* **2013**, *20*, 1047–1055.
8. Leifert, J.A.; Harkins, S.; Whitton, J.L. Full-length proteins attached to the HIV tat protein transduction domain are neither transduced between cells, nor exhibit enhanced immunogenicity. *Gene Ther.* **2002**, *9*, 1422–1428. [CrossRef]
9. Cedeño, C.; Pauwels, K.; Tompa, P. Protein delivery into plant cells: Toward in vivo structural biology. *Front. Plant Sci.* **2017**, *8*, 519. [CrossRef]

plants MDPI

Article

Integrated Full-Length Transcriptome and Metabolome Profiling Reveals Flavonoid Regulation in Response to Freezing Stress in Potato

Zhiguo Zhu [1,†], Lingling Wei [1,†], Lei Guo [1], Huihui Bao [1], Xuemei Wang [1], Philip Kear [2], Zhen Wang [3] and Guangtao Zhu [1,*]

1 Yunnan Key Laboratory of Potato Biology, Engineering Research Center of Sustainable Development and Utilization of Biomass Energy, Ministry of Education, School of Life Sciences, Yunnan Normal University, Kunming 650500, China
2 School of Life Sciences, Anhui Agricultural University, Hefei 230036, China
3 International Potato Center (CIP), CIP China Center for Asia Pacific, Beijing 100081, China
* Correspondence: zhuguangtao@ynnu.edu.cn
† These authors contributed equally to this work.

Abstract: Cold stress impairs plant growth and development, resulting in crop failure. Cultivated potato (*Solanum tuberosum* L.) is sensitive to freezing, while its wild relative, *S. commersonii*, has a strong freezing tolerance. To decipher the anti-freezing mechanism of CM, we carried out a transcriptomic and metabolomic analysis of an anti-freezing variety of CM (a type of *S. commersonii*) and a freeze-sensitive variety of DM (a type of *Solanum tuberosum* L.). A total of 49,232 high-quality transcripts from 12,811 gene loci, including 46,772 coding sequences and 2018 non-coding RNAs, were identified. KEEG enrichment analysis of differentially expressed genes (DEGs) between the two varieties showed that the flavonoid biosynthesis pathway was strongly induced by freezing stress, which was proven by flavonoid metabolome analysis. Consistent with the accumulation of more flavonoids, nearly all the pathway genes were significantly upregulated in CM than those in DM. The transcript levels of two *chalcone synthase* (*CHS-1*) isoforms and four isoforms of *flavonoid 3′-hydroxylase* (*F3′H-1*) were confirmed by qRT-PCR. Co-expression analysis identified one *Myb-related* and three *UGTs* (*UDP-glycosyltransferase*) that were significantly upregulated in CM during freezing stress. Our findings support that the flavonoid pathway was significantly enhanced by freezing stress and the greater accumulation ofglycosylatedflavonoids in resistant types than that of sensitive types, maybe accounting for the increased freezing tolerance of freeze-resistant potato varieties.

Keywords: *S. commersonii*; freezing stress; full-length transcriptome; metabolome; flavonoid metabolic pathway; transcription factor

Citation: Zhu, Z.; Wei, L.; Guo, L.; Bao, H.; Wang, X.; Kear, P.; Wang, Z.; Zhu, G. Integrated Full-Length Transcriptome and Metabolome Profiling Reveals Flavonoid Regulation in Response to Freezing Stress in Potato. *Plants* **2023**, *12*, 2054. https://doi.org/10.3390/plants12102054

Academic Editors: Zanmin Hu, Han Xiao, Yi Ren and Chengming Fan

Received: 19 March 2023
Revised: 10 May 2023
Accepted: 14 May 2023
Published: 22 May 2023

1. Introduction

The potato (*Solanum tuberosum* L.) is one of the most important tuber crops in the world, playing an important role in ensuring food security. The cultivated potato variety is sensitive to frost, exhibiting chilling damage at −0.8 °C and frost damage at −2 °C [1,2]. Even a brief exposure to cold stress can substantially reduce potato production [3]. The genetic improvement of the plant to improve freezing tolerance is the most effective way to address this problem.

Plants respond to environmental stimuli by regulating gene expression, and transcriptome analysis is an important method to investigate this process. Therefore, the acquisition of intact transcripts is a prerequisite for functional genomic [4]. Single-molecule real-time (SMRT) sequencing, a third-generation sequencing technique, can read >10 kb sequence in a single pass, so it adequately covers the majority of transcripts and provides the full-length nucleotide information without assembly [5,6]. SMRT presents huge advantages in transcriptomic analysis, such as the identification of novel transcripts, detection of alternatively spliced isoforms and non-coding RNA, especially in non-reference species [7]. In contrast, second-generation sequencing can read short sequences with high throughput and low cost, and it has an advantage in determining the accurate expression profiles of genes [8]. Furthermore, integrating second- and third-generation sequencing can produce comprehensive information to explore complex biological questions [7].

Environmental stimuli in plants produce various specialized metabolites. Higher levels of glucosinolates were produced in moderate and pronounced chilling conditions in *Arabidopsis* [9]. A high spermidine content was produced in chilling-treated cucumber; this metabolite also contributes to chilling tolerance [10]. Metabolites can bridge genotypes and phenotypes, and their presence and levels produce important clues to explore complex biological processes [11]. Multi-omics analysis revealed the regulatory network of steroidal glycoalkaloids in tomato fruit over the course of tomato domestication [12]. Metabolome-associated transcriptome analyses identified the putative terpene synthase genes affecting the biosynthesis of aroma-related volatile organic compounds in litchi flower [13]. The integration of transcriptome and metabolome revealed the defensive role of the spermidine decarboxylase gene *ADC1* (*Arginine decarboxylase gene*) and its associated pathway in response to cold stress in potato [14].

In contrast to the freezing sensitivity of cultivated potato, the wild relative *S. commersonii* possesses superior freezing tolerance (semi-lethal temperature is −11.5 °C or −4.5 °C with or without cold acclimation, respectively) [15]. Two major QTLs conferring freezing tolerance for this species were mapped to two intervals on chr02 and chr12 through a BC1 segregation population [16]. Furthermore, the overexpression of *ScGolS1* (*Galactinol Synthase 1*) derived from *S. commersonii* significantly increased freezing tolerance (−2 °C) in transgenic potato [15]. These findings expanded our understanding of freezing tolerance, illustrating that multiple genes and related pathways regulate it. Therefore, a comprehensive analysis of different potato varieties with contrasting traits based on the transcriptome and metabolome will provide new insights into freezing tolerance. In this study, we analyzed *S. tubersosum* (DM for short, doubled monoploid line) and *S. commersonii* (CM for short). The flavonoid pathway responded more strongly to freezing in CM than in DM. We confirmed this result by metabolic profiling and qRT-PCR verification of major alternative splicing isoforms of *CHS-1* and *F3′H-1*. In addition, three novel transcription factors and three GTs (*Gal-GT*, *Man-GT*, *Glu-GT*) in the flavonoid pathway, potentially contributing to freezing tolerance, were identified.

2. Results

2.1. Full-Length Transcriptome Analysis of S. commersonii

A total of 71.27 Gb of data was obtained using the Pacbio Sequel platform, containing 46.94 Mb subreads with an average length of 1518 bp (Figure 1a). The self-corrected redundant subreads generated circular consensus sequences (CCS). A total of 878,975 CCS with an average length of 1711 bp were generated by merging subreads (Figure 1b). After correction and clustering, 868,272 full-length non-chimeric reads (FLNCs) with an average length of 1672 bp were obtained from the CCS reads (Figure 1c). Subsequently, FLNCs were used to generate 62,511 high-quality and 13 low-quality isoforms. The high-quality isoforms with a length in the range of 58–5506 bp were used for further study. BUSCO (Benchmarking Universal Single-Copy Orthologs) evaluation showed that 96.2% of the isoforms were complete (Figure 1d), indicating the high quality of this transcriptome data. The isoforms generated 49,232 high-quality transcripts through matching and folding, representing 12,811 gene loci.

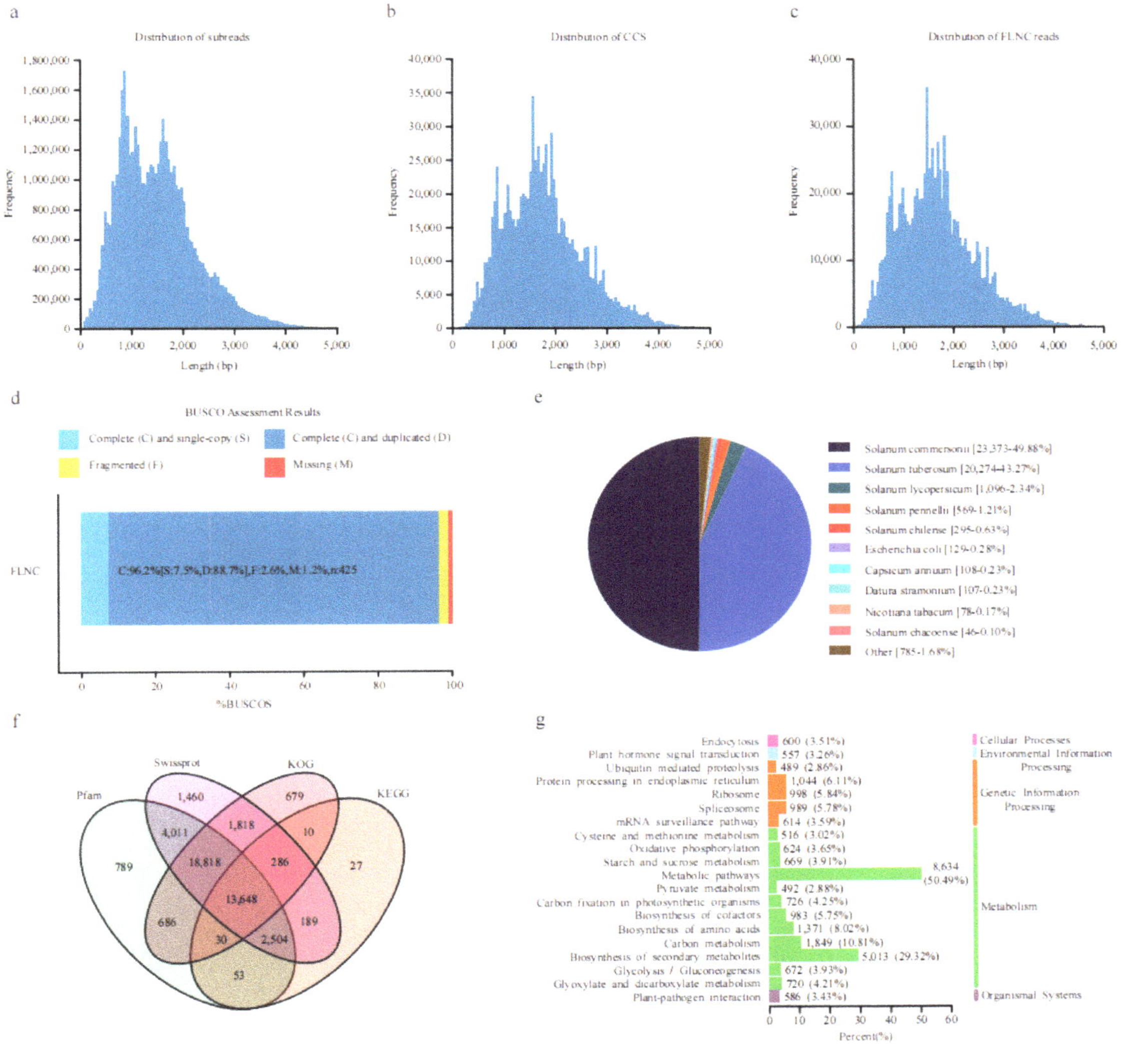

Figure 1. Overview of the full-length transcripts of *S. commersonii*. (**a**) Length distribution of subreads. (**b**) Length distribution of circular consensus sequences (CCS). (**c**) Length distribution of full-length non-chimeric reads (FLNC). (**d**) BUSCO evaluation of the transcript completeness. (**e**) Annotation of the full-length non-redundant isoforms from four databases. (**f**) Transcript annotation in Nr database. (**g**) KEGG functional annotation of the transcript.

2.2. Functional Annotation of High-Quality Transcripts from S. commersonii

Among all the transcripts, 46,857 (95.18%) were annotated in the NR protein database. The top three best BLAST hits were *S. commersonii* (49.88%), *S. tuberosum* (43.27%), and *S. lycopersicum* (Figure 1e). Using Swissprot, KEGG (Kyoto Encyclopedia of Genes and Genomes), Pfam (The Pfam protein families database), and KOG (Clusters of orthologous groups for complete eukaryotic genomes) for further annotation, a total of 2955, 6767, and 21,638 occurred in one, two, and three shared databases (Figure 1f). These 29,468 isoforms were grouped into three major GO categories (biological process, cellular component, and molecular function). For biological processes, the top three subgroups were the cellular process (20,875), metabolic process (18,618), and response to stimulus (12,777). For the cellular component, the top three subgroups were cell (26,433), cell part (26,433), and organelle (21,755). For molecular function, the top three subgroups were catalytic activity (13,374), binding (10,926), and transporter activity (1923) (Figure S1). In the KEGG classification, 29,451 isoforms were annotated in five branches (cellular processes, environmental information processing, genetic information processing, metabolism, and organismal systems), of which a high percentage of the isoforms (50.49%) were included in the metabolic pathways, with the second highest in the biosynthesis of secondary metabolites (Figure 1g), thus indicating that metabolic regulation plays an essential role in potato freezing responses (Table S1).

2.3. The Classification of Non-Redundant Transcripts

The transcripts were classified into non-coding and coding sequences (CDS) according to their coding ability (Figure 2a). A total of 2018 shared transcripts were supported as non-coding sequences by four databases, including 2008 long non-coding RNA (lncRNA, length > 200 bp) and 10 small non-coding RNA (sncRNA, length < 200 bp) (Figure 2b). A total of 86.25% (1732) lncRNA sequences overlapped with the 2245 unannotated transcripts.

In addition, 46,772 transcripts were predicted to possess coding ability with lengths ranging from 100 to 1719 bp, with 99% of CDS being longer than 100 bp (Figure 2c). The exon number of all expressed genes was also investigated, and 75.4% of genes had less than ten exons (Figure 2d). Furthermore, the CDSs were blasted against PlantRegMap (plant regulation data and analysis platform) for transcription factor (TF) prediction. Finally, 44,812 and 1910 CDSs were annotated as structural genes and TFs, accounting for 91.02% and 3.88% of all these transcripts, respectively (Figure 2a). These putative TFs could be categorized into 54 families; the top 15 families account for 65.62% of all TFs. The most abundant family was *bHLH*, which contained 44 members, followed by *MYB-related* (n = 40), *ERF* (n = 36), *C3H* (n = 33), *C2H2* (n = 33), *bZIP* (n = 32), and *MYB* (n = 32) (Figure S2).

Alternative splicing (AS) is an important way for the post-transcriptional regulation of genes. Therefore, we comprehensively surveyed AS events in response to freezing stress. We found that 70.14% and 50.06% of gene loci transcripts harbor more than one and two isoforms, respectively (Figure 2f). A total of 6633 AS events corresponding to 3045 genes loci were identified. These ASs were divided into intron retention (IR, 2379), alternative 3′ splice site (A3SS, 1614), alternative 5′ splice site (A5SS, 990), skipped exon (SE, 890), alternative first exon (AF, 566), alternative last exon (AL, 175), and mutually exclusive exon (MX, 19) types, which account for 35.87%, 24.33%,14.93%, 13.42%, 8.53%, 2.64%, and 0.29% of all events, respectively (Figure 2e). Full-length transcripts were also used to identify transcript fusion events. A total of 104 were found, including 74 inter-chromosomal and 30 intra-chromosomal fusion transcripts (Figure 2g). These results revealed that full-length transcriptomic analysis could identify a high degree of alternative splicing in potato during freezing stress.

2.4. Differentially Expressed Genes in Response to Cold Stress

The two potato varieties were treated in the freezing condition (−3 °C) for 12 h and showed distinct phenotypic responses to freezing treatment. At the 3rd hour, both varieties did not show freezing damage. At the 6th hour, DM showed obvious freezing damage with wilting leaves, while the leaves of CM were entirely frozen but not wilting. At the 12th hour, severe damage occurred on DM with a wilting stem, while mild damage occurred on CM with slightly curled leaves. After a 12 h recovery period at room temperature (25 °C), DM was completely dead, while CM could recover to a normal condition with minor injuries (Figure 3a). A comprehensive transcriptome analysis for the two varieties during this freezing stress time course was performed to reveal the regulatory mechanism of potato coping with freezing stress. Compared to the 0th hour, a total of 1320 (829 up and 491 down), 1895 (1180 up and 715 down), and 2258 (1544 and 714 down) differentially expressed genes (DEGs) were identified at the 3rd, 6th, and 12th hour in DM, respectively. A total of 426 (344 up and 82 down), 678 (411 up and 267 down), and 1405 (748 and 657 down) differentially expressed genes (DEGs) were identified at the 3rd, 6th, and 12th hour in CM, respectively (Figure 3b). Collectively, 1519 DEGs were found between any two time points in DM, and 24.22% of genes produced 957 AS events. A total of 1271 DEGs were found between any two time points in CM, and 20.38% of genes produced 589 AS events (Table S2).

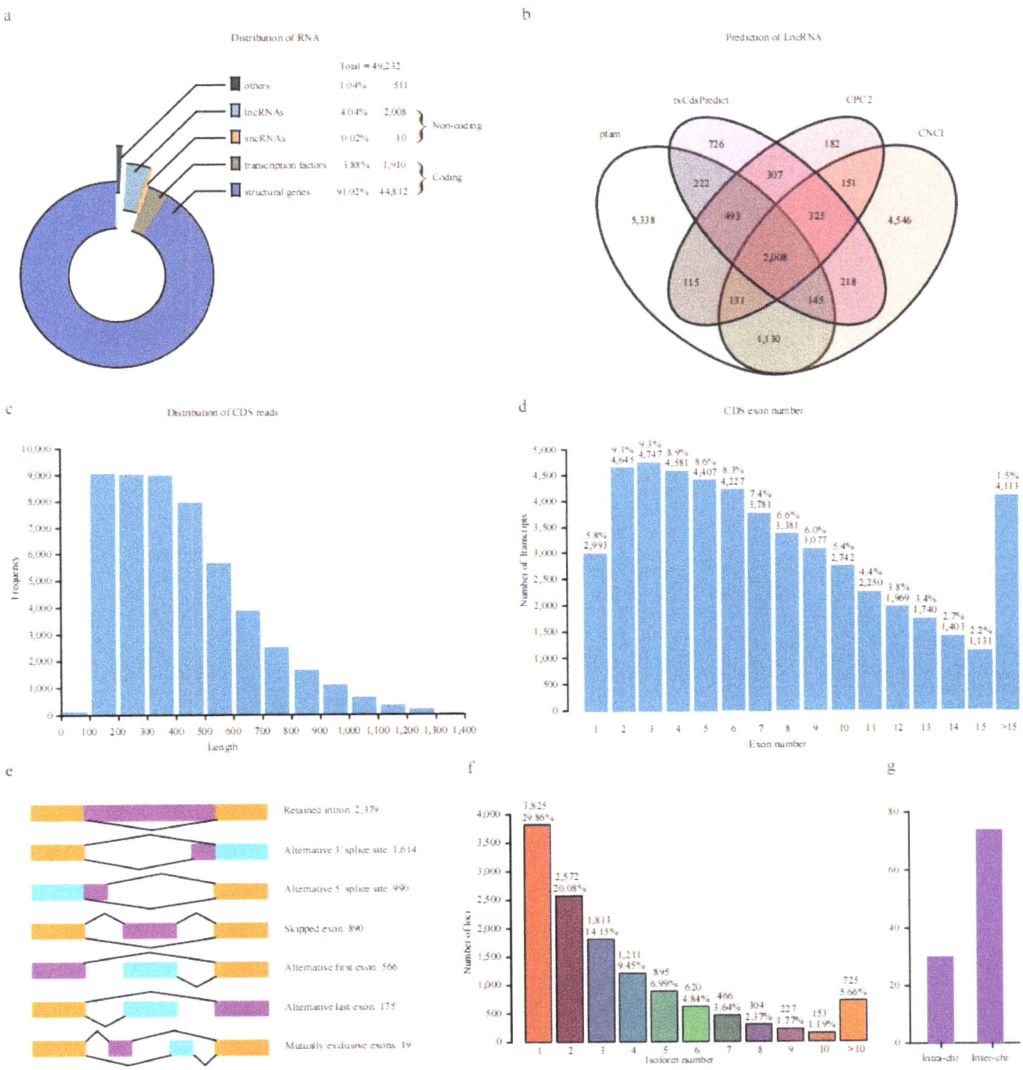

Figure 2. Characterization of non-redundant transcripts. (**a**) Categories of transcripts based on coding ability. (**b**) Venn diagram of the predicted lncRNA using four databases. (**c**) Length distribution of CDS reads length. (**d**) Statistic of exon numbers from one locus. The number and percentage were marked above the bars. (**e**) Visualization of seven types of alternative splicing events. The number was marked after the type. (**f**) Statistic of the isoform number produced by one locus. The number and percentage were marked above the bars. (**g**) The number of inter-/intra-chromosomal fusion transcripts.

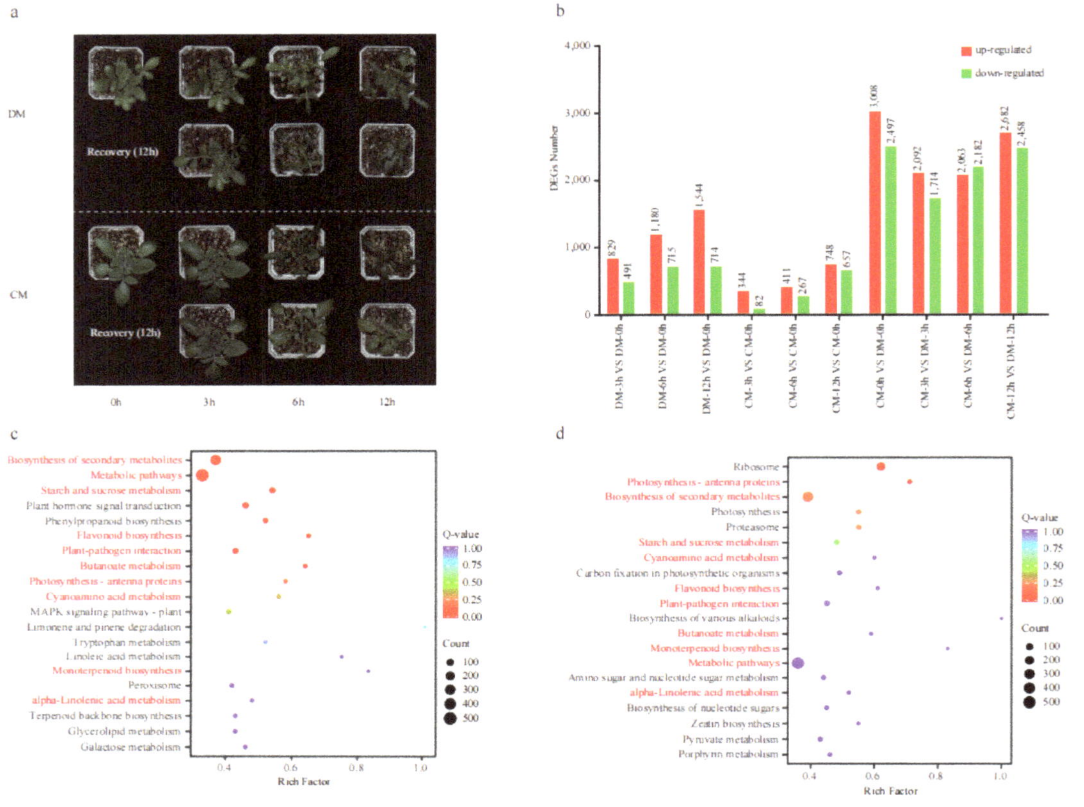

Figure 3. The phenotype of two varieties (DM and CM) under freezing treatment and transcriptome analysis. (**a**) The phenotype of two varieties under freezing treatment (0th, 3rd, 6th, and 12th) and recovery. (**b**) The number of differentially expressed genes (DEGs). The red and blue columns represent upregulated and downregulated genes with the number marked above, respectively. (**c**) KEGG enrichment analysis of the DEGs between two varieties at the 3rd hour. (**d**) KEGG enrichment analysis of the DEGs between two varieties at the 6th hour. Compared pathways sharing two pairs were marked in red.

Coincidentally, the number of upregulated DEGs is larger than that of downregulated DEGs in all three pairs for both varieties. This may indicate that many more positive regulatory genes than negative regulatory genes were activated in response to freezing and may also indicate that both varieties responded to freezing by activating a similar or the same mechanism. We analyzed the DEGs from each variety using KEGG, and the top 20 pathways ($p < 0.05$) were investigated. A total of 25 enriched pathways were shared between the two varieties, such as starch and sucrose metabolism, glutathione metabolism, galactose metabolism, zeatin biosynthesis, flavonoid biosynthesis, and fatty acid metabolism (Figure S3).

When we compared the DEGs between the two varieties, a total of 5505 (3008 up and 2497 down), 3806 (2092 up and 1714 down), 4245 (2063 and 2182 down), and 5140 (2682 up and 2458 down) DEGs were identified at the 0th, 3rd, 6th, 12th hour, respectively (Figure 3b). As seen in the 6th hour, DM showed obvious phenotype damage; the DEGs between the two varieties at the 3rd and 6th hour were further analyzed using KEGG (Figure 3c,d). Among the top 20 pathways, 7 were found in both time points, including starch and sucrose metabolism, flavonoid biosynthesis, plant pathogen interaction photosynthesis, cyanoamino acid metabolism, peroxisome, and alpha-linolenic acid metabolism. These results suggested that these pathways were strongly responsive to freezing and might play important roles in freezing tolerance. Given that flavonoids are reported to contribute to cold tolerance [17], we detected the contents of flavonoids in these two varieties.

2.5. *Differentially Accumulated Metabolites in Response to Freezing Stress*

A total of 100 flavonoid metabolites were identified, involving 9 species; they are flavone (n = 38), flavonol (n = 31), flavanone (n = 11), isoflavone (n = 8), anthocyanins (n = 5), dihydrochalcones (n = 4), chalcone (n = 1), flavanonol (n = 1), and proanthocyanidins (n = 1). Principal component analysis showed that all time points of DM were clustered together, indicating that the changes of metabolites in DM are not significant during freezing stress. However, the samples of CM at the 0th and 3rd hour diverged from the samples of CM at 6th and 12th, indicating that many more changed metabolites occurred. The flavonoid accumulation patterns of the resistant and sensitive potato varieties differed, and between 3 and 6 h of freezing CM illustrated a strong metabolic change, while DM did not (Figure 4a). This result indicates that the time the period between the 3rd and 6th hours is important for activating CM's metabolic-related pathway. Furthermore, we found that 49 and 68 metabolites were significantly affected by freezing stress in DM and CM, respectively (Tables S3 and S4). The number of differentially accumulated metabolites (DAM) was 14 (10 up and 4 down), 9 (9 up), and 26 (18 up and 8 down) after freezing treatment for 3, 6, and 12 h in DM, respectively. The DAM was 29 (2 up and 27 down), 51 (33 up and 18 down), and 46 (30 up and 16 down) after freezing stress treatment for the 3, 6, and 12 h in CM, respectively (Figure 4b). In summary, many more metabolites presented significant changes in CM than those in DM. In addition, except for the 3rd vs. 0th hour, the number of accumulated metabolites was much more than the decreased metabolites in CM. This indicated that more active flavonoid regulation occurred in CM than in DM.

K-means clustering analysis was used to group the changing metabolites, producing seven subclasses. The five metabolites in Subclass1 increased to the highest level at the 3rd hour and then decreased in DM. However, in CM, they sharply increased by more than six-fold at the 6th hour and then slowly increased to the highest level at the 12th hour. The 11 metabolites in Subclass 2 steadily increased to the highest level at the 12th hour in DM, while they first decreased and then increased in CM. The 21 metabolites in Subclass 3 changed in a narrow range in DM but sharply increased to a high level at the 6th hour in CM. These metabolites mainly belong to the quercetin and kaempferol pathway. Subclass 4 includes 26 metabolites, and these metabolites first increased and then decreased in DM but essentially stayed unchanged at a low level in CM. The 20 metabolites in Subclass 5 mostly stayed unchanged at a low level in DM, but in CM, they first decreased and then increased. The nine metabolites in Subclass 6 first increased then decreased in DM but kept dropping to their lowest level in CM at the 12th hour. Finally, the eight metabolites in Subclass 7 continuously increased in DM, while in CM, they first increased to the highest level at the 3rd hour and then decreased during further freezing stress (Figure 4c).

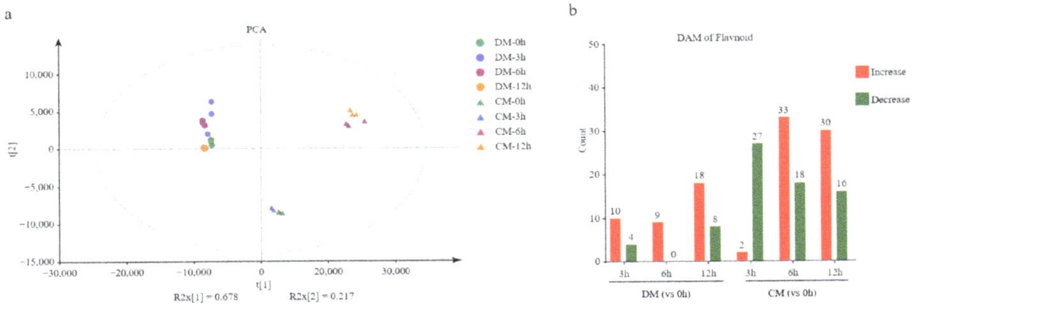

Figure 4. *Cont.*

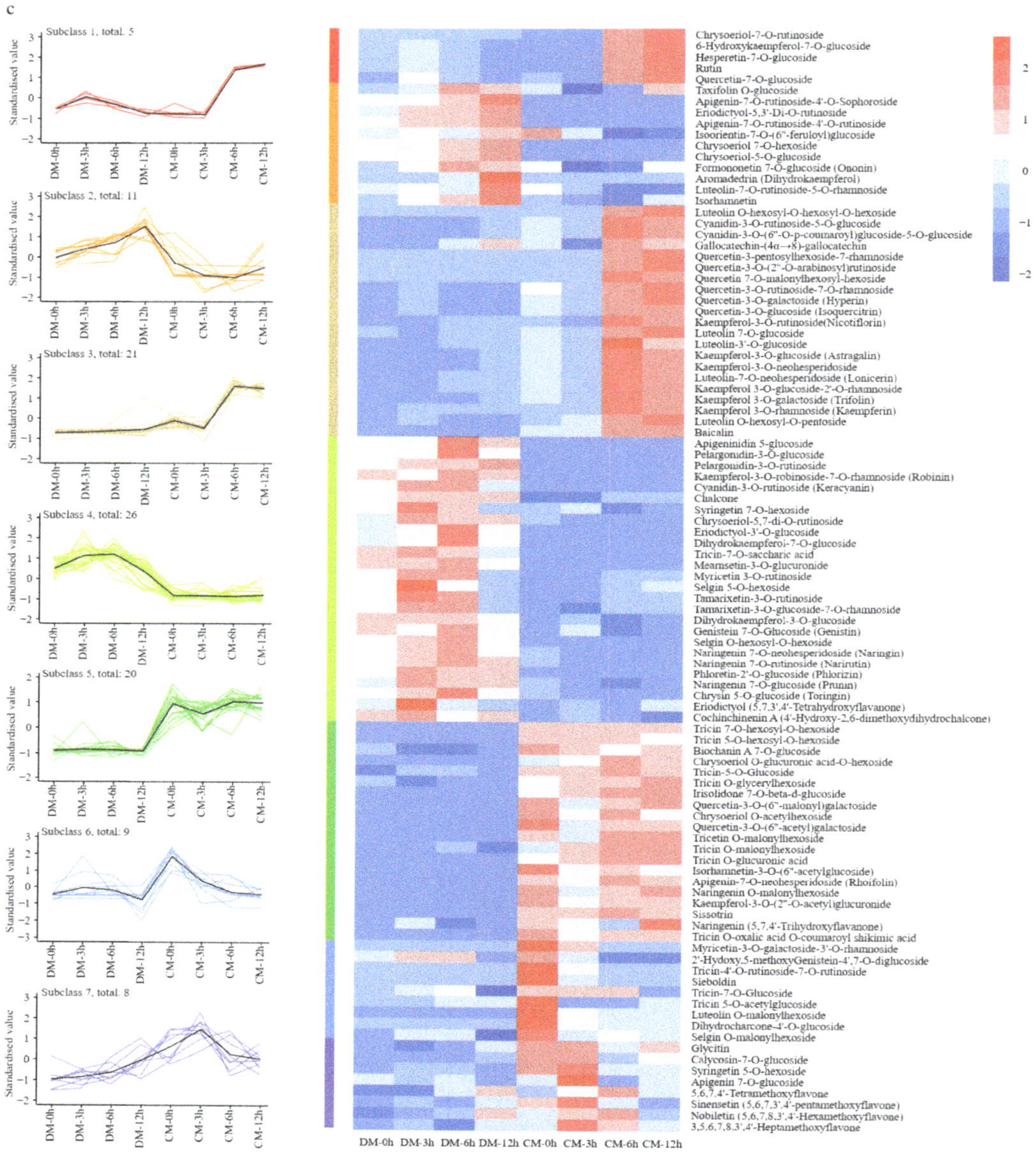

Figure 4. The analysis of the flavonoid metabolome. (**a**) Principal component analysis (PCA) of the flavonoid metabolites in two varieties during freezing treatment. (**b**) Differentially accumulated metabolite (DAM) analysis of two varieties during freezing treatment. (**c**) The K-means clustering analysis of DAMs and heatmap of metabolite content. The number of flavonoids in each subclass was marked with *n*.

2.6. The Flavonoid Pathway Was Significantly Upregulated in CM, More So than in DM

Five key genes in the flavonoid biosynthesis pathway from 4-coumaryoyl-CoA to kaempferol and quercetin via naringenin were comprehensively investigated based on transcriptomics (Figure 5a). All these genes showed increased expression in response to freezing in both varieties based on FPKM (Table S5). Specifically, the expression level of all the genes increased to the ma ximum values under freezing stress at the 6th hour. At the start of the pathway, *CHS-1* strongly responded to the freezing stress in both varieties. Compared to the 0th hour, the FPKM of *CHS-1* increased up to 12- and 3-fold in DM and CM, respectively. *CHS-2* increased by 3.6-fold but at a low level (FPKM < 10) in DM, while remaining relatively unchanged in CM during stress. These results indicate that *CHS-1* should be the predominant functional homolog, as *CHI-1* and *CHI-2* responded at different levels. *CHI-1* only increased by 1.5-fold in the two varieties, while *CHI-2* increased roughly by 6- and 2-fold in DM and CM, respectively. These results indicate that *CHI-2* should be the predominant functional homolog. The significantly upregulated levels of these two genes enable much more metabolic flow into the flavonoid pathway. Downstream components *F3′H-1* showed a maximum changing pattern, having more than 15- and 6-fold increases in DM and CM during freezing stress, respectively. *F3′H-2* increased 1.7-fold but at a low level (FPKM < 10) in DM, while maintaining relatively unchanged levels in CM during stress. *FLS* increased by 3- and 2-fold in DM and CM, respectively. Comparing the two varieties simultaneously, all assigned genes (*CHS-1*, *CHS-2*, *CHI-1*, *CHI-2*, *F3H*, *F3′H-1*, *F3′H-2* and *FLS*) (Table S5) in flavonoid pathway showed a much higher expression level in CM than those in DM.

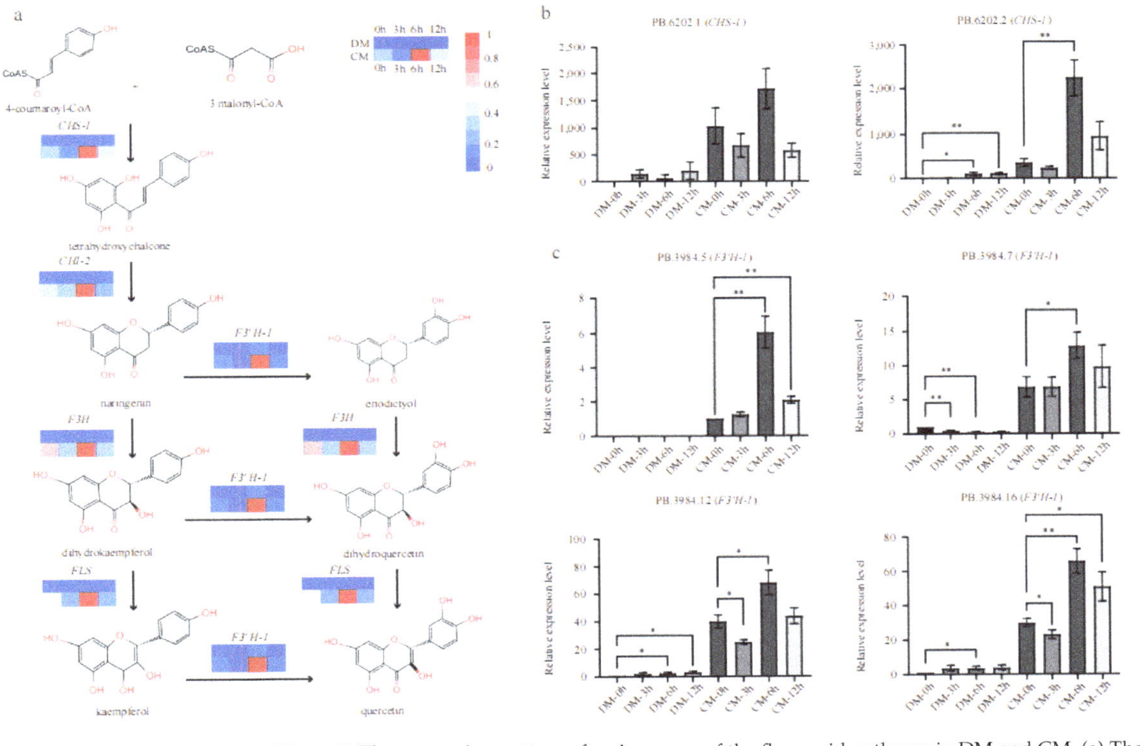

Figure 5. The expression pattern of major genes of the flavonoid pathway in DM and CM. (**a**) The pathway of flavonoid and the expression levels of *CHS*, *CHI*, *F3H*, *F3′H*, and *FLS* based on FPKM. (**b**) qRT-PCR identification of two *CHS* isoforms. Error bars indicate the mean ± standard error (SE) (n = 3). (**c**) qRT-PCR identification of four *F3′H* isoforms. Error bars indicate the mean ± standard error (SE) (n = 3). A two-sided t-test determined statistical significance. One and two stars indicate $p < 0.05$ and $p < 0.01$, respectively.

2.7. *qRT-PCR Confirmed the Expression Levels of CHS-1 and F3′H-1*

Two isoforms (PB.6202.1 and PB.6202.2) of *CHS-1* were identified from the AS data, and both showed significantly upregulated expression patterns in CM more so than in DM (Table S5). qRT-PCR verified their transcript levels, and the results showed that the two transcripts were upregulated during freezing treatment in both varieties, while PB.6202.1 and PB.62020.2 showed a higher level in CM than DM at each time point (Figure 5b). qRT-PCR also indicated that PB.62020.2 might primarily regulate the flavonoid metabolome under freezing stress. Based on the AS data, *F3′H-1* produced seven major isoforms (filtered with FPKM > 2). Four of them, PB3984.5, PB3984.7, PB3984.10, and PB3984.16, were unique in CM, and the other three isoforms showed a much higher expression in CM than in DM (Table S5). These four transcripts, PB3984.5, PB3984.7, PB3984.12, and PB3984.16, were verified by qRT-PCR (Figure 5c; Table S6). In agreement with the unique expression of PB3984.5 in CM based on FPKM, its expression was only detected in CM and increased to the max level under freezing stress at the 6th hour. All the rest of the three transcripts showed a much lower relative expression level in DM, while in CM, they showed high relative expression levels, especially at the 6th hour. The transcriptome analysis and qRT-PCR results showed that the major structural genes in the flavonoid pathway were significantly upregulated in CM under freezing stress. These extremely higher expression levels might contribute to the strong frost tolerance of *S. commersonii*.

2.8. *Multi-Omics Reveals New Functional Genes in the Flavonoid Pathway*

In order to comprehensively reveal the regulatory network of flavonoids subjected to freezing stress (Figure 6a), 1271 DEGs from CM were used for co-expression analysis. A total of 10 modules (except for M11) were identified, containing from 38 to 496 genes. To bridge the transcriptome and metabolome, we performed a correlation analysis between the gene modules and DAMs. Two modules, M1 and M4, were identified based on the correlation coefficients ($r > 0.7$ and $p < 0.05$) (Figure 6b). Interestingly, both *CHS-1* and *F3′H-1* were grouped in module M1. We next used these two genes as baits to search for possible regulators and isolated one *Myb-related* and three *ERFs* (*DREB2*, *ERF054*, and *ERF4*) (Figure 6c). Their expression levels during freezing treatment were confirmed by the FPKM value (Figure 6d). They were strongly responsive to freezing treatments. Compared to the 0th hour in CM, the expression of *Myb-related* reached the highest value at the 6th hour, while *DREB2*, *ERF054*, and *ERF4* were upregulated at the 3rd hour and then downregulated at the 6th hour. Using these TFs as baits, three UDP-transferases, *β-1,3-galactosyltransferase* (*Gal-GT*), *4-beta-mannosyltransferase* (*Man-GT*), and *O-glucosyltransferase* (*Glu-GT*), were identified. *Gal-GT* was simultaneously highly related to *MYB-related*, *ERF054*, and *DREB2*. *Man-GT* and *Glu-GT* were highly negatively related to *DREB2* and *ERF054*, respectively (Figure 6c). Compared to the normal condition, in CM, the expression levels of *Gal-GT*, *Man-GT*, and *Glu-GT* were upregulated by 3.7-fold, 3.8-fold, and 1.6-fold at the 6th hour, respectively (Figure 6e; Table S5). Using correlation analysis between these three GTs and all the glycosylated flavonoids, we found a total of 8, 4, and 5 glycosylated flavonoids, which are highly correlated ($|R| > 0.8$, $p < 0.05$) with these three genes, respectively, indicating that these identified GTs very likely play important roles in the glycosylation of flavonoids to protect against plant freezing stress (Table S7).

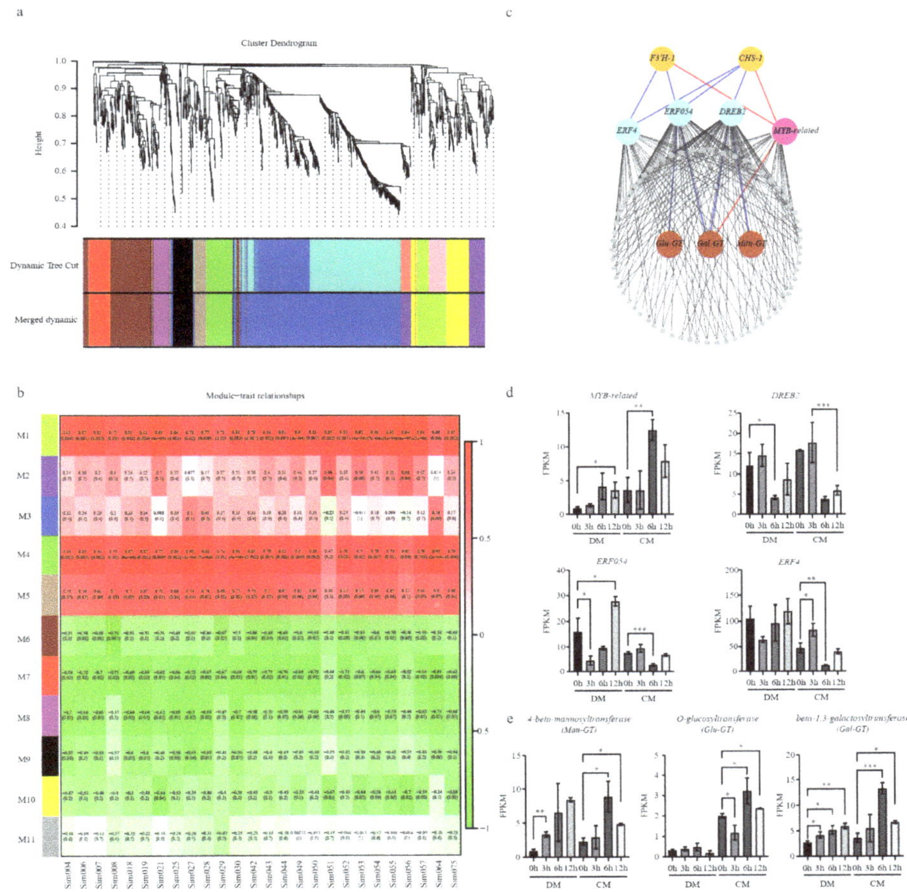

Figure 6. Co-expression network analysis identified new freezing-responsive genes in the flavonoid pathway. (**a**) Hierarchical clustering tree of the modules. (**b**) Module–trait relationships calculated by Pearson correlation coefficients. (**c**) The co-expression network of M1. The triangle and circle strands for TFs and structural genes, respectively. The red and blue lines indicate the positive and negative correlation, respectively; (**d**). Expression levels of *MYB-related*, *DREB2*, *ERF054*, and *ERF4*. (**e**) *Man-GT*, *Glu-GT*, and *Gal-GT* in CM and DM under freezing stress at different time points. A two-sided *t*-test determined statistical significance. One, two and three stars indicate $p < 0.05$, $p < 0.01$ and $p < 0.001$, respectively.

3. Discussion

During the long process of evolution, plants have generated multiple strategies to cope with stimuli in cold environments. The transcriptional regulation of co-responding genes is one important way to deal with cold stress. Multiple pathways are activated or suppressed in this process to alleviate cold damage. Here, we viewed that many genes and pathways dramatically respond to freezing treatment. A total of 1519 and 1271 genes were responsive to freezing stress in the two varieties examined. These genes were enriched in 25 shared pathways for both varieties (Figure S3). For example, starch and sucrose metabolism were found in three pairs of time-point comparisons, indicating its important role in coping with freezing in potato.

As an osmotic regulator and penetration protectant, sugar can reduce the cell's water potential and protect and maintain membrane stability. In chill-treated maize, *raffinose synthase* (*RAFS*), the key enzyme for raffinose biosynthesis, was significantly upregulated, and its photosynthetic product sugar was significantly accumulated. Seedings with high raffinose accumulation exhibited better chilling stress tolerance [18]. In *Camellia sinensis*, the content of soluble sugars in leaves rose under low-temperature stress [19]. Both the biosynthesis of unsaturated fatty acids and fatty acid elongation pathways were found in two pairs of time-point comparations in chickpea. The unsaturated fatty acid accumulation of membrane lipids was correlated with cold stress resistance. As well, after cold acclimation, there was an increase in the unsaturated fatty acid ratio compared with the saturated ones [20]. A study of winter wheat showed that linolenic acid and palmitic acid were closely related to cold resistance, and the index of unsaturated fatty acid variation is increased in a cold tolerance genotype [21]. In the cold-tolerant genotype *S. commersonii*, stearoyl-ACP (acyl carrier protein) desaturase was activated after cold acclimation and high polyunsaturated fatty acids were highly accumulated in the leaf [22]. Plant hormone signal transduction occurred at five time-point comparisons. Plants can regulate hormones in answer to cold stress, and these hormones can also aid in stress protection. Maize seedlings pretreated with ABA present greater cold tolerance than untreated seedlings [23]. Undoubtedly, these identified enriched pathways provide many clues for unraveling the freezing tolerance mechanism in potato.

Low temperatures can produce ROS (reactive oxygen species) accumulation, which affects many cellular functions [24]. Plants can protect against outer stress by activating antioxidant mechanisms for scavenging ROS [25], including enzymatic and non-enzymatic antioxidants. Flavonoids are among the most bioactive plant non-enzymatic antioxidants and are reported to perform better than other antioxidants, such as ascorbate [26]. *S. commersonii* presents superior freezing tolerance and could contribute excellent alleles for anti-frost breeding in potato. A few studies have been conducted to explore its freezing tolerance mechanism under different aspects, including soluble carbohydrate synthesis and *CBF* (*C-repeat binding factor*) regulons expression [27,28]. However, the flavonoid pathway contribution to the anti-frost trait of CM was never explored. In this study, we found that flavonoid biosynthesis genes were strongly responsive to freezing in both varieties, and correspondingly, the flavonoid content was affected by freezing treatment.

Compared with DM, more (68% vs. 49%) flavonoids showed significant change in CM, and these metabolites presented different, changing patterns, such as the Subclasses 1, 3 and 5, including 46 metabolites, which present high accumulation in CM but not in DM. Coincidentally, all detected quercetin ($n = 9$) and 7 out of 8 Kaempferol-3-glycosyl were among them. In a previous study, the effect of flavonoid content on freezing tolerance was carried out using 20 *Arabidopsis* mutant lines. All the knock-out mutants of flavonoid biosynthesis genes had reduced flavonoid content and showed impaired leaf-freezing tolerance. In contrast, the *MYB* activation mutant line pap1-D accumulated a high content of quercetin-3Glc, Kaempferol-3Rha-7Rha, and three anthocyanins, and showed improved freezing tolerance [24]. This study demonstrates the contribution of flavonoids to freezing tolerance, especially certain flavonoids such as glycosylated flavonoids.

Corresponding to the accumulation pattern of metabolites in CM, nearly all the major structural genes were significantly upregulated in CM, especially at the 6th hour after freezing (Figure 5), while these genes only showed a mild upregulation in DM. In addition, all the AS isoforms in CM showed an overwhelming expression superiority to that in DM. An upregulated pathway means an upstream regulator was also activated or suppressed. Four TFs were identified based on gene expression correlation. The *MYB-related* gene showed a high positive correlation (R > 0.9) with *CHS-1* and *F3′H-1*, indicating it may activate these structural genes. *FtMYB12* in *F. tataricum* could be greatly induced by low temperatures, and its overexpression in *Arabidopsis* resulted in enhanced cold tolerance [29]. *AtMYB12* could upregulate a related gene and lead to the production of rutin in transgenic tobacco callus cultures [30]; rutin is also a glycosylated flavonol. Meanwhile, three *ERF* genes were also identified. Multiple studies found that *ERFs* could be induced by cold and were key transcription factors involved in cold-responsive gene expressions [31–33]. In addition, several studies revealed that some *ERFs* could regulate flavonoid biosynthesis. In citrus, *CitERF32* and *CitERF33* could activate the transcription of *CHI*, affecting the accumulation of flavonoid production [34]. In tomato, the overexpressing *SlERF.G3-like* could increase the expression of structural genes and lead to the accumulation of flavonoid compounds [35]. In addition, multiple studies found that cold could induce *ERFs* and were the key transcription factors involved in cold-responsive gene expressions. Although the three identified *ERFs* negatively correlated with *CHS-1* and *F3′H-1* from the network, their expression level increased (3rd vs. 0th) in CM.

In plants, flavonoids usually exist in their "decorated" forms catalyzed by different enzymes, such as *glycosyltransferases* or *methyltransferases*. The glycosylation of flavonoids usually confers the biological activity of corresponding derivatives [36]. For example, in *Arabidopsis*, two *UDP-glycosyltransferases* could be induced by cold, salt, and drought stresses, and they could modulate the accumulation of modified anthocyanin, contributing to enhanced stress tolerance [37]. In rice, two *UDP-glycosyltransferases* affect the natural variation in flavones, and product accumulation contributes to UV-B tolerances [36]. Here, three UGTs were found to be related to flavonoid biosynthesis and freezing stress. However, different UGTs show different substrate specificity, and the end products can also present different biological activity. Therefore, further experimental studies are needed to clarify their exact functions.

4. Materials and Methods

4.1. Plant Material and Sample Preparation

The plant varieties used in this study were *Solanum tubersosum* (DM) and *Solanum commersonii* (CM). The plants were grown for 30 days under a 16 h light/8 h dark photoperiod, placed for 3 days at 3 °C followed by −3 °C treatment, and leaf samples were harvested at 0 h, 3 h, 6 h, and 12 h for transcriptome and metabolome assays. Three biological replicates for each time point were prepared. Total RNA was isolated from the above samples using RNAprep Pure Polyphenol Plant TOTAL RNA Extraction Kit (DP441) and treated with DNase I (Fermentas, Carlsbad, CA, USA) according to the manufacturer's instructions. The RNA from each sample was sequenced using 150-bp paired-end Illumina sequencing with libraries of 350-bp insert sizes. In addition, one bulked RNA sample of *S. commersonii*, generated from all time points, was sequenced using the Pacbio Sequel platform.

4.2. Illumina Data Analysis

Fragments were mapped to the reference genome sequence (DM_v6.1) using hisat2. The gene expression levels were estimated by FPKM (fragments per thousand bases of transcript per million mapped reads). Differential expression analysis was performed using the DESeq R package. Genes identified by DESeq with *padj* (q) < 0.05 and Log2 (Fold-Change) ≥ 1 or Log2 (Fold-Change) ≤ -1 were defined as differentially expressed. The gene expression profiles were calculated using StringTie v.1.3.3.

4.3. PacBio Data Analysis

The Iso-Seq data were processed using SMRT-Analysis software package v3.0 (https://github.com/ben-lerch/IsoSeq-3.0/blob/master/README.md, accessed on 19 January 2022). A circular consensus sequence (CCS) was generated from subread BAM files with the following parameters: skip-polish, min-passes 2, and min-rq 0.99. CCS reads were classified into full-length and non-full-length reads using lima. Full-length FASTA files were fed into the cluster and polish step with default parameters. The polished reads were aligned to the DM genome (v6.1) using the software pbmm2 (https://github.com/PacificBiosciences/pbmm2, accessed on 19 January 2022). Finally, the aligned BAM files were fed into the collapse step with default parameters. BUSCO was used to evaluate the quality of full-length transcripts. AS events were extracted and quantified using SUPPA2 (https://github.com/comprna/SUPPA, accessed on 21 January 2022) based on a GFF containing long-read isoforms. TransDecoder (https://github.com/

TransDecoder/TransDecoder/releases, accessed on 22 January 2022) was used to identify CDS regions within the transcript sequences. For transcription factor (TF) prediction, we first detected the ORFs of isoforms by TransDecoder. Then, the ORFs were mapped onto PlantTFDB (http://planttfdb.gao-lab.org/prediction.php, accessed on 23 January 2022).

4.4. *Identification of the lncRNA*

CNCI (Coding-Non-Coding Index), CPC (Coding Potential Calculator), txCdsPredict, and Pfam databases were used to predict the coding potential of transcripts. Transcripts with coding potential were filtered out, and those without coding potential were candidate sets of lncRNAs according to the threshold: CPC threshold < 0, CNCI threshold < 0, and txCdsPredict threshold < 500.

4.5. *Identification of Fusion Transcripts*

The Python script (fusion_finder.py) in the PBTRANSCRIPT-TOFU package (http://github.com/PacificBiosciences/cDNA_primer/, accessed on 25 January 2022) was used to identify fusion transcripts. Each candidate fusion transcript must be mapped to two or more loci with a distance >= 10 kb. Each locus must cover at least 5% of the transcripts, and the total coverage of fusion transcripts is at least 99%.

4.6. *Functional Annotation of Genes*

Genes were annotated by public databases, including the NCBI Non-Redundant Protein Database (Nr), NCBI Non-Redundant Nucleotide Database (Nt), SwissProt, Protein Family (Pfam), Gene Ontology (GO), Kyoto Encyclopedia of Genes and Genomes (KEGG), and Clusters of Orthologous Groups of proteins (KOG/COG/eggNOG). The similarity was filtered with an E-value threshold of 10^{-5}.

4.7. *Analysis of Differentially Expressed Transcripts*

All Illumina reads were mapped to the SMRT full-length transcripts to identify fragments per kilobase of exon per million fragments mapped (FPKM) values. DESeq2v1.4.5 was used for differentially expressed genes (DEGs) analysis. The *p*-value could be assigned to each gene and adjusted by the Benjamini and Hochberg approach for controlling the false discovery rate. Genes with $q \leq 0.05$ and $|\text{Log2 (Fold-Change)}| \geq 1$ were identified as DEGs.

4.8. *Co-Expression Analysis*

A weighted gene co-expression network analysis was performed using the WGCNA package (v1.69) with the default parameters. An adjacency matrix was constructed and converted into a topological overlap matrix (TOM). A dynamic hybrid tree-cut algorithm (the R package dynamicTreeCut, v.1.63) was used to detect the modules. A correlation matrix between phenotypes and gene modules was constructed using the WGCNA package. Finally, the relationship between genes was visualized using Cytoscape (v.3.8.2).

4.9. *Quantitative Real-Time PCR (qRT-PCR) Analysis*

For RNA extraction, we used the Plant Total RNA Kit (Tiangen, Beijing, China), and PrimeScript™ RT Reagent Kit with gDNA Eraser (TaKaRa, Kusatsu, Japan) was used for the synthesis of the first-strand cDNA. The ABI Step, One Plus instrument was used for qRT-PCR experiments. Experiments were repeated 3 times. The comparative 2−ΔΔCT (A method of qRT-PCR fluorescence quantitative data analysis, ΔΔCt = ΔCt experimental group − ΔCt control group) method was used for gene expression level analysis. For quantitative analysis, Exocyst complex component gene (*Sec3A*) was used as the internal reference. Reactions contained the following: 5 μL of 2x × TB Green Premix Ex Taq II, 1 μL of template cDNA, 0.4 μL of forward and Universal miRNA qPCR Primer, 0.2 μL of Passive Reference Dye and water to 10 μL. PCR amplification was performed as follows: 95 °C for 30 s, 40 cycles of 95 °C for 5 s, 58 °C for 30 s.

5. Conclusions

In this study, we reconstructed a multi-omics database for freezing research in potato. KEGG enrichment analysis of DEGs revealed that some shared pathways were activated in freezing sensitive and tolerant varieties, indicating they launched similar mechanisms to cope with freezing. The transcriptome and metabolome integration illustrated that the flavonoid pathway was strongly responsive to freezing stimuli in both genotypes but to different degrees. The four newly identified transcription factors (one *Myb-related* and three *ERF*) and their associated UGTs are potential contributors to the freezing tolerance of CM in concert with other structural genes in the flavonoid pathway. All the results provide insights to explore the freezing tolerance mechanism mediated by the flavonoid pathway and other pathways. These results will also provide new genetic modules from *S.commersonii* for potato anti-freezing breeding.

Supplementary Materials: The following supporting information can be downloaded at: https://www.mdpi.com/article/10.3390/plants12102054/s1, Figure S1: GO functional annotation of all the transcripts; Figure S2: Transcription factor family classification; Figure S3: KEGG pathway enrichment analysis of the DEGs identified in DM and CM; Table S1: Functional annotation of all transcripts; Table S2: DEGs in DM and CM under freezing treatment; Table S3: Detection of flavonoid metabolites in DM and CM; Table S4: Differential accumulated metabolites at different time points; Table S5: The expression level of *CHI, CHS, F3H, F3′H, FLS*, three *UDP—glycosyltransferases*, and four transcriptional; Table S6: Primers for quantitative PCR in DM and CM; Table S7: Correlation between different *glycosyltransferases* and glycosylated flavonoid derivatives.

Author Contributions: G.Z. designed the project and revised the manuscript. Z.Z. performed all the data analysis. L.W. and P.K. wrote the paper with contributions from other authors. L.G. performed the qRT-PCR assay, and H.B. prepared plant material. X.W. performed phenotypic identification. Z.W. contributed to the experimental design. All authors have read and agreed to the published version of the manuscript.

Funding: This research was funded by the China National Key Research and Development Program (2022YFF1002500), the National Natural Science Foundation of China (U2002204 and 32272725), and Yunnan Fundamental Research Project (202001AV070003), and the Guangdong Major Project of Basic and Applied Basic Research (2021B0301030004).

Data Availability Statement: Data recorded in the current study are available in all tables of the manuscript.

Conflicts of Interest: The authors declare no competing financial interests.

Abbreviations

SMRT	Single-molecule real-time
CCS	Circular consensus sequences
FLNC	The full length is not chimeric
AS	Alternatively spliced
TF	Transcription factor
ADC1	Arginine decarboxylase gene
DEG	Differentially expressed gene
DAM	Differentially accumulated metabolite
GO	Gene ontology
KEGG	Kyoto Encyclopedia of Genes and Genomes
Pfam	The Pfam protein family database
KOG	Clusters of orthologous groups for complete eukaryotic genomes
F3′H	Flavonoid 3′-hydroxylase
F3H	Flavanone 3-hydroxylase
CHS	Chalcone synthase
CHI	Chalcone isomerase
FLS	Flavonol synthase

References

1. Men, F.; Liu, M. *Potato Cultivation Physiology*; China Agriculture Press: Beijing, China, 1995; pp. 65–80.
2. Li, P.; Huner, N.; Toivio-Kinnucan, M.; Chen, H.; Palta, J. Potato freezing injury and survival, and their relationships to other stress. *Am. Potato J.* **1981**, *58*, 15–29. [CrossRef]
3. Chang, D.; Sohn, H.; Cho, J.; Im, J.; Jin, Y.; Do, G.; Kim, S.; Cho, H.; Lee, Y. Freezing and frost damage of potato plants: A case study on growth recovery, yield response, and quality changes. *Potato Res.* **2014**, *57*, 99–110. [CrossRef]
4. Waadt, R.; Seller, C.; Hsu, P.; Takahashi, Y.; Munemasa, S.; Schroeder, J. Plant hormone regulation of abiotic stress responses. *Nat. Rev. Mol. Cell Biol.* **2022**, *23*, 680–694. [CrossRef] [PubMed]
5. Zhou, C.; Zhu, C.; Tian, C.; Xu, K.; Huang, L.; Shi, B.; Lai, Z.; Lin, Y.; Guo, Y. Integrated volatile metabolome, multi-flux full-length sequencing, and transcriptome analyses provide insights into the aroma formation of postharvest *jasmine* (*Jasminum sambac*) during flowering. *Postharvest Biol. Technol.* **2022**, *183*, 111726. [CrossRef]
6. Hu, T.; Chitnis, N.; Monos, D.; Dinh, A. Next-generation sequencing technologies: An overview. *Hum. Immunol.* **2021**, *82*, 801–811. [CrossRef]
7. Wong, L.; Razali, S.; Deris, Z.; Danish-Daniel, M.; Tan, M.; Nor, S.; Ma, H.; Min, W.; Yantao, L.; Asaduzzaman, M. Application of second-generation sequencing (SGS) and third-generation sequencing (TGS) in aquaculture breeding program. *Aquaculture* **2022**, *548*, 737633. [CrossRef]
8. Cao, H.; Wang, Y.; Zhang, W.; Chai, X.; Zhang, X.; Chen, S.; Yang, F.; Zhang, C.; Guo, Y.; Liu, Y. A short-read multiplex sequencing method for reliable, cost-effective and high-throughput genotyping in large-scale studies. *Hum. Mutat.* **2013**, *34*, 1715–1720. [CrossRef]
9. Kissen, R.; Eberl, F.; Winge, P.; Uleberg, E.; Martinussen, I.; Bones, A. Effect of growth temperature on glucosinolate profiles in *Arabidopsis thaliana* accessions. *Phytochemistry* **2016**, *130*, 106–118. [CrossRef] [PubMed]
10. Shen, W.; Nada, K.; Tachibana, S. Involvement of polyamines in the chilling tolerance of cucumber cultivars. *Plant Physiol.* **2000**, *124*, 431–440. [CrossRef]
11. Li, H.; Yang, X.; Shang, Y.; Zhang, Z.; Huang, S. Vegetable biology and breeding in the genomics era. *Sci. China Life Sci.* **2023**, *66*, 226–250. [CrossRef] [PubMed]
12. Zhu, G.; Wang, S.; Huang, Z.; Zhang, S.; Liao, Q.; Zhang, C.; Lin, T.; Qin, M.; Peng, M.; Yang, C. Rewiring of the fruit metabolome in tomato breeding. *Cell* **2018**, *172*, 249–261.e212. [CrossRef]
13. Abbas, F.; Guo, S.; Zhou, Y.; Wu, J.; Amanullah, S.; Wang, H.; Shen, J. Metabolome and transcriptome analysis of terpene synthase genes and their putative role in floral aroma production in *Litchi chinensis*. *Physiol. Plant.* **2022**, *174*, e13796. [CrossRef] [PubMed]
14. Kou, S.; Chen, L.; Tu, W.; Scossa, F.; Wang, Y.; Liu, J.; Fernie, A.; Song, B.; Xie, C. The *Arginine decarboxylase* gene *ADC1*, associated to the putrescine pathway, plays an important role in potato cold-acclimated freezing tolerance as revealed by transcriptome and metabolome analyses. *Plant J.* **2018**, *96*, 1283–1298. [CrossRef] [PubMed]
15. He, F.; Xu, J.; Jian, Y.; Duan, S.; Hu, J.; Jin, L.; Li, G. Overexpression of *Galactinol synthase 1* from *Solanum commersonii* (*ScGolS1*) confers freezing tolerance in transgenic potato. *Hortic. Plant J.* **2022**, *in press*. [CrossRef]
16. Tu, W.; Dong, J.; Zou, Y.; Zhao, Q.; Wang, H.; Ying, J.; Wu, J.; Du, J.; Cai, X.; Song, B. Interspecific potato somatic hybrids between *Solanum malmeanum* and *S. tuberosum* provide valuable resources for freezing-tolerance breeding. *Plant Cell Tissue Organ Cult.* **2021**, *147*, 73–83. [CrossRef]
17. Peng, Z.; Wang, Y.; Zuo, W.; Gao, Y.; Li, R.; Yu, C.; Liu, Z.; Zheng, Y.; Shen, Y.; Duan, L. Integration of metabolome and transcriptome studies reveals flavonoids, abscisic acid, and nitric oxide comodulating the freezing tolerance in *Liriope spicata*. *Front. Plant Sci.* **2022**, *12*, 764625. [CrossRef] [PubMed]
18. Han, Q.; Qi, J.; Hao, G.; Zhang, C.; Wang, C.; Dirk, L.; Downie, A.; Zhao, T. *ZmDREB1A* regulates *Raffinose synthase* controlling raffinose accumulation and plant chilling stress tolerance in maize. *Plant Cell Physiol.* **2019**, *61*, 331–341. [CrossRef]
19. Zhu, Z.; Jiang, J.; Jiang, C.; Li, W. Effects of low-temperature stress on SOD activity, soluble protein content and soluble sugar content in *Camellia sinensis* leaves. *J. Anhui Agric. Univ.* **2011**, *38*, 24–26.
20. Kazemi Shahandashti, S.; Maali Amiri, R.; Zeinali, H.; Ramezanpour, S. Change in membrane fatty acid compositions and cold-induced responses in chickpea. *Mol. Biol. Rep.* **2013**, *40*, 893–903. [CrossRef]
21. Özcan, M.; Rosa, A.; Dessi, M.; Marongiu, B.; Piras, A.; Al Juhaimi, F. Quality of wheat germ oil obtained by cold pressing and supercritical carbon dioxide extraction. *Czech J. Food Sci.* **2013**, *31*, 236–240. [CrossRef]
22. Vega, S.; del Rio, A.; Bamberg, J.; Palta, J. Evidence for the up-regulation of *stearoyl-ACP (Δ9) desaturase* gene expression during cold acclimation. *Am. J. Potato Res.* **2004**, *81*, 125–135. [CrossRef]
23. Janowiak, F.; Maas, B.; Dörffling, K. Importance of abscisic acid for chilling tolerance of maize seedlings. *J. Plant Physiol.* **2002**, *159*, 635–643. [CrossRef]
24. Schulz, E.; Tohge, T.; Zuther, E.; Fernie, A.; Hincha, D. Flavonoids are determinants of freezing tolerance and cold acclimation in *Arabidopsis thaliana*. *Sci. Rep.* **2016**, *6*, 34027. [CrossRef] [PubMed]
25. Dvořák, P.; Krasylenko, Y.; Zeiner, A.; Samaj, J.; Takac, T. Signaling toward reactive oxygen species-scavenging enzymes in plants. *Front. Plant Sci.* **2021**, *11*, 618835. [CrossRef]
26. Hernández, I.; Alegre, L.; Van Breusegem, F.; Munné-Bosch, S. How relevant are flavonoids as antioxidants in plants? *Trends Plant Sci.* **2009**, *14*, 125–132. [CrossRef]

27. Kopeć, P.; Rapacz, M.; Arora, R. Post-translational activation of *CBF* for inducing freezing tolerance. *Trends Plant Sci.* **2022**, *27*, 415–417. [CrossRef]
28. Chen, L.; Zhao, H.; Chen, Y.; Jiang, F.; Zhou, F.; Liu, Q.; Fan, Y.; Liu, T.; Tu, W.; Walther, D. Comparative transcriptomics analysis reveals a *calcineurin B—like* gene to positively regulate constitutive and acclimated freezing tolerance in potato. *Plant Cell Environ.* **2022**, *45*, 3305–3321. [CrossRef] [PubMed]
29. Zhou, M.; Wang, C.; Qi, L.; Yang, X.; Zhanmin, S.; Yu, T.; Tang, Y.; Shao, J.; Wu, Y. Ectopic expression of *Fagopyrum tataricum FtMYB12* improves cold tolerance in *Arabidopsis thaliana*. *J. Plant Growth Regul.* **2015**, *34*, 362–371. [CrossRef]
30. Pandey, A.; Misra, P.; Chandrashekar, K.; Trivedi, P. Development of *AtMYB12*-expressing transgenic tobacco callus culture for production of rutin with biopesticidal potential. *Plant Cell Rep.* **2012**, *31*, 1867–1876. [CrossRef]
31. Zhang, Y.; Ming, R.; Khan, M.; Wang, Y.; Dahro, B.; Xiao, W.; Li, C.; Liu, J. *ERF9* of *Poncirus trifoliata* (L.) Raf. undergoes feedback regulation by ethylene and modulates cold tolerance via regulating a *glutathione S-transferase U17* gene. *Plant Biotechnol. J.* **2021**, *20*, 183–200. [CrossRef]
32. Khan, M.; Hu, J.; Dahro, B.; Ming, R.; Zhang, Y.; Wang, Y.; Alhag, A.; Li, C.; Liu, J. *ERF108* from *Poncirus trifoliata* (L.) Raf. functions in cold tolerance by modulating raffinose synthesis through transcriptional regulation of *PtrRafS*. *Plant J.* **2021**, *108*, 705–724. [CrossRef] [PubMed]
33. Sun, X.; Zhao, T.; Gan, S.; Ren, X.; Fang, L.; Karungo, S.; Wang, Y.; Chen, L.; Li, S.; Xin, H. Ethylene positively regulates cold tolerance in grapevine by modulating the expression of ETHYLENE RESPONSE FACTOR 057. *Sci. Rep.* **2016**, *6*, 24066. [CrossRef]
34. Zhao, C.; Liu, X.; Gong, Q.; Cao, J.; Shen, W.; Yin, X.; Grierson, D.; Zhang, B.; Xu, C.; Li, X. Three *AP2/ERF* family members modulate flavonoid synthesis by regulating type IV *Chalcone isomerase* in *citrus*. *Plant Biotechnol. J.* **2020**, *19*, 671–688. [CrossRef] [PubMed]
35. Huang, Y.; Zhang, B.; Sun, S.; Xing, G.; Wang, F.; Li, M.; Tian, Y.; Xiong, A. *AP2/ERF* transcription factors involved in response to tomato yellow leaf curly virus in tomato. *Plant Genome* **2016**, *9*, 1–15. [CrossRef] [PubMed]
36. Peng, M.; Shahzad, R.; Gul, A.; Subthain, H.; Shen, S.; Lei, L.; Zheng, Z.; Zhou, J.; Lu, D.; Wang, S. Differentially evolved *glucosyltransferases* determine natural variation of rice flavone accumulation and UV-tolerance. *Nat. Commun.* **2017**, *8*, 1975. [CrossRef]
37. Li, P.; Li, Y.; Zhang, F.; Zhang, G.; Jiang, X.; Yu, H.; Hou, B. The *Arabidopsis UDP-glycosyltransferases UGT79B2* and *UGT79B3*, contribute to cold, salt and drought stress tolerance via modulating anthocyanin accumulation. *Plant J.* **2017**, *89*, 85–103. [CrossRef]

plants

MDPI

Article

A *SUPERMAN-like* Gene Controls the Locule Number of Tomato Fruit

Mi Zhang [1,2,†], Enbai Zhou [1,2,†], Meng Li [2], Shenglan Tian [1,2] and Han Xiao [2,*]

[1] University of Chinese Academy of Sciences, 19A Yuquan Rd, Beijing 100049, China; zhangmi@cemps.ac.cn (M.Z.); zhouenbai@cemps.ac.cn (E.Z.); tianshenglan@sippe.ac.cn (S.T.)

[2] National Key Laboratory of Plant Molecular Genetics, CAS Center for Excellence in Molecular Plant Sciences, Institute of Plant Physiology and Ecology, Chinese Academy of Sciences, 300 Fenglin Rd., Shanghai 200032, China; limeng22387@126.com

* Correspondence: hanxiao@cemps.ac.cn; Tel.: +86-21-5492-4301

† These authors contribute equally to this work.

Abstract: Tomato (*Solanum lycopersicum*) fruits are derived from fertilized ovaries formed during flower development. Thus, fruit morphology is tightly linked to carpel number and identity. The *SUPERMAN* (*SUP*) gene is a key transcription repressor to define the stamen–carpel boundary and to control floral meristem determinacy. Despite *SUP* functions having been characterized in a few plant species, its functions have not yet been explored in tomato. In this study, we identified and characterized a *fascinated and multi-locule fruit* (*fmf*) mutant in *Solanum pimpinellifolium* background harboring a nonsense mutation in the coding sequence of a zinc finger gene orthologous to *SUP*. The *fmf* mutant produces supersex flowers containing increased numbers of stamens and carpels and sets malformed seedless fruits with complete flowers frequently formed on the distal end. *fmf* alleles in cultivated tomato background created by CRISPR-Cas9 showed similar floral and fruit phenotypes. Our results provide insight into the functional conservation and diversification of *SUP* members in different species. We also speculate the *FMF* gene may be a potential target for yield improvement in tomato by genetic engineering.

Keywords: tomato (*Solanum lycopersicum*); *Solanum pimpinellifolium*; fruit morphology; flower development; locule number; *SUPERMAN*

Citation: Zhang, M.; Zhou, E.; Li, M.; Tian, S.; Xiao, H. A *SUPERMAN-like* Gene Controls the Locule Number of Tomato Fruit. *Plants* **2023**, *12*, 3341. https://doi.org/10.3390/plants12183341

Academic Editor: Hisato Kunitake

Received: 3 September 2023
Revised: 19 September 2023
Accepted: 20 September 2023
Published: 21 September 2023

1. Introduction

Tomato is one of major vegetable fruit crops cultivated worldwide, providing a rich resource of nutrients for human diets. As a typical type of fleshy fruit, tomato fruits develop from fertilized ovaries undergoing substantial post-fertilization growth, but many features of fruit morphology are defined during flower development. For example, the elongated and pear fruit shapes found in some cultivars are mainly defined by ovary shape before fertilization. Indeed, the two major fruit shape genes *SUN* and *OVATE* set ovary shape [1,2]. In cultivated tomato, the fruits have multiple locules derived from carpels. Thus, the locule number is largely attributed to the number of carpels formed during flower development, which is mainly controlled by *LOCULE NUMBER* (*LC*, ortholog of *WUS*) and *FASCIATED* (*FAS*, ortholog of *CLV3*) [3–6].

The bisexual tomato flower, like most angiosperm flowers, consists of four distinctive whorls of floral organs: sepals in the outmost whorl, petals in the second whorl, and stamens and carpels in the respective third and fourth whorls. The number of individual floral organs are varied among cultivated tomato accessions, but the wild ancestor *S. pimpinellifolium* flower has five sepals, five petals, five stamens fused at the base to form a cone-like structure, and a single pistil formed by two fused carpels [7]. The number and position of floral organs in each whorl is governed by floral homeotic genes and cadastral genes controlling floral organ boundary [8,9]. *SUPERMAN* (*SUP*) is the first identified boundary gene playing a crucial role in maintaining the boundary between stamen and carpel whorls in Arabidopsis flowers because the typic floral phenotype observed in *sup* mutants is the increased number of stamens at the expense of carpels [10,11]. The extra stamens in *sup* mutants likely resulted from an identity change in some whorl 4 cells, through which their fate was changed from female to male [12]. *SUP* encodes a transcription factor containing a C_2H_2 type zinc finger DNA binding domain and an EAR-like repressor domain [11,13,14]. Its specific expression at the boundary between whorl 3 and 4 prevents the expression of *APETALA3* (*AP3*) and *PISTILLATA* (*PI*) reaching the central region of the floral meristem [11,12].

Characterization of *SUP* orthologs in a few plant species, including rice (*Oryza sativa*), *Petunia hybrida*, and *Medicago truncatula*, has revealed functional conservation and divergence. The *sup* mutants of *Petunia hybrida* (*phsup1*) also show an increased stamen number at the expense of carpels, but form extra tissues connecting the inner three whorls [15]. Rice *small reproductive organs* (*sro*) harboring a mutation in a *SUP-like* gene does not show defects in floral organ number and organ identity, instead the mutant has smaller stamens and pistils [16]. Mutations in the Medicago *SUPERMAN* (*MtSUP*) gene cause an increased number of the inner three whorls and more flowers formed on individual inflorescence, indicating the *MtSUP* gains novel functions in the legume species [17]. It has been proposed that the transcriptional divergences detected in the flowers between Arabidopsis and Medicago underlie the functional diversification [18]. Similarly, this kind of expression divergence may be also applicable to the rice *SUP* ortholog, since *SRO* is specifically expressed in stamen filaments, not at the stamen–carpel boundary [16]. Because *SUP* orthologs have been only characterized in a few plant species, it remains to determine whether expression divergence explains the evolution of *SUP* functions. Nevertheless, the phenotypic variations observed in these *sup* mutants highlight the need to explore the functions of *SUP* orthologs in more plant species.

As shape and size are important fruit quality traits in tomato [19–22], understanding how key regulatory genes control the number of floral organs, especially the carpel number, not only sheds new light on gene conservation and diversification but also provides new strategies for breeding new varieties to meet the increasing demand for diverse fruit quality. In this study, we report the characterization of *FMF*, orthologous to *SUP*, regulating floral organ number, and ovary development in tomato.

2. Results

2.1. Phenotypic Analysis of the Fmf Mutant

Through screening for mutants showing fruit phenotypes including shape and size, we identified one such mutant in an Ethyl Methane Sulphonate (EMS)-mutagenized S. pimpinellifolium (accession LA1781) population. The mutant named *fasciated and multi-locule fruit* (*fmf*) had abnormal fruit morphology. The *fmf* mutant also exhibited abnormal flower development. Compared with the wild type (LA1781), *fmf* flowers had apparently normal sepals and petals, but its stamen cones were cracked and styles often split and shorter (Figure 1A–H). Dissection of floral organs revealed that *fmf* flowers had more than five stamens, in contrast to the five stamens constantly observed in the wild type flower (Figure 1I,J). *fmf* carpels had either unfused multiple ovaries connected at the base or multiple styles failed to be fused together at the distal end (Figure 1K–P). fmf ovaries contained multiple locules (carpels) and in most cases without ovule developed (Figure 1Q–V). Quantification of floral organs at anthesis showed that the numbers of stamens and carpels had significantly increased in *fmf*, on average 8.9 stamens and 6.2 carpels were formed (Figure 1W). *fmf* styles were dramatically shortened (Figure 1X). In addition, *fmf* mutation had a weak impact on inflorescence development: *fmf* inflorescences were shorter, containing slightly fewer flowers (Supplemental Figure S1).

Though *fmf* set fruits poorly (Supplemental Figure S1A), multiple-locule parthenocarpic fruits were formed occasionally (Figure 2A–D). Strikingly, complete flowers were often formed at the distal ends of these seedless *fmf* fruits (Figure 2E). To test the functionality of *fmf* stamens and carpels, we either pollinated *fmf* pistils with wild type pollens or wild type pistils with *fmf* pollens. *fmf* flowers pollinated with wild type pollens produced seedless fruits, whereas wild type plants produced seeded fruits when pollinated with *fmf* pollens (Figure 2F,G). These results suggest that *fmf* is female-sterile but male-fertile.

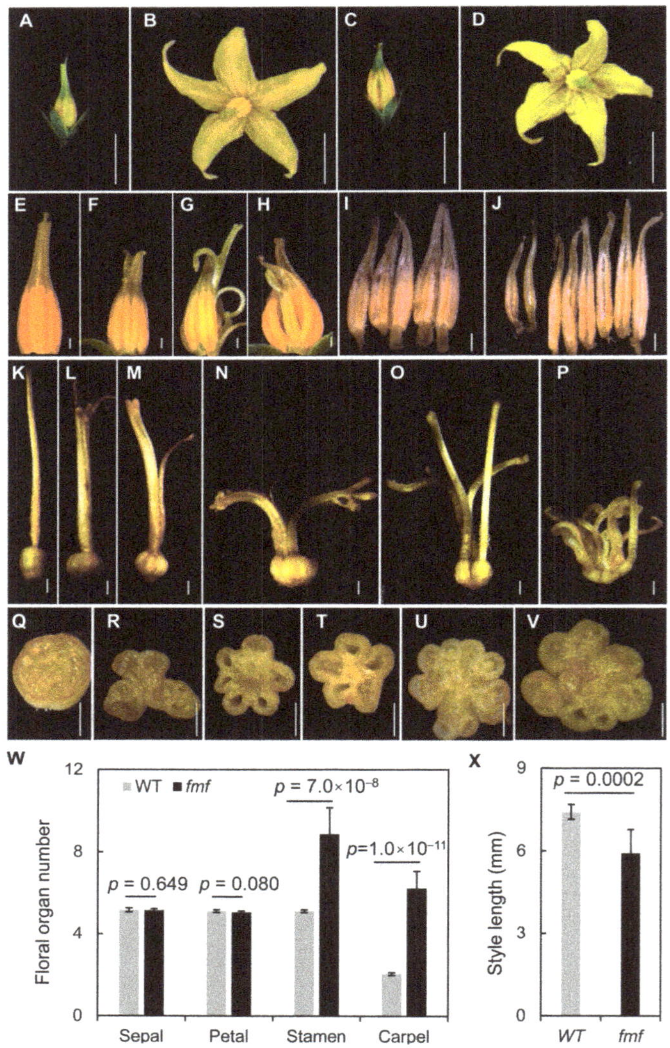

Figure 1. Floral phenotypes of *fmf* and wild type. (**A–D**) Unopened (**A**,**C**) and anthesis (**B**,**D**) flowers of the *fmf* mutant (**C**,**D**) and wild type ((**A**,**B**), LA1781). (**E–H**) Anther cones of wild type (**E**) and *fmf* (**F–H**). (**I**,**J**) Dissected anthers from a single flower of wild type (**I**) or *fmf* (**J**). (**K–P**) Dissected carpels of wild type (**K**) and *fmf* (**L–P**). (**Q–V**) Transversely spliced ovaries of wild type (**Q**) and *fmf* (**R–V**). (**W**) Quantification of the numbers of floral organs of wild type and *fmf*. (**X**) Style length of wild type and *fmf*. Measurements in (**W**,**X**) were conducted on 100 flowers from 10 plants for each genotype. Data represent means ± SD. p values were calculated using Student's t-test. Scale bar: 5 mm (**A–D**), 500 mm (**E–V**).

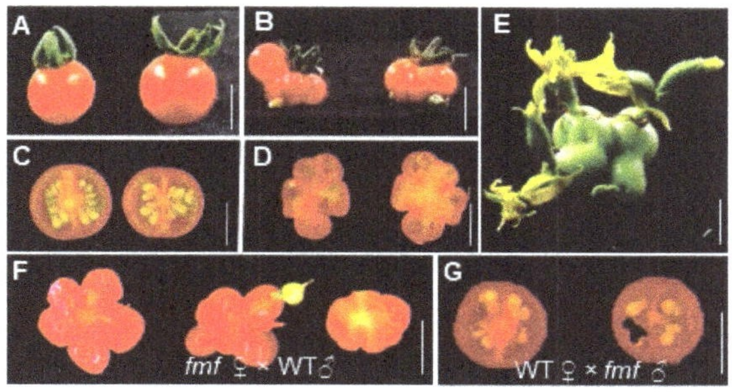

Figure 2. Fruit development of *fmf* and wild type. (**A**,**B**) Representative images of mature fruits of wild type (**A**) and *fmf* (**B**). (**C**,**D**) Transversely sliced fruits of wild type (**C**) and *fmf* (**D**). (**E**) A representative *fmf* fruit showing extra flowers developed on the distal parts of the fruit. (**F**) Images of parthenocarpic fruits developed from *fmf* flowers pollinated with wild type pollens. (**G**) Seeded fruits developed from wild type flowers pollinated with *fmf* pollens. Scale bar, 1 cm.

We further investigated the development of *fmf* flowers by histological analysis and scanning electronic microscopy (SEM). Histological analysis revealed that the *fmf* mutation did not affect sepal and petal formation (Figure 3A–E). The first noticeable difference between wild type and *fmf* flowers was observed after stamen primordia were formed, which meant that in *fmf* flowers an additional whorl of stamen primordia was developed in between the third and fourth whorls (Figure 3C,F). Later, *fmf* flowers developed more carpels (Figure 3G–L). In the center of the *fmf* flowers, there were often flower structures (Figure 3J). SEM analysis also confirmed that the timing and position of sepal and petal primordia were not impacted by *fmf* mutation (Figure 4A–D), while extra stamen and carpel primordia were developed afterward (Figure 4E–H). When wild type flowers reached anthesis stage, their pistils were almost closed, like a closed mouth (Figure 4I,J). In contrast, the pistils failed to be enclosed in *fmf* flowers (Figure 4K,L). Within the *fmf* flowers, new complete flower-like structures were also observed (Figure 4K). These results indicate that mutation in the *FMF* gene mainly affects the development of the two innermost whorls, stamens, and carpels.

2.2. Molecular Cloning of the FMF Gene

To identify the causal mutation underlying the *fmf* phenotypes, we generated an F_2 population by crossing the *fmf* mutant to cv. Moneymaker (LA2706). Rough mapping was conducted by BSA-seq using pooled genomic DNA isolated from wild type and *fmf* plants. The *FMF* locus was placed on a region around 66 Mb on chromosome 9 (ITAG4.0) showing maximal difference in the SNP index (Figure 5A). Then, we performed fine mapping using 384 *fmf* plants from the segregation population. Using Indel markers, the *FMF* locus was further narrowed down to an 83.85 kb interval between markers xps2515 and xps2452. The interval contains four annotated genes: *Solyc09g089580*, *Solyc09g089590*, *Solyc09g089600*, and *Solyc09g089610*, which encode 2-oxoglutarate (2OG) and the Fe(II)-dependent oxygenase superfamily protein, transcriptional regulator TAC1, the zinc finger protein, and the ethylene receptor-like protein (ETR6), respectively (Figure 5B). After sequencing the four candidate genes, the *fmf* mutant only contains a nonsense mutation in *Solyc09g089590*, which the cytosine at position 306 of the coding sequence was changed to adenine and the mutation introduced a stop codon after translation of 101 amino acids (102S*). Solyc09g089590 is a putative C_2H_2-type transcription factor containing transcription repression domain—the EAR-motif [13,14]. The deduced fmf protein was truncated, lacking the C-terminal containing the EAR-motif, suggesting that the *fmf* mutation may impair its transcription in regulation activity.

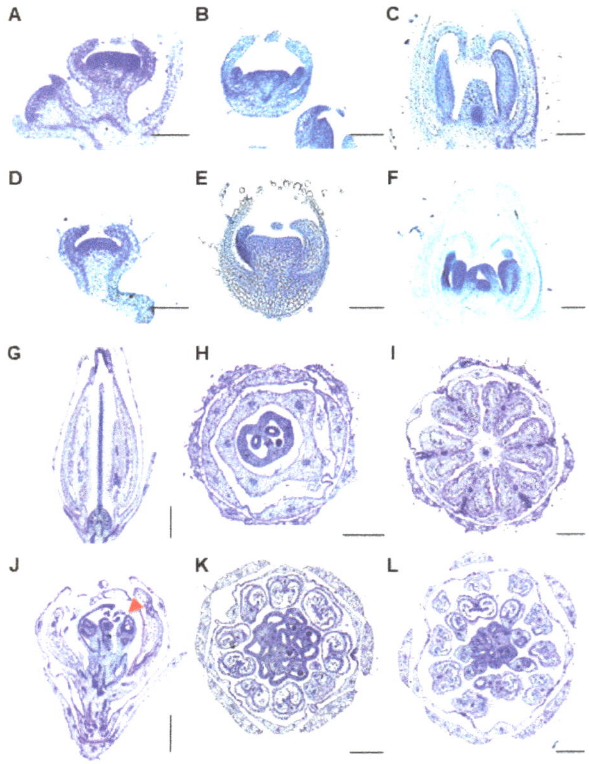

Figure 3. Histological analysis of *fmf* and wild type flowers. (**A**,**C**) Parafilm sections of type flower buds at the stages when sepal (**A**), petal (**B**), and stamen (**C**) was formed, respectively. (**D**–**F**) Parafilm sections of *fmf* flower buds at stages similar to wild type showing in (**A**–**C**). (**G**–**I**) Parafilm sections of mature wild type flowers (one day before anthesis). (**J**–**L**) Parafilm sections of mature *fmf* flowers. Longitudinal (**G**,**J**) and transverse (**H**,**I**,**K**,**L**) sections were made to reveal the structures of inner floral organs. The red arrowhead in (**J**) indicates a flower-like structure. Scale bar: 100 μm (**A**–**F**), 500 μm (**H**,**I**,**K**,**L**), 1 mm (**G**,**J**).

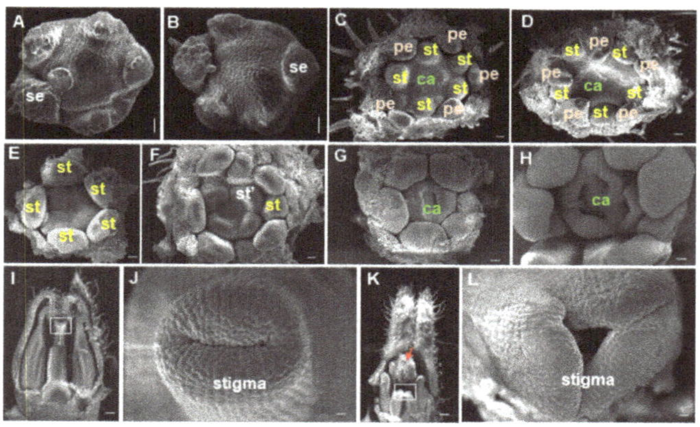

Figure 4. SEM analysis of *fmf* and wild type flowers. (**A**,**B**) Flower buds of wild type (**A**) and *fmf* (**B**) at sepal initiation stage. (**C**,**D**) Flower buds of wild type (**C**) and *fmf* (**D**) in which petal primordia

initiated. (**E**,**F**) Second whorl of stamen primordia formed in *fmf* flower buds. (**G**,**H**) Carpel fusion just started in wild type (**G**) and *fmf* (**H**) flower buds. Multi-carpels were observed in *fmf*, contrasting to two carpels in wild type. (**I–L**) Overview of mature wild type (**I**,**K**) and *fmf* (**K**,**L**) flowers and close examination of their stigma (**J**,**L**). The white boxes in (**I**,**K**) indicate stigma parts observed in (**J**,**L**) and the red arrow in (**K**) points to a flower-like tissue developed within the *fmf* flower. se, sepal; pe, petal., st, stamen; ca, carpel; st', extra stamen. Scale bar: 20 μm (**A–H**), 100 μm (**I**,**K**), 10 μm (**J**,**L**).

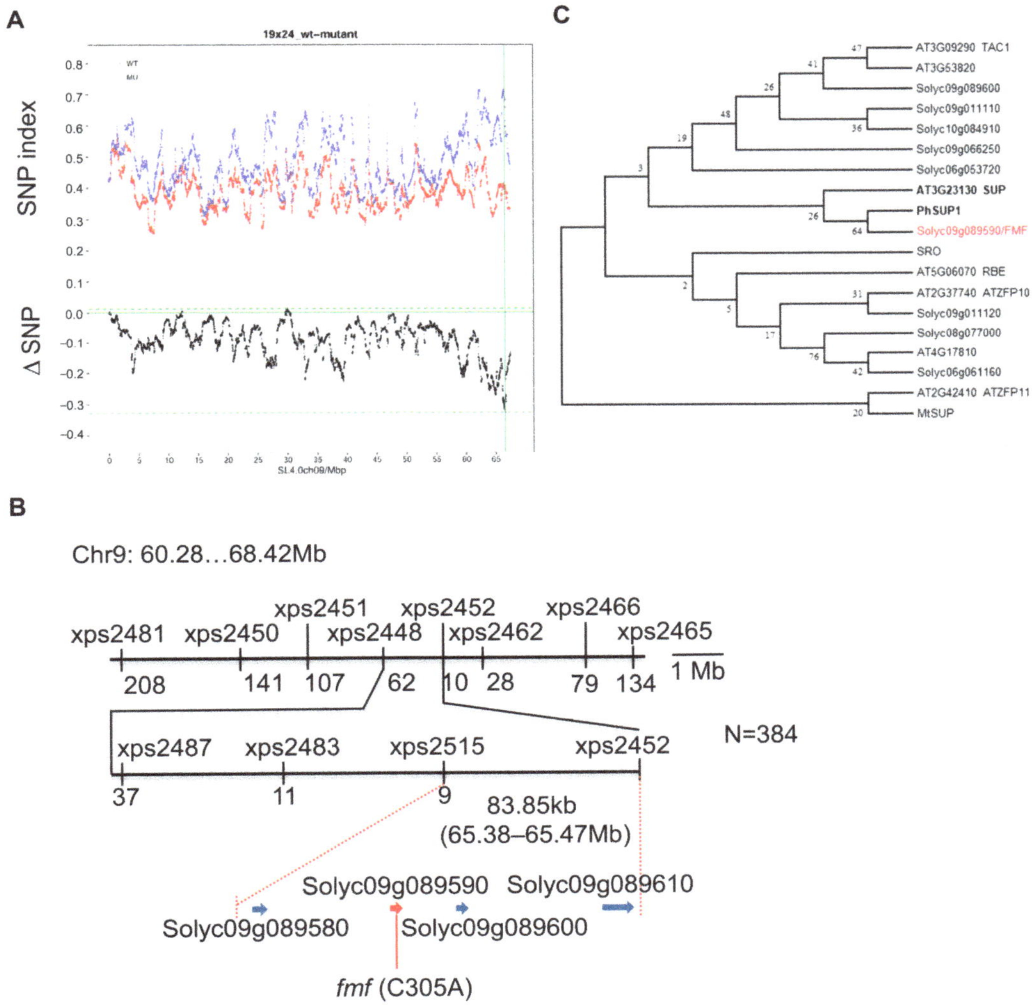

Figure 5. Cloning of the *FMF* gene. (**A**) SNP index of bulked DNA samples of wild type and *fmf* plants from a F2 population derived from a cross between *fmf* (in LA1781 background) and Moneymaker (LA2706). The graph shows that maximal SNP deviation between *fmf* and wild type was detected around 67 Mb on chromosome 9. (**B**) Fine mapping of the *FMF* locus. The numbers beneath the markers with the names starting with xps are the numbers of recombinants. *FMF* was placed on an interval of 83.85 kb region containing four predicted genes. (**C**) Phylogenetic tree of FMF and its close homologs in tomato and Arabidopsis. The tree was generated using MEGA7 with bootstrap of 1000.

Sequence and phylogenetic analysis of Solyc09g089590 and its close homologs from tomato and other species showed that Solyc09g089590 shares the highest similarity (e-value = 2×10^{-24}) and is grouped with Arabidopsis SUP and Petunia PhSUP1. SUP is thought to function as a cadastral gene to maintain the floral organ boundaries, in which its mutation causes formation of extra stamens at the expense of carpels [10]. Given *fmf* shows similar stamen phenotype with Arabidopsis *sup* mutants, Solyc09g089590 is likely orthologous to SUP and the missense mutation in this gene is responsible for the observed *fmf* floral morphology. However, tomato likely have a paralog Solyc06g053720 close to SUP; it also shares the highest sequence identity with SUP (e-value = 2×10^{-25}). Moreover, on the inferred phylogenetic tree, Solyc09g089590/FMF was not grouped with Arabidopsis TELOMERASE ACTIVATOR1 (TAC1), suggesting that it is unlikely orthologous to TAC1 as annotated by international tomato genome sequencing project (version ITAG4.0).

2.3. Creation of New *fmf* Alleles by CRISPR-Cas9

To further confirm that *Solyc09g089590* underlies the *FMF* locus, we generated *Solyc09g089590* mutants in Moneymaker background by CRISPR-Cas9. We obtained two different alleles *fmf-cr2* and *fmf-cr4* that showed *Solyc09g089590* was successfully edited; *fmf-cr2* had 4 bp deletion (+182~+185 start from start codon ATG) and *fmf-cr4* had 9 bp deletion (+175~+183) (Supplemental Figure S2A). The deletion in *fmf-cr2* introduced premature translation stop codon after 63 amino acids, which disrupted the conserved C_2H_2 zinc finger DNA binding domain encoding by *Solyc09g089590*. The *fmf-cr4* mutation resulted in loss of the first three amino acids of the invariant QALGGH motif in the C_2H_2 domain, which also likely disrupted this functional domain. *fmf-cr2* and *fmf-cr4* showed identical flower phenotype (Supplemental Figure S2B–D, Figure 6). Thus, a more detailed phenotypic analysis was conducted just on the *fmf-cr2* allele. Like the *fmf* mutant in LA1781 background, *fmf-cr2* in Moneymaker background also had increased numbers of stamens and carpels, shorter and unfused styles (Figure 6A–S). *fmf-cr2* was also male fertile and set seedless fruits, but unlike *fmf* fruits in LA1781 background, *fmf-cr2* fruits only had cracks, no extra flower structure was formed in the fruits (Figure 6T–X), suggesting that mutations in the *FMF* gene have different impacts on cell proliferation and differentiation in whorl 4 between cultivated tomato and its wild relatives. Nevertheless, both in LA1781 and Moneymaker backgrounds, mutations in the *FMF* gene caused very similar defects in stamen and carpel development.

2.4. FMF Expression during Flower Development

We investigated *FMF* expression in developing flowers of wild type by in situ hybridization. *FMF* transcripts were detected in stamen and carpel primordia at the early flower stage, then in stamens when all floral organs were formed (Figure 7A–D). FMF was also expressed weakly in petals and carpels, but not in sepals. Because *WUS* is a crucial regulator of floral meristem identity and its ortholog *SlWUS* controls locule number in tomato [3,6,23], we also compared its expression in wild type and *fmf* flowers. Compared with wild type, *SlWUS* expression was much stronger in stamen and carpel primordia of *fmf* flowers (Figure 7E,F).

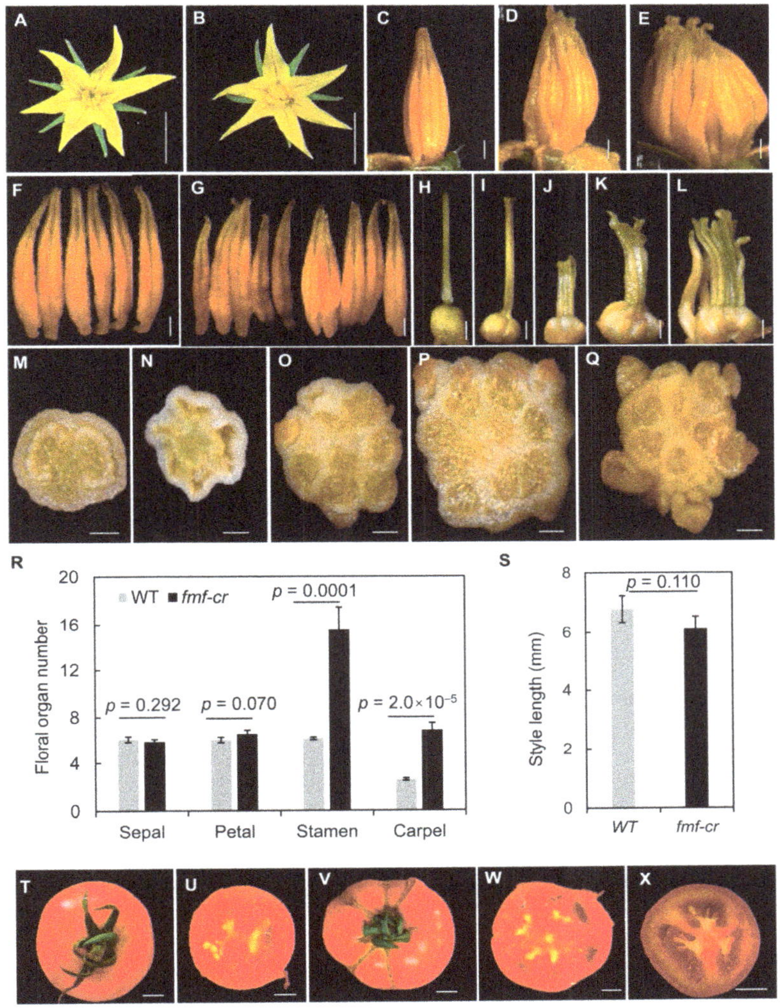

Figure 6. Flower and fruit phenotypes of *fmf-cr2* in cultivated tomato background. (**A**,**B**) Anthesis flowers of the *fmf-cr2* mutant (**B**) and wild type ((**A**) Moneymaker). (**C**–**E**) Anther cones of wild type (**C**) and *fmf-cr2* (**D**,**E**). (**F**,**G**) Dissected anthers from a single flower of wild type (**F**) or *fmf-cr2* (**G**). (**H**–**L**) Dissected carpels of wild type (**H**) and *fmf-cr2* (**I**–**L**). (**M**–**Q**) Transversely spliced ovaries of wild type (**M**) and *fmf* (**N**–**Q**). (**R**) Quantification of the numbers of floral organs of wild type and *fmf*. (**S**) Style length of wild type and *fmf*. Measurements in (**R**,**S**) were conducted on 40 flowers from 4 plants for each genotype. Data represent means ± SD. *p* values were calculated using Student's *t*-test. (**T**–**W**) Mature fruits of wild type (**T**,**U**) and *fmf-cr2* (**V**,**W**). No seed was observed in *fmf-cr2* fruits. (**X**) Seeded fruits developed from wild type flowers pollinated with *fmf-cr2* pollens. Scale bar: 1 cm (**A**–**E**,**T**–**X**), 1 mm (**F**–**L**), 500 μm (**M**–**Q**).

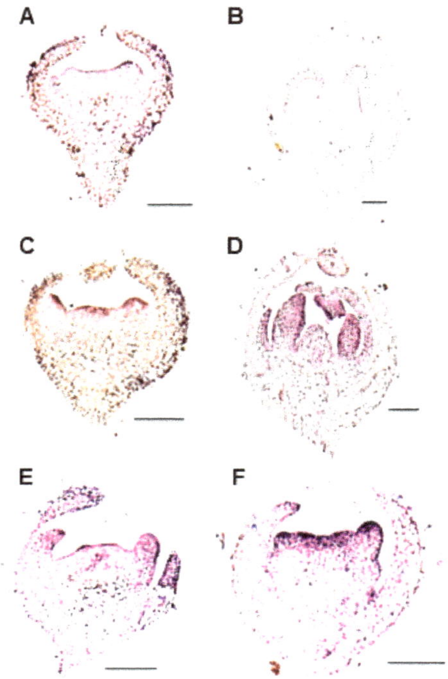

Figure 7. *FMF* expression in wild type flowers. (**A–D**) *FMF* expression wild type flowers revealed by in situ hybridization. (**A**,**B**) Flower sections were probed by *FMF* sense probe. (**C**,**D**) Flower sections were probed by *FMF* antisense probe. (**E**) *SlWUS* expression in wild type flowers. (**F**) *SlWUS* expression in *fmf* flowers. Scale bar, 100 μm.

3. Discussion

In many cases, the fruit develops from the ovary after fertilization. Thus, flower development has considerable impact on fruit morphology. For example, *SUN*, *OVATE*, *LC/SlWUS*, and *FAS/SlCLV3* control tomato fruit shape and size through their early actions on ovary development [1–4,6,23,24]. In this study, we identified *FMF* that is likely orthologous to the Arabidopsis *SUP* gene controlling tomato fruit morphology. The *fmf* mutant identified in EMS-mutagenized *S. pimpinellifolium* LA1781populations and the *fmf-cr* mutants in *S. lycopersicum* cv. Moneymaker created by CRISPR-Cas9 showed almost identical flower and fruit phenotypes except no flower structures were observed in the *fmf-cr* fruits (Figures 1, 2 and 6). All *fmf* mutants contained more stamens and carpels, shorter and often unfused styles, and fewer if not absent ovules. Moreover, the *fmf* mutants produced functional pollens. These morphological features make the *FMF* gene a potential target for genetic manipulation to create desirable fruit traits in tomato and other Solanaceous crops, i.e., creating weak *sup* alleles to increase locule number.

Different *sup* alleles exhibit a wide spectrum of flower abnormality in Arabidopsis and *M. truncatula*, ranging from superman to superwoman to supersex [8,17,18,25–28]. Compared with the *sup* mutants identified in other plant species, *fmf* mutants only had supersex phenotype-producing bisexual flowers with more stamens and carpels, sharing higher similarity with Arabidopsis *sup-5* and *M. truncatula mtsup-2* [8,17]. No *fmf* flower showed superman (having extra stamens and carpelloid carpels) or superwoman (bisexual flowers with all stamens on a single whorl and an indeterminate whorl 4) morphology as observed in some alleles of Arabidopsis and *M. truncatula sup* mutants. Similarly, low spectrum of phenotypic variations was also detected in the rice *sro* mutant and the *sup* mutants of *P. hybrida* [15,16]. The differences in phenotypic variations observed in allelic *sup* mutants across plant species imply that *SUP* regulation may be integrated into species–specific genetic networks controlling floral development.

Characterization of *sup* mutants in several plant species including Arabidopsis, *M. truncatula*, *P. hybrida*, and rice has revealed the conservation and diversification of *SUP* functions across plant species [8,10–12,15–17,25,27–42]. *fmf*, *sup*, *phsup1* and *mtsup* mutants have increased numbers of stamens and/or carpels in their respective flowers. Therefore, the four eudicot species have conserved *SUP* functions in regulation of floral organ numbers. However, rice *rso* flowers have reduced-size stamens and carpels but no changes in the number and identity of floral organs [16], suggesting that *SUP* functions are not well conserved between eudicots and monocots. It is also possible that SRO is not the true ortholog of Arabidopsis SUP because it is grouped with RABBIT EAR (RBE), a close homolog of SUP, though their relationship is not well supported by phylogenetic analysis. SUP belongs to a subclade of the zinc finger protein family, containing a single C_2H_2 zinc finger DNA binding domain and an EAR-repressor domain. It is notable that members in this subclade share low sequence similarity in the regions beyond the two domains as indicated by their phylogenetic relationships (Figure 5C). However, FMF is closer to Petunia PhSUP1 and Arabidopsis SUP, providing additional evidence to support that FMF is orthologous to SUP.

The differences in flower phenotypes among *sup* mutants of the five species may be explained by different spatiotemporal expression patterns of *SUP* orthologs. *FMF* is expressed in whorl 2 and 3 at early stage and mainly expressed in stamens at late stage. *SUP* is expressed on the two sides of the stamen–carpel boundary [10–12,27,31,32]. However, such an expression pattern is not observed for *SRO*, *MtSUP*, and *PhSUP1* [15–17]. For example, *MtSUP* is expressed in carpel primordia and the common primordia where petals and stamens developed later, and *SRO* is specifically expressed in stamen filaments.

fmf mutants have a distinctive floral phenotype not observed in any other *sup* mutants: complete flowers formed in the *fmf* ovaries/fruits in the genetic background of *S. pimpinellifolium* LA1781 (Figures 2 and 6). Such a phenotype indicates that floral meristem is not terminated in a timely manner in *fmf* flowers. It remains to further explore whether additional genetic components are involved and what caused the incomplete penetration in the cultivar Moneymaker. Given *WUS* is required for floral meristem identity [43], we speculate that the *SlWUS* activity in LA1781 and Moneymaker flowers may be somewhat different because the increased locule number in cultivated tomato is highly associated with DNA variations in the *SlWUS* gene. In addition, *FMF* mutations also slightly affected inflorescence length and the number of flowers formed on individual inflorescence. Nevertheless, our results from characterization of the *FMF* gene in tomato reveal an undescribed *SUP* function.

4. Materials and Methods

4.1. Materials and Plant Growth Conditions

S. pimpinellifolium accession LA1781 and cultivated tomato cv. Moneymaker (LA2706) were grown in phytotrons or plastic greenhouse. When grown in phytotron, plant pots in blonde peat (Pindstrup Mosebrug A/S, Ryomgaard, Denmark) were grown under the condition of temperatures of 18 to 25 °C with a relative humidity of 70–80% and illuminated by 150 $mE \cdot m^{-2} \cdot s^{-1}$ light intensity for 16 h. Plants grown in a plastic greenhouse were under natural solar radiation, and the during the growth season (late March to middle of July) the temperature ranged from 15 to 29 °C and relative humidity from 32 to 90%.

4.2. Bulked-Segregant Analysis-Seq (BSA-Seq) and Map-Based Cloning

After examination of fruit morphology, individual *fmf* plants segregating in a F2 population derived from a cross between *fmf* (pollen donor) and Moneymaker were sampled for DNA extraction using method previously described [2].

For BSA-seq, additional sampling was conducted, which leaves from around 100 plants for each genotype were pooled for the mutant and wild type before DNA extraction.

The two pooled DNA samples were sequenced by Hiseq 2500 (Illumina). Raw reads were quality checked by *fastqc* (version v0.11.9, http://www.bioinformatics.babraham.ac.uk/projects/fastqc/, accessed on 7 September 2019), followed by the removal of low-quality reads and adaptor sequences by *Trimmomatic* (version 0.39, accessed on 22 February 2021) [44]. The processed clean reads were mapped to the tomato reference genome ITAG4.0 downloaded from Sol Genomics Network (https://solgenomics.net/, accessed on 22 February 2021) using BWA-MEM (version BWA-0.7.17, 15 August 2021) [45]. Mapping results were sorted by *samtools faidx* (version 1.13, accessed on 9 August 2021) [46]. We used the functions AddOrReplaceReadGroups, MarkDuplicates, BamIndexStats in the Picard software (release 1.119. http://broadinstitute.github.io/picard/, accessed on 16 January 2022) to further remove duplicates and index the mapped reads. SNPs were called using the HaplotypeCaller program in *GATK* (v4.2.4.0, accessed on 16 January 2022) with following parameters: -stand-call-conf 30 --native-pair-hmm-threads 15 -mbq 20 [47]. The SNP index was then calculated for each allele as described previously [48]. Sliding window analysis on SNP-index plots was carried out with 1 Mb window size and 10 kb increment and the plot graph was draw by *ggplot2* (https://github.com/tidyverse/ggplot2, accessed on 20 February 2022).

After rough mapping, fine mapping was focused on an 8.14 Mb region (60.28–68.42 Mb) using developed Indel and CAPS markers. Marker information can be found in Supplemental Table S1.

4.3. Generation of *fmf-cr* Mutants by CRISPR-Cas9

The design of target oligoes and vector construction were previously described [48]. Briefly, target oligoes containing AAATCAGCTCAAGCTCTTGG (start at 166 after ATG) were designed using the online tool CRISPR-P (http://cbi.hzau.edu.cn/crispr/, accessed on 6 March 2022) [49]. Oligoes after annealing were cloned into the psgR-Cas9-At vector and further assembled into pCambia1300 [50]. The agrobacterium tumefaciens strain GV3101 harboring the plasmid was used for plant transformation using Moneymaker cotyledons as explants according to the method previously described [51].

4.4. Phylogenetic Analysis of FMF Homologs

Homologous proteins of Arabidopsis SUP and tomato FMF were identified by BlastP using FMF protein sequence as query on Araport11 protein sequences of Arabidopsis (TAIR, https://www.arabidopsis.org/, accessed on 7 December 2022) and ITAG4.0 protein sequences of tomato (SGN, https://solgenomics.net/, accessed on 7 December 2022). The sequences of SUP orthologs in Petunia, rice, and Medicago were retrieved from NCBI (https://www.ncbi.nlm.nih.gov/protein/, accessed on 7 December 2022). Phylogenetic analysis was conducted using the MEGA7.0 software [52]. The original phylogenetic tree was constructed using the neighbor-joining method and then the bootstrap consensus tree was inferred from 1000 replicates.

4.5. Microscopy

Flower buds at different developmental stages were collected and fixed in FAA fixative solution at 4 °C overnight. The samples were dehydrated through ethanol series (50% –70%–85%–95%–100%). For histological analysis, the flower samples were embedded in Paraplast® (P3558, Sigma-Aldrich, St. Louis, MO, USA) and 10 μm sections were made by a Leica microtome using a Leica microtome (Leica) and were briefly stained by 0.05% toluidine blue. For SEM analysis, the flower samples after dehydration were subjected to critical point drying in liquid nitrogen and coated with gold, then were examined under an electronic microscope (Zeiss Merlin Compact, Oberkochen, Germany).

4.6. In Situ Hybridization

Flower buds were fixed in FAA overnight at 4 °C and embedded in Paraplast (Sigma-Aldrich, St. Louis, MO, USA) after dehydration. The samples were then sliced into 8 μm sections using a microtome (Leica, Wetzlar, Germany). After dewaxing and rehydration, the sections were probed by digoxigenin-labeled sense and antisense riboprobes as previously described [53].

Supplementary Materials: The following supporting information can be downloaded at: https://www.mdpi.com/article/10.3390/plants12183341/s1, Figure S1: Inflorescence phenotype of *fmf* and wild type; Figure S2: Generations of *fmf* mutants by CRISPR-CAS9; Table S1: Primers used in this study.

Author Contributions: Conceptualization, H.X.; validation, M.Z. and E.Z.; formal analysis, M.Z. and E.Z.; investigation, M.Z., E.Z., M.L. and S.T.; data curation, M.Z. and H.X.; writing, H.X.; supervision and project administration, H.X.; funding acquisition, H.X. All authors have read and agreed to the published version of the manuscript.

Funding: This research was funded by Strategic Priority Research Program from the Chinese Academy of Sciences (grant no. XDA24030404), The Science and Technology Commission of Shanghai Municipality (grant no. 23ZR1470500) and National Natural Science Foundation of China (grant no. 31672164 and 32072577).

Data Availability Statement: Materials generated in this study will be available upon request to the corresponding author (H.X.).

Acknowledgments: The authors would like to thank the Tomato Genetics Resource Center (TGRC) at University of California, Davis for kindly providing the seeds of LA2706 (Moneymaker) and LA1781.

Conflicts of Interest: The authors declare no conflict of interest.

References

1. Liu, J.; Van Eck, J.; Cong, B.; Tanksley, S.D. A new class of regulatory genes underlying the cause of pear-shaped tomato fruit. *Proc. Natl. Acad. Sci. USA* **2002**, *99*, 13302–13306. [CrossRef]
2. Xiao, H.; Jiang, N.; Schaffner, E.; Stockinger, E.J.; Van Der Knaap, E. A Retrotransposon-Mediated Gene Duplication Underlies Morphological Variation of Tomato Fruit. *Science* **2008**, *319*, 1527–1530. [CrossRef]
3. Chu, Y.; Jang, J.; Huang, Z.; van der Knaap, E. Tomato locule number and fruit size controlled by natural alleles of *lc* and *fas*. *Plant Direct* **2019**, *3*, e00142. [CrossRef]
4. Cong, B.; Barrero, L.S.; Tanksley, S.D. Regulatory change in YABBY-like transcription factor led to evolution of extreme fruit size during tomato domestication. *Nat. Genet.* **2008**, *406*, 800–804. [CrossRef]
5. Munos, S.; Ranc, N.; Botton, E.; Bérard, A.; Rolland, S.; Duffé, P.; Carretero, Y.; Le Paslier, M.C.; Delalande, C.; Bouzayen, M.; et al. Increase in tomato locule number is controlled by two single-nucleotide polymorphisms located near WUSCHEL. *Plant Physiol.* **2011**, *156*, 2244–2254. [CrossRef]
6. Xu, C.; Liberatore, K.L.; MacAlister, C.A.; Huang, Z.; Chu, Y.-H.; Jiang, K.; Brooks, C.; Ogawa-Ohnishi, M.; Xiong, G.; Pauly, M.; et al. A cascade of arabinosyltransferases controls shoot meristem size in tomato. *Nat. Genet.* **2015**, *47*, 784–792. [CrossRef]
7. Xiao, H.; Radovich, C.; Welty, N.; Hsu, J.; Li, D.; Meulia, T.; van der Knaap, E. Integration of tomato reproductive developmental landmarks and expression profiles, and the effect of SUN on fruit shape. *BMC Plant Biol.* **2009**, *9*, 49. [CrossRef]
8. Breuil-Broyer, S.; Trehin, C.; Morel, P.; Boltz, V.; Sun, B.; Chambrier, P.; Ito, T.; Negrutiu, I. Analysis of the Arabidopsis superman allelic series and the interactions with other genes demonstrate developmental robustness and joint specification of male-female boundary, flower meristem termination and carpel compartmentalization. *Ann. Bot.* **2016**, *117*, 905–923. [CrossRef]
9. Monniaux, M.; Vandenbussche, M. How to Evolve a Perianth: A Review of Cadastral Mechanisms for Perianth Identity. *Front. Plant Sci.* **2018**, *9*, 1573. [CrossRef]
10. Bowman, J.L.; Sakai, H.; Jack, T.; Weigel, D.; Mayer, U.; Meyerowitz, E.M. SUPERMAN, a regulator of floral homeotic genes in Arabidopsis. *Development* **1992**, *114*, 599–615. [CrossRef]
11. Sakai, H.; Medrano, L.J.; Meyerowitz, E.M. Role of SUPERMAN in maintaining Arabidopsis floral whorl boundaries. *Nature* **1995**, *378*, 199–203. [CrossRef]
12. Prunet, N.; Yang, W.; Das, P.; Meyerowitz, E.M.; Jack, T.P. *SUPERMAN* prevents class B gene expression and promotes stem cell termination in the fourth whorl of *Arabidopsis thaliana* flowers. *Proc. Natl. Acad. Sci. USA* **2017**, *114*, 7166–7171. [CrossRef]
13. Hiratsu, K.; Mitsuda, N.; Matsui, K.; Ohme-Takagi, M. Identification of the minimal repression domain of SUPERMAN shows that the DLELRL hexapeptide is both necessary and sufficient for repression of transcription in Arabidopsis. *Biochem. Biophys. Res. Commun.* **2004**, *321*, 172–178. [CrossRef]
14. Hiratsu, K.; Ohta, M.; Matsui, K.; Ohme-Takagi, M. The SUPERMAN protein is an active repressor whose carboxy-terminal repression domain is required for the development of normal flowers. *FEBS Lett.* **2002**, *514*, 351–354. [CrossRef]
15. Nakagawa, H.; Ferrario, S.; Angenent, G.C.; Kobayashi, A.; Takatsuji, H. The Petunia Ortholog of Arabidopsis *SUPERMAN* Plays a Distinct Role in Floral Organ Morphogenesis. *Plant Cell* **2004**, *16*, 920–932. [CrossRef]

16. Xu, W.; Zhu, W.; Yang, L.; Liang, W.; Li, H.; Chen, M.; Luo, Z.; Huang, G.; Duan, L.; Dreni, L.; et al. SMALL REPRODUCTIVE ORGANS, a SUPERMAN-like transcription factor, regulates stamen and pistil growth in rice. *New Phytol.* **2021**, *233*, 1701–1718. [CrossRef]
17. Rodas, A.L.; Roque, E.; Hamza, R.; Gómez-Mena, C.; Minguet, E.G.; Wen, J.; Mysore, K.S.; Beltrán, J.P.; Cañas, L.A. *MtSUPERMAN* plays a key role in compound inflorescence and flower development in *Medicago truncatula. Plant J.* **2020**, *105*, 816–830. [CrossRef]
18. Rodas, A.L.; Roque, E.; Hamza, R.; Gómez-Mena, C.; Beltrán, J.P.; Cañas, L.A. SUPERMAN strikes again in legumes. *Front. Plant Sci.* **2023**, *14*, 1120342. [CrossRef]
19. Maurya, D.; Mukherjee, A.; Akhtar, S.; Chattopadhyay, T. Development and validation of the OVATE gene-based functional marker to assist fruit shape selection in tomato. *3 Biotech* **2021**, *11*, 474. [CrossRef]
20. Rodríguez, G.R.; Muños, S.; Anderson, C.; Sim, S.C.; Michel, A.; Causse, M.; Gardener, B.B.; Francis, D.; van Der Knaap, E. Distribution of SUN, OVATE, LC, and FAS in the tomato germplasm and the relationship to fruit shape diversity. *Plant Physiol.* **2011**, *156*, 275–285. [CrossRef]
21. Sierra-Orozco, E.; Shekasteband, R.; Illa-Berenguer, E.; Snouffer, A.; van der Knaap, E.; Lee, T.G.; Hutton, S.F. Identification and characterization of GLOBE, a major gene controlling fruit shape and impacting fruit size and marketability in tomato. *Hortic. Res.* **2021**, *8*, 138. [CrossRef]
22. van der Knaap, E.; Chakrabarti, M.; Chu, Y.H.; Clevenger, J.P.; Illa-Berenguer, E.; Huang, Z.; Keyhaninejad, N.; Mu, Q.; Sun, L.; Wang, Y.; et al. What lies beyond the eye: The molecular mechanisms regulating tomato fruit weight and shape. *Front. Plant Sci.* **2014**, *5*, 227. [CrossRef]
23. Li, H.; Qi, M.; Sun, M.; Liu, Y.; Liu, Y.; Xu, T.; Li, Y.; Li, T. Tomato Transcription Factor SlWUS Plays an Important Role in Tomato Flower and Locule Development. *Front. Plant Sci.* **2017**, *8*, 457. [CrossRef]
24. Fletcher, J.C. The CLV-WUS Stem Cell Signaling Pathway: A Roadmap to Crop Yield Optimization. *Plants* **2018**, *7*, 87. [CrossRef]
25. Bondada, R.; Somasundaram, S.; Marimuthu, M.P.; Badarudeen, M.A.; Puthiyaveedu, V.K.; Maruthachalam, R. Natural epialleles of Arabidopsis SUPERMAN display superwoman phenotypes. *Commun. Biol.* **2020**, *3*, 772. [CrossRef]
26. Cao, X.; Jacobsen, S.E. Role of the Arabidopsis DRM Methyltransferases in De Novo DNA Methylation and Gene Silencing. *Curr. Biol.* **2002**, *12*, 1138–1144. [CrossRef]
27. Gaiser, J.C.; Robinson-Beers, K.; Gasser, C.S. The Arabidopsis SUPERMAN gene mediates asymmetric growth of the outer integument of ovules. *Plant Cell* **1995**, *7*, 333–345. [CrossRef]
28. Jacobsen, S.E.; Meyerowitz, E.M. Hypermethylated SUPERMAN epigenetic alleles in Arabidopsis. *Science* **1997**, *277*, 1100–1103. [CrossRef]
29. Bereterbide, A.; Hernould, M.; Farbos, I.; Glimelius, K.; Mouras, A. Restoration of stamen development and production of functional pollen in an alloplasmic CMS tobacco line by ectopic expression of the *Arabidopsis thaliana SUPERMAN* gene. *Plant J.* **2002**, *29*, 607–615. [CrossRef]
30. Dinkins, R.; Pflipsen, C.; Thompson, A.; Collins, G.B. Ectopic Expression of an Arabidopsis Single Zinc Finger Gene in Tobacco Results in Dwarf Plants. *Plant Cell Physiol.* **2002**, *43*, 743–750. [CrossRef]
31. Huang, H.; Ma, H. FON1, an Arabidopsis Gene That Terminates Floral Meristem Activity and Controls Flower Organ Number. *Plant Cell* **1997**, *9*, 115. [CrossRef]
32. Ito, T.; Sakai, H.; Meyerowitz, E.M. Whorl-Specific Expression of the SUPERMAN Gene of Arabidopsis Is Mediated by cis Elements in the Transcribed Region. *Curr. Biol.* **2003**, *13*, 1524–1530. [CrossRef]
33. Kater, M.M.; Franken, J.; Van Aelst, A.; Angenent, G.C. Suppression of cell expansion by ectopic expression of the Arabidopsis SUPERMAN gene in transgenic petunia and tobacco. *Plant J.* **2000**, *2*, 407–413. [CrossRef]
34. Kazama, Y.; Fujiwara, M.T.; Koizumi, A.; Nishihara, K.; Nishiyama, R.; Kifune, E.; Abe, T.; Kawano, S. A SUPERMAN-like Gene is Exclusively Expressed in Female Flowers of the Dioecious Plant Silene latifolia. *Plant Cell Physiol.* **2009**, *50*, 1127–1141. [CrossRef]
35. Meister, R.J.; Kotow, L.M.; Gasser, C.S. SUPERMAN attenuates positive INNER NO OUTER autoregulation to maintain polar development of Arabidopsis ovule outer integuments. *Development* **2002**, *129*, 4281–4289. [CrossRef]
36. Nandi, A.K.; Kushalappa, K.; Prasad, K.; Vijayraghavan, U. A conserved function for Arabidopsis SUPERMAN in regulating floral-whorl cell proliferation in rice, a monocotyledonous plant. *Curr. Biol.* **2000**, *10*, 215–218. [CrossRef]
37. Nibau, C.; Di Stilio, V.S.; Wu, H.-M.; Cheung, A.Y. Arabidopsis and Tobacco SUPERMAN regulate hormone signalling and mediate cell proliferation and differentiation. *J. Exp. Bot.* **2010**, *62*, 949–961. [CrossRef]
38. Rohde, A.; Grunau, C.; Beck, D.L.; Montagu, M.V.; Rosenthal, A.; Boerjan, W. carpel, a New Arabidopsis Epi-Mutant of the SUPERMAN Gene: Phenotypic Analysis and DNA Methylation Status. *Plant Cell Physiol.* **1999**, *40*, 961–972. [CrossRef]
39. Sakai, H.; Krizek, B.A.; Jacobsen, S.E.; Meyerowitz, E.M. Regulation of SUP Expression Identifies Multiple Regulators Involved in Arabidopsis Floral Meristem Development. *Plant Cell* **2000**, *12*, 1607. [CrossRef]
40. Xu, K.; Wang, L.; Liu, N.; Xie, X.; Zhu, Y. Characterization of a SUPERMAN-like Gene, MdSUP11, in apple (Malus × domestica Borkh.). *Plant Physiol. Biochem.* **2018**, *125*, 136–142. [CrossRef]
41. Yun, J.-Y.; Weigel, D.; Lee, I. Ectopic Expression of SUPERMAN Suppresses Development of Petals and Stamens. *Plant Cell Physiol.* **2002**, *43*, 52–57. [CrossRef] [PubMed]
42. Zhao, J.; Liu, M.; Jiang, L.; Ding, L.; Yan, S.S.; Zhang, J.; Dong, Z.; Ren, H.; Zhang, X. Cucumber SUPERMAN Has Conserved Function in Stamen and Fruit Development and a Distinct Role in Floral Patterning. *PLoS ONE* **2014**, *9*, e86192. [CrossRef] [PubMed]

43. Laux, T.; Mayer, K.F.X.; Berger, J.; Jürgens, G. The *WUSCHEL* gene is required for shoot and floral meristem integrity in *Arabidopsis*. *Development* **1996**, *122*, 87–96. [CrossRef]
44. Bolger, A.M.; Lohse, M.; Usadel, B. Trimmomatic: A flexible trimmer for Illumina sequence data. *Bioinformatics* **2014**, *30*, 2114–2120. [CrossRef]
45. Li, H.; Durbin, R. Fast and accurate short read alignment with Burrows—Wheeler transform. *Bioinformatics* **2009**, *25*, 1754–1760. [CrossRef] [PubMed]
46. Li, H.; Handsaker, B.; Wysoker, A.; Fennell, T.; Ruan, J.; Homer, N.; Marth, G.; Abecasis, G.; Durbin, R.; 1000 Genome Project Data Processing Subgroup. The Sequence Alignment/Map format and SAMtools. *Bioinformatics* **2009**, *25*, 2078–2079. [CrossRef] [PubMed]
47. McKenna, A.; Hanna, M.; Banks, E.; Sivachenko, A.; Cibulskis, K.; Kernytsky, A.; Garimella, K.; Altshuler, D.; Gabriel, S.; Daly, M.; et al. The Genome Analysis Toolkit: A MapReduce framework for analyzing next-generation DNA sequencing data. *Genome Res.* **2010**, *20*, 1297–1303. [CrossRef]
48. Xu, Q.; Li, R.; Weng, L.; Sun, Y.; Li, M.; Xiao, H. Domain-specific expression of meristematic genes is defined by the LITTLE ZIPPER protein DTM in tomato. *Commun. Biol.* **2019**, *2*, 134. [CrossRef]
49. Lei, Y.; Lu, L.; Liu, H.-Y.; Li, S.; Xing, F.; Chen, L.-L. CRISPR-P: A Web Tool for Synthetic Single-Guide RNA Design of CRISPR-System in Plants. *Mol. Plant* **2014**, *7*, 1494–1496. [CrossRef]
50. Liu, W.; Zhu, X.; Lei, M.; Xia, Q.; Botella, J.R.; Zhu, J.-K.; Mao, Y. A detailed procedure for CRISPR/Cas9-mediated gene editing in Arabidopsis thaliana. *Sci. Bull.* **2015**, *60*, 1332–1347. [CrossRef]
51. McCormick, S.; Niedermeyer, J.; Fry, J.; Barnason, A.; Horsch, R.; Fraley, R. Leaf disc transformation of cultivated tomato (*L. esculentum*) using Agrobacterium tumefaciens. *Plant Cell Rep.* **1986**, *5*, 81–84. [CrossRef] [PubMed]
52. Kumar, S.; Stecher, G.; Tamura, K. MEGA7: Molecular evolutionary genetics analysis version 7.0 for bigger datasets. *Mol. Biol. Evol.* **2016**, *33*, 1870–1874. [CrossRef] [PubMed]
53. Weng, L.; Zhao, F.; Li, R.; Xu, C.; Chen, K.; Xiao, H. The Zinc Finger Transcription Factor *SlZFP2* Negatively Regulates Abscisic Acid Biosynthesis and Fruit Ripening in Tomato. *Plant Physiol.* **2015**, *167*, 931–949. [CrossRef] [PubMed]

plants

MDPI

Review

Guidelines for Performing CRISPR/Cas9 Genome Editing for Gene Validation and Trait Improvement in Crops

Nikolaos Tsakirpaloglou, Endang M. Septiningsih and Michael J. Thomson *

Department of Soil and Crop Sciences, Texas A&M University, College Station, TX 77843, USA; n.tsakirpaloglou@ag.tamu.edu (N.T.); eseptiningsih@tamu.edu (E.M.S.)
* Correspondence: m.thomson@tamu.edu

Abstract: With the rapid advances in plant genome editing techniques over the past 10 years, more efficient and powerful crop genome editing applications are now possible. Candidate genes for key traits can be validated using CRISPR/Cas9-based knockouts and through the up- and down-regulation of gene expression. Likewise, new trait improvement approaches can take advantage of targeted editing to improve stress tolerance, disease resistance, and nutritional traits. However, several key steps in the process can prove tricky for researchers who might be new to plant genome editing. Here, we present step-by-step guidelines and best practices for a crop genome editing pipeline that should help to improve the rate of success. Important factors in the process include proper target sequence analysis and single guide RNA (sgRNA) design, sequencing of the target site in the genotypes of interest, performing an in vitro CRISPR/Cas9 ribonucleoprotein (RNP) assay to validate the designed sgRNAs, preparing the transformation constructs, considering a protoplast editing step as further validation, and, finally, stable plant transformation and mutation detection by Sanger and/or next-generation sequencing. With these detailed guidelines, a new user should be able to quickly set up a genome editing pipeline in their crop of interest and start making progress with the different CRISPR/Cas-based editing variants for gene validation and trait improvement purposes.

Keywords: genome editing; crop improvement; high-throughput pipeline; CRISPR/Cas9

Citation: Tsakirpaloglou, N.; Septiningsih, E.M.; Thomson, M.J. Guidelines for Performing CRISPR/Cas9 Genome Editing for Gene Validation and Trait Improvement in Crops. *Plants* **2023**, *12*, 3564. https://doi.org/10.3390/plants12203564

Academic Editors: Zanmin Hu, Han Xiao, Yi Ren and Chengming Fan

Received: 14 September 2023
Revised: 10 October 2023
Accepted: 11 October 2023
Published: 13 October 2023

1. Introduction

Looming threats to global food security and a growing need for improved human nutrition justify increasing investments in next-generation technologies to accelerate progress in crop improvement. However, new paradigms are needed to efficiently translate genomic research discoveries into accelerated crop improvement programs to address future challenges. A new generation of plant breeding technologies, including genomic selection, genome editing, and, ultimately, controlled genetic recombination, will be required to address future challenges in developing more nutritious foods and climate-resilient crops [1]. Of these new technologies, genome editing offers an unprecedented opportunity to rapidly characterize and deploy key genes and alleles for crop improvement [2].

In 2012, a landmark discovery demonstrated that Clustered Regularly Interspaced Short Palindromic Repeat (CRISPR)-based bacterial defense systems can be used for precise genome editing based on a single guide RNA (sgRNA) and the nuclease Cas9 [3]. Since then, CRISPR technologies have begun to transform the fields of medicine, microbiology, and plant and animal research [4]. Massive investments in CRISPR technologies for medical research have begun to spill over into the plant sciences, and new insights, discoveries, and applications are being published weekly. Genome editing offers a promising approach to rapidly accelerate plant breeding efforts by precisely modifying genes or genome areas of interest [5]. The regulatory policies for genome-edited products vary by jurisdiction but may be straightforward in some cases [6–11].

The identification of causal genes controlling key traits in crop species often begins with quantitative trait locus (QTL) mapping and genome-wide association studies (GWASs).

Subsequently, a list of candidate genes is refined, and functional validation is performed through complementation and/or genome editing [12]. Recent advances in genome editing techniques have begun to greatly accelerate progress in the genetic dissection of key traits in plants but have not yet been thoroughly optimized for crop improvement [13]. Although the greatest impact of CRISPR/Cas-based techniques currently lies in functional gene validation, more opportunities will arise for trait improvement as the technology matures and the regulatory framework moves towards facilitating the commercialization of gene-edited products in certain jurisdictions [8–11,14]. The past 10 years have seen rapid progress in CRISPR/Cas-based genome editing in plants, and it is a good time to take stock of the lessons learned and to review the latest advances for efficient genome editing in crop plants. Here, we present guidelines and recommended best practices for a step-by-step approach for the development and optimization of a genome editing pipeline for different crop species (Figure 1). Although these guidelines are currently best suited for gene validation studies, in the future, we expect that the outputs of this pipeline could fit into the conventional breeding pipeline to further advance and improve various crops for different traits [15].

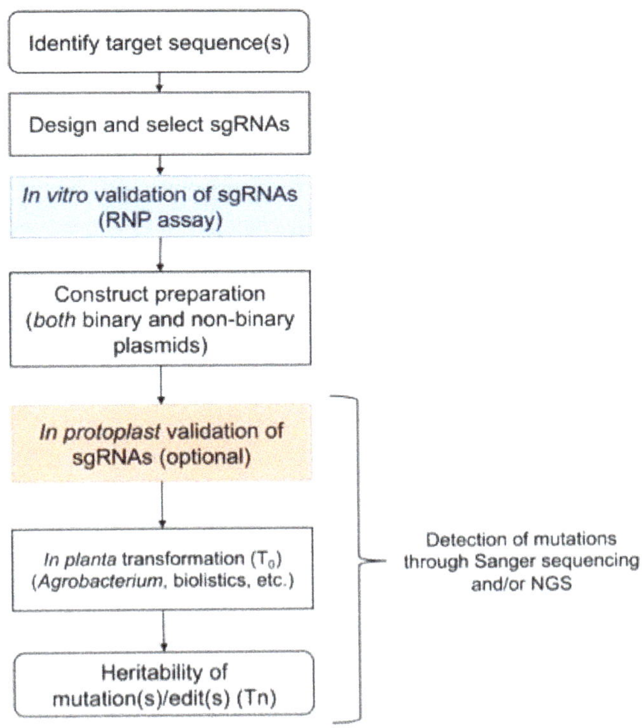

Figure 1. A stepwise approach for genome editing in various crop species. A series of different steps is followed for the selection of target sequences, the design of sgRNAs, and their validation in vitro or with protoplasts before stable transformation. Detection of the putative CRISPR/Cas9-mediated edits can be performed using Sanger sequencing or next-generation sequencing in T_0 and subsequent generations. It is important to note that as the development pipeline primarily uses transgenic modification as a step in the production of genome-edited plants, the need to remove such modifications is a universal prerequisite for the different jurisdictions [10,16,17].

2. Procedure

2.1. In Silico Sequence Analysis

After selecting the target candidate genes for editing (as previously discussed in [7]), the first steps are to obtain the target sequences, including genomic DNA, mRNA, and coding sequence (CDS), from online sources and to confirm the target gene structure through multiple sequence alignments. Although predicted gene annotations can be useful to get started, they are not always reliable, and confirming the gene structure manually helps to provide better results. The precise locations of the start codon and the gene's introns and exons are required to help ensure the successful selection of the desired target sequences and the subsequent design of sgRNAs, depending on the goals of the experiment. For example, disrupting (knocking out) gene function is most efficiently achieved by designing sgRNAs in an exon near the 5′-end of the gene. This increases the likelihood of generating frameshift mutations that lead to premature stop codons and truncated proteins. It is important to note that the inclusion of alternative splicing variants is important at the transcript and/or coding sequence level since this will impact exon/intron boundaries.

1. Identify the target sequences of the crop species of interest from their respective databases (Table 1).
2. Download the genomic sequence of interest and the respective transcript and coding sequences.

Note: Ensure that the transcript and coding sequences of all the predicted splicing variants are downloaded.

3. Map the obtained transcript and coding sequences to the genomic sequence to prepare the gene structure for the respective gene, including intron/exon locations for all transcript variants.

Note: In our case, we used MAFFT Multiple Sequence Alignment Software Version 7 [18] through Benchling (https://www.benchling.com/ (accessed on 1 September 2023)).

4. Annotate the target genomic sequence manually or confirm the gene annotation found in various sequence databases.

Table 1. A list of online annotated databases for different crop species that can serve as repositories for identifying and obtaining the required sequence information for the respective target(s).

Crop Species	Web Links (All Links Were Accessed on 1 September 2023)
Rice	• http://rice.uga.edu/ • https://rapdb.dna.affrc.go.jp/
Wheat	• http://wheat-urgi.versailles.inra.fr/Seq-Repository/Assemblies • https://plants.ensembl.org/Triticum_aestivum/Info/Index • https://wheat.pw.usda.gov/GG3/
Barley	• https://wheat.pw.usda.gov/GG3/ • https://ics.hutton.ac.uk/barleyrtd/index.html • http://plants.ensembl.org/Hordeum_vulgare/Info/Index
Maize	• https://www.maizegdb.org/ • https://www.plantgdb.org/ZmGDB/
Sorghum	• https://www.plantgdb.org/SbGDB/ • https://plants.ensembl.org/Sorghum_bicolor/Info/Index • https://www.sorghumbase.org/
Solanaceae	• https://solgenomics.net/
Cotton	• https://cottonfgd.net/ • https://www.cottongen.org/
Soybean	• https://soybase.org/
Legumes	• https://www.legumeinfo.org/

2.2. Design of Guide RNAs

The purpose of this step is to design and select sgRNAs that will create the desirable edits in the target areas of the genome with high efficiency. Currently, there are many online software tools (online software without the need to be installed) for the design of sgRNAs for different crop species (Figure 2). All these tools enable the design of sgRNAs using different parameters and algorithms. Interestingly, there is significant variation between the output of these tools for the same input sequence, leading to questions regarding which tool or algorithm provides the most accurate classification for the design and selection of sgRNAs. Hence, the best practice is to compare the output of multiple online tools to identify the sgRNAs present in all outputs, assuming that the selection of the specific sequences using multiple tools/algorithms implies optimal design and efficiency and therefore a greater chance of success. Mapping the "common" sgRNAs onto the genomic sequence often reveals that most of these sgRNAs are clustered in specific areas of the target gene. In addition to selecting sgRNAs that are common across the design tools, additional characteristics of the obtained sgRNAs, such as predicted efficiency, a lower probability for off-target modifications, and the sequence position for the creation of mutations, help to increase the chance of successful sgRNAs. Note that a previous study suggested that the sgRNA rankings using the current set of online design tools are not entirely predictive in plants, since most of the algorithms were trained using animal systems [19]. Therefore, the best practice would be to design multiple gRNAs per target, so that there is a backup if one of the sgRNA designs fails. This also allows for simple size selection when screening PCR amplicons for a double mutant event, which would lead to a deletion of the segment between the two sgRNAs.

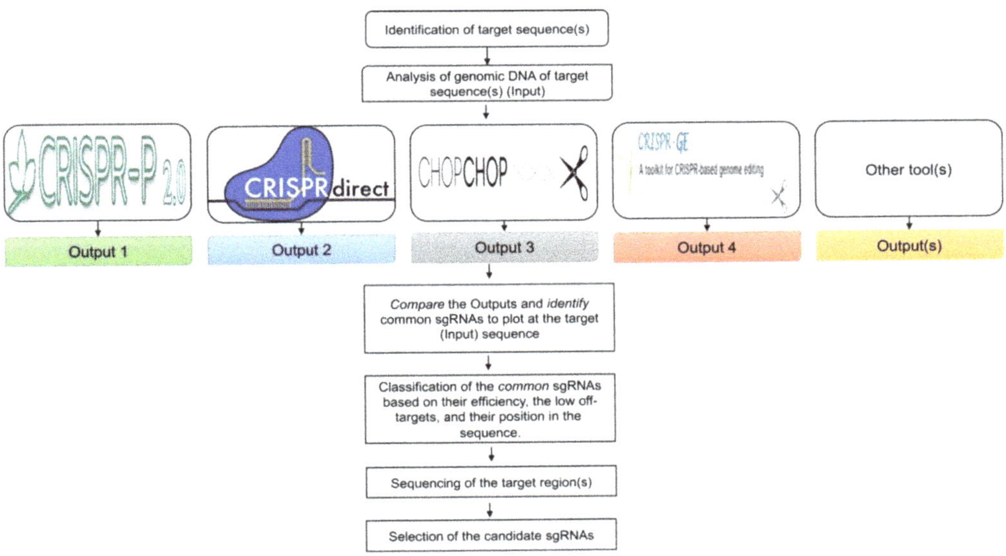

Figure 2. Design of sgRNAs. Analysis of the genomic DNA sequence of the respective target(s) can be performed through different web tools that are commonly used for the design of sgRNAs in plants. Comparison of the output of these databases results in the identification and selection of sgRNAs present in all the outputs. Classification of the common sgRNAs can be performed based on their efficiency, fewer potential off-target sites, and their position in the sequence. Before the final selection of the sgRNAs, sequencing of the target regions in the sgRNA region is recommended to ensure consistency between the in-silico and the actual cultivar sequence.

5. Use the genomic sequence of the target region as an "input sequence" for the analysis in online databases commonly used for the design of sgRNAs in plants (Table 2), such as CRISPR-P 2.0 [20], CRISPR-direct [21], CHOPCHOP [22], target design [23], CRISPR-PLANT v2 [24], and/or other available web tools [25], focusing on the design of sgRNAs.

Note: In our experiments, we observed that the output of this analysis from the respective databases differed substantially for the same target region(s).

Table 2. A list of online databases that are commonly used for the design of sgRNAs in different crop species.

Online Databases	Web Links (All Links Were Accessed on 1 September 2023)
CRISPR-P 2.0	http://crispr.hzau.edu.cn/CRISPR2/
CRISPR-direct	https://crispr.dbcls.jp
Chopchop	https://chopchop.cbu.uib.no
Target design	http://skl.scau.edu.cn/targetdesign/
CRISPR-PLANT v2	http://omap.org/crispr2/

All these databases integrate genome data from several crop species for the design of sgRNAs; however, if the desired species is not available, a close relative species could be utilized as the template.

6. Identify the "common" sgRNAs that are present in the output of most, if not all, databases.

Note: In our case, we found that most of the common sgRNAs were often clustered at certain, potentially conserved, regions of the target sequence.

7. Map the selected sgRNAs on the gene structure alignments and classify them based on the following criteria:
   i. Targeting all the transcript variants (pay attention to alternative splicing);
   ii. Predicted high efficiency;
   iii. Fewer off-targets;
   iv. For knockout mutations, create the intended mutations near the beginning of the coding sequence to generate premature stop codons and truncated peptides.

2.3. Primer Design and Sequencing of Target Regions

It is well documented that allelic variations within a genome of the same species, such as insertions/deletions (InDels) and single nucleotide polymorphisms (SNPs), occur quite frequently [26]. It is therefore imperative to ensure that the actual target sequences of the selected sgRNA regions (see Section 2.2) are identical to those obtained electronically based on the reference genome, before sgRNA selections are finalized, to ensure their proper functioning in the respective target region(s). For this purpose, sequencing of the target region is recommended, preferably using PCR with proofreading Taq polymerase and TOPO cloning.

8. Design primers that flank the target regions using NCBI Primer Blast [27] and Oligo Analyzer (IDT) (https://www.idtdna.com/calc/Analyzer/Home/Instructions (accessed on 1 September 2023)).

Note: The ideal target amplicon size should fall between 500 bp and 1200 bp to yield the best results from Sanger sequencing. However, the size of the amplicon may vary, especially in cases of polyploid crops, where it is necessary to design genome-specific primers.

The target site(s) of the sgRNA(s) should be positioned at various locations within the target region, with a minimum distance of ≥150 bp from the respective primers, avoiding the exact center of the amplicon so that multiple bands can be easily visualized when cleaving the target site with the in vitro RNP assay (see Section 2.4 below).

9. Amplify the target region using proofreading Taq polymerase.

Note: Various providers have different enzymes with proofreading activities. Some of the commonly used enzymes include Phusion High-Fidelity DNA Polymerase (2 U/μL) (Thermo Fisher Scientific, Waltham, MA, USA), Q5 High-Fidelity DNA Polymerase (NEB), and KAPA HiFi DNA Polymerase (Roche, Indianapolis, IN, USA).

10. Agarose electrophoresis and excision of the target fragments from the gel, followed by purification with a gel extraction kit, enables the collection of clean amplicons with minimal impurities.
11. Sanger sequencing of the obtained amplicon can be performed at this point; however, TOPO-cloning (for example using the Zero Blunt™ TOPO™ PCR Cloning Kit, Thermo Fisher Scientific, Waltham, MA, USA) of the fragments and submission of 3–5 clones derived from each amplicon for sequencing is highly recommended, especially when sequencing is intended for genotyping the target region. This ensures that all possible sequence variants in the amplicon are detected.

2.4. In Vitro CRISPR/Cas9 Ribonucleoprotein (RNP) Assay Validation Step 1

The in vitro cleavage of DNA amplicons using the CRISPR/Cas9 RNP complex can serve as a useful validation step to assess the functionality and relative efficiency of the CRISPR system for the desirable targets [28,29]. For this purpose, the procedure described on the NEB website (https://www.neb.com/protocols/2014/05/01/in-vitro-digestion-of-dna-with-cas9-nuclease-s-pyogenes-m0386, accessed on 1 September 2023) and the associated protocol from protocols.io (https://dx.doi.org/10.17504/protocols.io.be6fjhbn, accessed on 1 September 2023) for the in vitro digestion of DNA with Cas9 nuclease can be followed. In short, a purified/eluted PCR amplicon flanking the target site (used in Section 2.3. above) is incubated with synthetic sgRNAs and purified Cas9 nuclease. This is then run on gel electrophoresis to distinguish the cut (i.e., successfully cleaved by the sgRNA/Cas9 RNP complex) vs. uncut (i.e., not cleaved by the RNP complex) bands, which indicates the level of efficiency of Cas9 to cleave the target site with those specific sgRNA designs.

12. Perform the assay using the amplicon that was eluted from step 11 as substrate DNA.

Notes: In our experience ≥70 ng of substrate DNA is adequate for clear visualization in the agarose gel.

The sgRNAs can be synthesized and obtained from Synthego Corporation, Redwood City, CA, USA (https://www.synthego.com, accessed on 1 September 2023), whereas Cas9 nuclease can be obtained commercially from NEB (Catalog No.: M0386), Sigma-Aldrich, Burlington, MA, USA (Catalog No.: CAS9PROT), Thermo Fisher (Catalog No.: A36499), or other companies.

13. Visualization of the fragments can be performed using agarose electrophoresis (generally 1.5–2% (*w*/*v*)), at 100 V for 20–45 min, depending on the size of the expected fragments. A single uncut band indicates failure of the sgRNA with that target amplicon, whereas a larger uncut band combined with two smaller cut bands (often quite faint) indicates successful cleavage by Cas9 with those sgRNA designs.

2.5. Construct Preparation

Multiple modular approaches have been proposed by different research groups for plasmid preparation [30–33]. Most of them are based on the utilization of type IIS restriction enzymes and subsequent Golden Gate cloning for the assembly of the different modules. The flexibility that these systems offer in terms of combining different components enables the consideration of several approaches for construct design and development, depending on the target crop, as well as the overall size of the plasmid. Moreover, several RNA polymerase III (Pol-III)-driven and Pol-II-driven systems have been widely employed for efficient in planta expression of multiplexed sgRNA systems [5,14]. The utilization of the endogenous cell machinery for precise processing and the efficient production of multiple sgRNAs has also been quite successful [34,35].

14. Assemble binary and/or non-binary version(s) of the plasmids, utilizing the different components of the previously described modular systems.

Note: Consider the use of codon-optimized Cas9 nucleases for monocot or dicot species, depending on the target crop.

The utilization of tRNA-sgRNA and cys4 systems is recommended for the expression of multiple sgRNAs from a single promoter in different crop species [31]. All relevant plasmid constructs for CRISPR-Cas9 techniques in plants are publicly available, along with detailed information, from the Addgene repository (www.addgene.org, accessed on 1 September 2023).

2.6. Protoplast Transformation and Validation of sgRNAs (Validation step 2; Optional)

The development of a robust and reproducible protoplast isolation method enables the rigorous evaluation of genome editing components before stable transformation, as well as the potential for protoplast regeneration in several crop species [36–39]. For this purpose, we tried to establish a systematic approach in our lab for the isolation and transformation of protoplasts for both monocot and dicot crop species [40,41] to serve as a platform for assessing the effectiveness of our constructs to generate induced mutations at the target sites.

15. Isolate protoplasts from the respective crop species.

Note: A comprehensive list of protoplast isolation protocols for different plant species can be found in Reed and Bargmann, 2021 [38].

16. Optimize the length of the viability of the isolated protoplasts [42] at different temperatures (e.g., 4 °C, 13 °C, and 25 °C).

Note: To determine the viability of the protoplasts, we used the Plant Cell Viability Assay Kit (Sigma-Aldrich, Burlington, MA, USA, Cat. No.: PA0100-1KT).

17. Optimize the polyethylene glycol (PEG) concentration [43,44] to enable the efficient transformation of isolated protoplasts using a non-binary vector that contains a constitutive promoter to drive GFP expression.
18. Optimize the plasmid concentration required for efficient protoplast transformation by testing different concentrations of the previously utilized GFP plasmid (see step 17).
19. Assess the GFP expression levels when driven by promoters that are used to drive the expression of different biological components of the designed genome editing cassette (if needed).
20. Transform protoplasts using the non-binary version of the cassette containing Cas9 nuclease and the sgRNAs.

Perform DNA extraction and screening to detect mutations to determine the efficacy of the tested sgRNAs (see Section 2.8).

2.7. Stable Transformation of Crop Species for Efficient Genome Editing

For the efficient delivery and expression of biological and/or genome editing components, several plant transformation strategies have been considered [45–47]. Agrobacterium-mediated transformation remains the popular choice for plant transformation and the intact integration of large DNA fragments into plant chromosomes [48]. However, limitations in the range of plant host genotypes that are competent for Agrobacterium infection, combined with lengthy regeneration procedures for many crop species, create a bottleneck that often prohibits high-throughput transformation and genome editing [45]. Nonetheless, Agrobacterium-mediated transformation is still the method of choice for most genome editing experiments. Particle bombardment is also considered a popular approach for nuclear plant transformation; however, the detection of multiple integration sites and the high frequency of random rearrangements of the integrated copies, which can lead to unpredictable effects in the events generated using this approach, present further challenges for their downstream analysis and application [45]. The biolistic delivery of in vitro transcripts or ribonucleoprotein complexes of CRISPR/Cas9 has also been applied recently in several crop species to preclude genome integration effects and to potentially facilitate downstream commercial applications [49–52]. However, these alternate biolistic approaches have very low efficiencies and therefore require very large numbers of transformed explants to be successful.

Since *Agrobacterium*- and biolistic-based transformation methods require lengthy and labor-intensive in vitro tissue culture steps to regenerate whole plants from transformed explants or callus material, alternative approaches for in planta transformation have been explored. Nanoparticle-mediated gene transformation strategies for plant genetic engineering constitute an attractive strategy that could bypass the hurdles of tissue culture [53]. Recently, we were able to confirm carbon nanotube-mediated plasmid DNA delivery in rice leaves and seeds [54]. Despite the low efficiency detected in the editing of germinating rice seeds, the passive absorption and expression of plasmids attached to carbon nanotubes present an intriguing avenue for further exploration and optimization. At the same time, pollen magnetofection, a novel transformation methodology that involves the attachment of recombinant DNA to magnetic nanoparticles capable of penetrating pollen grains in a magnetic field [55,56], has not yet proven successful [57].

Virus-based RNA delivery systems have also attracted attention recently for the in planta delivery of genome editing components and the subsequent detection of mutations without the need for transgene integration and tissue culture. Multiplex genome editing has been achieved in tobacco using virus-based RNA delivery of the entire CRISPR/Cas9 cassette; however, the detected mutations were eliminated in subsequent generations of progeny lines, indicating a loss of transmissibility [58]. Alternatively, high editing efficiency was observed in wheat and maize when the viral system relied on infections of previously transgenic plants that overexpressed the Cas9 nuclease [59,60], further highlighting the potential of this approach despite limitations because of the narrow host range and other physical constraints.

Recent developments in the utilization of specific morphogenic factors for reprogramming somatic cells into embryonic ones hold the promise of improving crop transformation, particularly for monocot and recalcitrant species, or even eliminating the need for the tissue culture step for dicot species [45]. Moreover, the concurrent expression of a single growth regulator or more growth regulators, in combination with plasmids containing genome editing components, was shown to promote the development of stably transformed crops [61–67]. In addition, morphogenic factors improved the Agrobacterium-mediated transformation of maize and sorghum seedling leaf bases, enabling the direct and rapid formation of transformed plant embryos, which highlights the potential advantages of using leaf bases as the starting explant for maize and sorghum transformation and genome editing [68]. Nevertheless, despite recent advances in the utilization of morphogenic factors to accelerate crop transformation, there are still challenges in the development of efficient and reliable transformation systems that find widespread adoption, particularly for monocot species. Moreover, tissue and genotype dependency still hinder the widespread adoption of transgenic and genome editing technologies in many economically important species.

21. Stable transformation of the genome editing components using the desirable method.
22. Detection of mutations as described below (see Section 2.8).

2.8. Detection of Mutations

To readily detect the induced mutations generated using CRISPR/Cas application, several approaches could be considered. The restriction enzyme (RE) site loss method involves targeting a previously selected RE site within the sequence of interest that is proximal to the protospacer adjacent motif (PAM) [69]. It is well established that the Cas9 nuclease introduces a blunt cut three nucleotides from the PAM. Therefore, the selection of a DNA target with an overlapping restriction site near the PAM can result in disruption of the RE site during the repair process of the double-strand break via the error-prone non-homologous end joining pathway. Mutations can be detected by amplification of the target site and subsequent restriction digestion of the obtained amplicon with an appropriate RE, followed by gel electrophoresis (i.e., cut amplicons have an intact RE site and are not edited, whereas uncut bands indicate a disrupted RE site and therefore a successful edit). The sensitivity of this assay could be improved by digesting the template DNA restriction enzyme and subsequent PCR amplification [69,70]. The Surveyor assay, which identifies mismatches in the target amplicon due to edits, constitutes an alternative approach for the detection of induced mutations at the target site(s) due to genome editing [71–73]. Although it is theoretically applicable to any target sequence, the RE site loss approach is less sensitive because it requires a higher mutagenesis rate [69]. In any case, both approaches have prerequisites that make them cumbersome to fully utilize in high-throughput genome editing experiments, since the selection of sgRNAs is often designed in regions with rare or even no restriction sites, and the mutation rate can be limited. Therefore, it is important to consider alternative methods for the detection of induced mutations through genome editing, regardless of the target site and/or the mutation rate. For this reason, the utilization of Sanger and/or next-generation sequencing (NGS) approaches is preferred [74,75]. Sequencing the target site guarantees that any successful mutations will be detected.

23. Amplify the target sequence using primers flanking the induced mutation site using a proofreading polymerase.

Note: Perform agarose electrophoresis and elute the expected size amplicon from the gel using a gel extraction kit or perform a PCR clean-up if there are no other PCR byproducts.

24. Submit the eluted (or purified) amplicon(s) for Sanger sequencing.
25. Analyze the sequencing results using Synthego's free bioinformatics tool, Inference of CRISPR Edits (ICE) [76], to easily assess the possibility of putative CRISPR-induced edits at the target sites.
26. Further verification of the putative edits can be performed by TOPO-cloning of the eluted (or pure) amplicon(s) and subsequent submission of at least five clones for Sanger sequencing. This ensures that all possible variants are properly characterized, including heterozygous and biallelic loci.

Note: We recommend that submission of $\geq$ five clones per PCR amplicon suffice for the detection of potential edits using Sanger sequencing. For more accurate results, it is advisable to submit more clones [69] or to conduct multiplex target amplicon sequencing (see step 28).

27. Compare the sequencing results by mapping them to the template sequence (see step 3).
28. (Optional) Perform next-generation sequencing (NGS) of the target amplicons using the eluted (or pure) amplicon obtained previously (step 27) as the initial template (nested PCR) to determine the rate of induced mutations generated by CRISPR/Cas and potential off-target edits (Figure 3). To be cost-effective, this requires many amplicon samples to be pooled together using barcoded DNA indices to label each sample during the sequencing library preparation.

Note: A three-stage target amplicon workflow for the next-generation sequencing of haploid and polyploid crop species, based on the 16S Metagenomic Sequencing Library Preparation, can be used [77,78].

In our case, a pair-end target amplicon analysis (one million read pairs (with a Read Length of 250 × 250)) was performed using Illumina MiSeq, with either 48 or 96 barcoded samples per run.

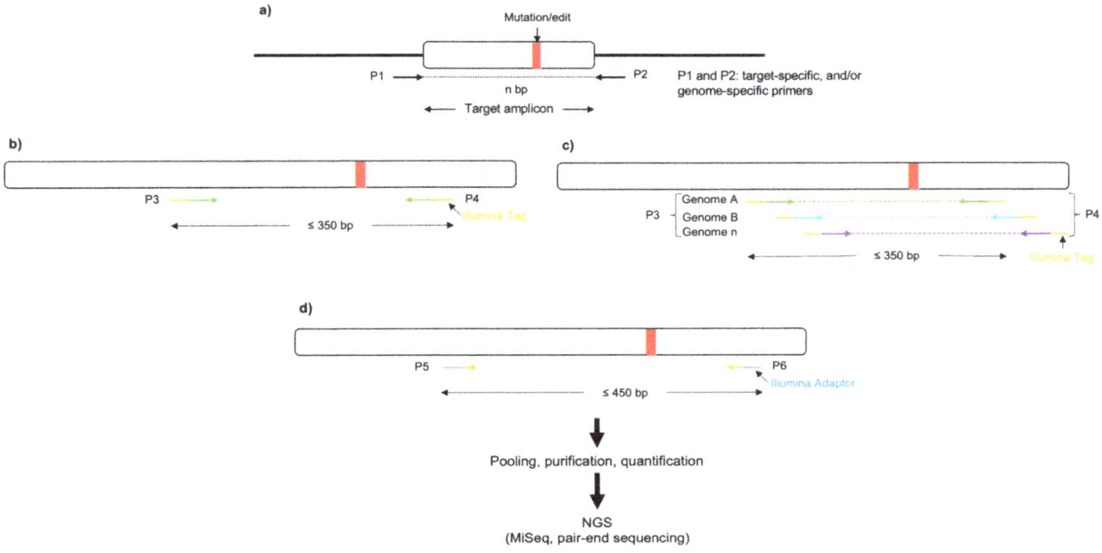

Figure 3. Schematic representation of a three-stage target amplicon workflow for the next-generation sequencing of haploid and polyploid crop species, based on the 16S Metagenomic Sequencing Library Preparation. (**a**) Forward and reverse primers (P1 and P2, respectively) flanking the region of interest are used to amplify templates from the genomic DNA (in polyploid crops P1 and P2 could serve as genome-specific primers). (**b**,**c**) Nested PCR is performed, using the eluted/purified amplicon from the previous step as a template, to incorporate Illumina overhang adapters into the primers (P3 and P4) for the respective genomes. (**d**) Additionally, nested PCR enables the addition of multiplexing indices and Illumina (San Diego, CA, USA) sequencing adapters (P5 and P6; https://support-docs.illumina.com/SHARE/AdapterSeq/illumina-adapter-sequences.pdf (accessed on 1 September 2023)). The generated libraries are then normalized, pooled, and sequenced on the MiSeq system.

All the PCR steps involved in the preparation of NGS libraries were performed using KAPA HiFi DNA Polymerase (Roche, Indianapolis, IN, USA), as has been suggested previously [79]. However, it has come to light recently that Quantabio RepliQa Hifi Toughmix produces the best outcomes for a variety of genomes and applications [80].

This approach is mainly suitable for the identification of unintended on-target modifications that have been introduced within the immediate vicinity of the target site [81]. A comprehensive assessment of off-target edits and other unintended modifications that could be present in genome-edited plants would require a different approach [82].

29. Analysis of the obtained Illumina MiSeq results using CRIS.py [83], a versatile and high-throughput analysis program for CRISPR-based genome editing.

3. Conclusions and Future Perspectives

Implementing these step-by-step guidelines should improve the rate of success for routine CRISPR/Cas9 (or other Cas nuclease) genome editing experiments in most major crops. Devoting time in the initial stages to carefully guide RNA design and validation is essential to ensure good results, especially for crops that have a lengthy in vitro tissue culture and regeneration process, where taking shortcuts early on can lead to disappointment at the end. At the same time, sgRNAs can fail for various reasons, so it is still prudent to have backup designs, such as targeting two nearby sgRNAs to knock out a target gene, so if one fails, there is still a good chance of knocking out the gene function with the single remaining guide RNA. Another bottleneck is *Agrobacterium*-mediated transformation and in vitro regeneration [84], since several crop species are difficult to transform and are often very slow to regenerate, such as cotton, which can take 18–24 months. As several research groups around the world are exploring alternative *in planta* transformation approaches, it is hoped that this bottleneck will be resolved over the next few years. Once a truly high-throughput transformation system is in place, this will enable genome editing techniques that are currently too difficult to be performed routinely, such as allele replacements using homology-directed repair. Once genotype-independent transformation and regeneration become feasible, genome editing can extend beyond the use of transformation-ready model genotypes and can be more readily applied to elite breeding materials. Applying genome editing to elite breeding materials could accelerate the development of edited plants for agricultural use, but necessitates careful assessment of the off-target modifications and unintended transgenic insertions in the resulting plants [8–10].

Author Contributions: Conceptualization, N.T., M.J.T. and E.M.S.; methodology, N.T., M.J.T. and E.M.S.; software, N.T.; investigation, N.T.; resources, M.J.T. and E.M.S.; writing—original draft preparation, N.T., M.J.T. and E.M.S.; writing—review and editing, N.T., M.J.T. and E.M.S.; visualization, N.T.; supervision, M.J.T. and E.M.S.; project administration, M.J.T. and E.M.S.; funding acquisition, M.J.T. and E.M.S. All authors have read and agreed to the published version of the manuscript.

Funding: This research was funded by the National Institute of Food and Agriculture, U.S. Department of Agriculture, grant no. 2020-67013-31811 (M.J.T. and E.M.S.), Texas A&M AgriLife Research funds (M.J.T. and E.M.S.), and by the H.M. Beachell Endowed Chair for Rice Improvement (M.J.T.).

Data Availability Statement: The data presented in this study are available within the published article.

Conflicts of Interest: The authors declare no conflict of interest.

References

1. Barabaschi, D.; Tondelli, A.; Desiderio, F.; Volante, A.; Vaccino, P.; Valè, G.; Cattivelli, L. Next Generation Breeding. *Plant Sci.* **2016**, *242*, 3–13. [CrossRef]
2. Chen, K.; Wang, Y.; Zhang, R.; Zhang, H.; Gao, C. CRISPR/Cas Genome Editing and Precision Plant Breeding in Agriculture. *Annu. Rev. Plant Biol.* **2019**, *70*, 28.1–28.31. [CrossRef] [PubMed]
3. Jinek, M.; Chylinski, K.; Fonfara, I.; Hauer, M.; Doudna, J.A.; Charpentier, E. A Programmable Dual-RNA-Guided DNA Endonuclease in Adaptive Bacterial Immunity. *Science* **2012**, *337*, 816–821. [CrossRef] [PubMed]
4. Wang, J.Y.; Doudna, J.A. CRISPR Technology: A Decade of Genome Editing Is Only the Beginning. *Science* **2023**, *379*, eadd8643. [CrossRef] [PubMed]
5. Gao, C. Genome Engineering for Crop Improvement and Future Agriculture. *Cell* **2021**, *184*, 1621–1635. [CrossRef] [PubMed]
6. Jenkins, D.; Dobert, R.; Atanassova, A.; Pavely, C. Impacts of the Regulatory Environment for Gene Editing on Delivering Beneficial Products. *In Vitro Cell. Dev. Biol. Plant* **2021**, *57*, 609–626. [CrossRef] [PubMed]
7. Buchholzer, M.; Frommer, W.B. An Increasing Number of Countries Regulate Genome Editing in Crops. *New Phytol.* **2023**, *237*, 12–15. [CrossRef] [PubMed]
8. Eckerstorfer, M.F.; Grabowski, M.; Lener, M.; Engelhard, M.; Simon, S.; Dolezel, M.; Heissenberger, A.; Lüthi, C. Biosafety of Genome Editing Applications in Plant Breeding: Considerations for a Focused Case-Specific Risk Assessment in the EU. *Biotech* **2021**, *10*, 10. [CrossRef] [PubMed]
9. Kawall, K.; Cotter, J.; Then, C. Broadening the GMO Risk Assessment in the EU for Genome Editing Technologies in Agriculture. *Env. Sci. Eur.* **2020**, *32*, 106. [CrossRef]
10. Agapito-Tenfen, S.Z.; Okoli, A.S.; Bernstein, M.J.; Wikmark, O.-G.; Myhr, A.I. Revisiting Risk Governance of GM Plants: The Need to Consider New and Emerging Gene-Editing Techniques. *Front. Plant Sci.* **2018**, *9*, 1874. [CrossRef]
11. Turnbull, C.; Lillemo, M.; Hvoslef-Eide, T.A.K. Global Regulation of Genetically Modified Crops Amid the Gene Edited Crop Boom—A Review. *Front. Plant Sci.* **2021**, *12*, 630396. [CrossRef] [PubMed]
12. Thomson, M.J.; Biswas, S.; Tsakirpaloglou, N.; Septiningsih, E.M. Functional Allele Validation by Gene Editing to Leverage the Wealth of Genetic Resources for Crop Improvement. *Int. J. Mol. Sci.* **2022**, *23*, 6565. [CrossRef] [PubMed]
13. Cardi, T.; Murovec, J.; Bakhsh, A.; Boniecka, J.; Bruegmann, T.; Bull, S.E.; Eeckhaut, T.; Fladung, M.; Galovic, V.; Linkiewicz, A.; et al. CRISPR/Cas-Mediated Plant Genome Editing: Outstanding Challenges a Decade after Implementation. *Trends Plant Sci.* **2023**, *28*, 1144–1165. [CrossRef] [PubMed]
14. Zhu, H.; Li, C.; Gao, C. Applications of CRISPR-Cas in Agriculture and Plant Biotechnology. *Nat. Rev. Mol. Cell Biol.* **2020**, *21*, 661–677. [CrossRef] [PubMed]

15. Tsakirpaloglou, N.; Trijatmiko, K.R.; Septiningsih, E.; Thomson, M.J. Gene and Base Editing Tools for the Improvement of Cereal Traits. In *Developing Sustainable and Health Promoting Cereals and Pseudocereals: Conventional and Molecular Breeding*; Rakszegi, M., Papageorgiou, M., Rocha, J.M., Eds.; Elsevier: Amsterdam, The Netherlands, 2022.
16. Lema, M. Regulatory Assessment of Off-Target Changes and Spurious DNA Insertions in Gene-Edited Organisms for Agri-Food Use. *J. Regul. Sci.* **2021**, *9*, 1–15. [CrossRef]
17. Zhao, H.; Wolt, J.D. Risk Associated with Off-Target Plant Genome Editing and Methods for Its Limitation. *Emerg. Top. Life Sci.* **2017**, *1*, 231–240. [CrossRef] [PubMed]
18. Katoh, K.; Standley, D.M. MAFFT Multiple Sequence Alignment Software Version 7: Improvements in Performance and Usability. *Mol. Biol. Evol.* **2013**, *30*, 772–780. [CrossRef] [PubMed]
19. Naim, F.; Shand, K.; Hayashi, S.; O'Brien, M.; McGree, J.; Johnson, A.A.T.; Dugdale, B.; Waterhouse, P.M. Are the Current GRNA Ranking Prediction Algorithms Useful for Genome Editing in Plants? *PLoS ONE* **2020**, *15*, e0227994. [CrossRef] [PubMed]
20. Liu, H.; Ding, Y.; Zhou, Y.; Jin, W.; Xie, K.; Chen, L.-L. CRISPR-P 2.0: An Improved CRISPR-Cas9 Tool for Genome Editing in Plants. *Mol. Plant* **2017**, *10*, 530–532. [CrossRef]
21. Naito, Y.; Hino, K.; Bono, H.; Ui-Tei, K. CRISPRdirect: Software for Designing CRISPR/Cas Guide RNA with Reduced off-Target Sites. *Bioinformatics* **2015**, *31*, 1120–1123. [CrossRef]
22. Labun, K.; Montague, T.G.; Krause, M.; Torres Cleuren, Y.N.; Tjeldnes, H.; Valen, E. CHOPCHOP v3: Expanding the CRISPR Web Toolbox beyond Genome Editing. *Nucleic Acids Res.* **2019**, *47*, W171–W174. [CrossRef] [PubMed]
23. Xie, X.; Ma, X.; Zhu, Q.; Zeng, D.; Li, G.; Liu, Y.-G. CRISPR-GE: A Convenient Software Toolkit for CRISPR-Based Genome Editing. *Mol. Plant* **2017**, *10*, 1246–1249. [CrossRef] [PubMed]
24. Minkenberg, B.; Zhang, J.; Xie, K.; Yang, Y. CRISPR-PLANT v2: An Online Resource for Highly Specific Guide RNA Spacers Based on Improved off-Target Analysis. *Plant Biotechnol. J.* **2019**, *17*, 5–8. [CrossRef] [PubMed]
25. Uniyal, A.P.; Mansotra, K.; Yadav, S.K.; Kumar, V. An Overview of Designing and Selection of SgRNAs for Precise Genome Editing by the CRISPR-Cas9 System in Plants. *3 Biotech* **2019**, *9*, 223. [CrossRef] [PubMed]
26. Mammadov, J.; Aggarwal, R.; Buyyarapu, R.; Kumpatla, S. SNP Markers and Their Impact on Plant Breeding. *Int. J. Plant Genom.* **2012**, *2012*, 728398. [CrossRef] [PubMed]
27. Ye, J.; Coulouris, G.; Zaretskaya, I.; Cutcutache, I.; Rozen, S.; Madden, T.L. Primer-BLAST: A Tool to Design Target-Specific Primers for Polymerase Chain Reaction. *BMC Bioinform.* **2012**, *13*, 134. [CrossRef] [PubMed]
28. Mehravar, M.; Shirazi, A.; Mehrazar, M.M.; Nazari, M. In Vitro Pre-Validation of Gene Editing by CRISPR/Cas9 Ribonucleoprotein. *Avicenna J. Med. Biotechnol.* **2019**, *11*, 259–263. [PubMed]
29. Liu, Y.; Tao, W.; Wen, S.; Li, Z.; Yang, A.; Deng, Z.; Sun, Y. In Vitro CRISPR/Cas9 System for Efficient Targeted DNA Editing. *mBio* **2015**, *6*, e1714–e1715. [CrossRef]
30. Hahn, F.; Korolev, A.; Sanjurjo Loures, L.; Nekrasov, V. A Modular Cloning Toolkit for Genome Editing in Plants. *BMC Plant Biol.* **2020**, *20*, 179. [CrossRef]
31. Čermák, T.; Curtin, S.J.; Gil-Humanes, J.; Čegan, R.; Kono, T.J.Y.; Konečná, E.; Belanto, J.J.; Starker, C.G.; Mathre, J.W.; Greenstein, R.L.; et al. A Multipurpose Toolkit to Enable Advanced Genome Engineering in Plants. *Plant Cell* **2017**, *29*, 1196–1217. [CrossRef]
32. Lowder, L.G.; Zhang, D.; Baltes, N.J.; Paul, J.W.; Tang, X.; Zheng, X.; Voytas, D.F.; Hsieh, T.-F.; Zhang, Y.; Qi, Y. A CRISPR/Cas9 Toolbox for Multiplexed Plant Genome Editing and Transcriptional Regulation. *Plant Physiol.* **2015**, *169*, 971–985. [CrossRef]
33. Xing, H.-L.; Dong, L.; Wang, Z.-P.; Zhang, H.-Y.; Han, C.-Y.; Liu, B.; Wang, X.-C.; Chen, Q.-J. A CRISPR/Cas9 Toolkit for Multiplex Genome Editing in Plants. *BMC Plant Biol.* **2014**, *14*, 327. [CrossRef]
34. Tsai, S.Q.; Wyvekens, N.; Khayter, C.; Foden, J.A.; Thapar, V.; Reyon, D.; Goodwin, M.J.; Aryee, M.J.; Joung, J.K. Dimeric CRISPR RNA-Guided FokI Nucleases for Highly Specific Genome Editing. *Nat. Biotechnol.* **2014**, *32*, 569–576. [CrossRef] [PubMed]
35. Xie, K.; Minkenberg, B.; Yang, Y. Boosting CRISPR/Cas9 Multiplex Editing Capability with the Endogenous TRNA-Processing System. *Proc. Natl. Acad. Sci. USA* **2015**, *112*, 3570–3575. [CrossRef] [PubMed]
36. Davey, M.R.; Anthony, P.; Power, J.B.; Lowe, K.C. Plant Protoplast Technology: Current Status. *Acta Physiol. Plant* **2005**, *27*, 117–130. [CrossRef]
37. Yue, J.-J.; Yuan, J.-L.; Wu, F.-H.; Yuan, Y.-H.; Cheng, Q.-W.; Hsu, C.-T.; Lin, C.-S. Protoplasts: From Isolation to CRISPR/Cas Genome Editing Application. *Front. Genome Ed.* **2021**, *3*, 717017. [CrossRef] [PubMed]
38. Reed, K.M.; Bargmann, B.O.R. Protoplast Regeneration and Its Use in New Plant Breeding Technologies. *Front. Genome Ed.* **2021**, *3*, 734951. [CrossRef] [PubMed]
39. Yoo, S.D.; Cho, Y.H.; Sheen, J. Arabidopsis Mesophyll Protoplasts: A Versatile Cell System for Transient Gene Expression Analysis. *Nat. Protoc.* **2007**, *2*, 1565–1572. [CrossRef] [PubMed]
40. Biswas, S.; Wahl, N.J.; Thomson, M.J.; Cason, J.M.; McCutchen, B.F.; Septiningsih, E.M. Optimization of Protoplast Isolation and Transformation for a Pilot Study of Genome Editing in Peanut by Targeting the Allergen Gene Ara h 2. *Int. J. Mol. Sci.* **2022**, *23*, 837. [CrossRef] [PubMed]
41. Biswas, S.; Bridgeland, A.; Irum, S.; Thomson, M.J.; Septiningsih, E.M. Optimization of Prime Editing in Rice, Peanut, Chickpea, and Cowpea Protoplasts by Restoration of GFP Activity. *Int. J. Mol. Sci.* **2022**, *23*, 9809. [CrossRef]
42. Larkin, P.J. Purification and Viability Determinations of Plant Protoplasts. *Planta* **1976**, *128*, 213–216. [CrossRef]
43. Krens, F.A.; Molendijk, L.; Wullems, G.J.; Schilperoort, R.A. In Vitro Transformation of Plant Protoplasts with Ti-Plasmid DNA. *Nature* **1982**, *296*, 72–74. [CrossRef]

44. Mathur, J.; Koncz, C. PEG-Mediated Protoplast Transformation with Naked DNA. In *Arabidopsis Protocols*; Martinez-Zapater, J.M., Salinas, J., Eds.; Humana Press: Totowa, NJ, USA, 1998; Volume 82, pp. 267–276.
45. Chen, Z.; Debernardi, J.M.; Dubcovsky, J.; Gallavotti, A. Recent Advances in Crop Transformation Technologies. *Nat. Plants* **2022**, *8*, 1343–1351. [CrossRef]
46. Lee, K.; Wang, K. Strategies for Genotype-Flexible Plant Transformation. *Curr. Opin. Biotechnol.* **2023**, *79*, 102848. [CrossRef]
47. Gordon-Kamm, W.; Barone, P.; Svitashev, S.; Sander, J.D.; Kumar, S.; Jones, T. *Strategies for CRISPR/Cas9-Mediated Genome Editing: From Delivery to Production of Modified Plants*; Burleigh Dodds Science Publishing: Cambridge, UK, 2021.
48. Hamilton, C.M.; Frary, A.; Lewis, C.; Tanksley, S.D. Stable Transfer of Intact High Molecular Weight DNA into Plant Chromosomes. *Proc. Natl. Acad. Sci. USA* **1996**, *93*, 9975–9979. [CrossRef]
49. Svitashev, S.; Schwartz, C.; Lenderts, B.; Young, J.K.; Mark Cigan, A. Genome Editing in Maize Directed by CRISPR-Cas9 Ribonucleoprotein Complexes. *Nat. Commun.* **2016**, *7*, 13274. [CrossRef]
50. Liang, Z.; Chen, K.; Li, T.; Zhang, Y.; Wang, Y.; Zhao, Q.; Liu, J.; Zhang, H.; Liu, C.; Ran, Y.; et al. Efficient DNA-Free Genome Editing of Bread Wheat Using CRISPR/Cas9 Ribonucleoprotein Complexes. *Nat. Commun.* **2017**, *8*, 14261. [CrossRef]
51. Liang, Z.; Chen, K.; Zhang, Y.; Liu, J.; Yin, K.; Qiu, J.-L.; Gao, C. Genome Editing of Bread Wheat Using Biolistic Delivery of CRISPR/Cas9 in Vitro Transcripts or Ribonucleoproteins. *Nat. Protoc.* **2018**, *13*, 413–430. [CrossRef]
52. Park, J.; Choe, S. DNA-Free Genome Editing with Preassembled CRISPR/Cas9 Ribonucleoproteins in Plants. *Transgenic Res.* **2019**, *28*, 61–64. [CrossRef]
53. Lv, Z.; Jiang, R.; Chen, J.; Chen, W. Nanoparticle-Mediated Gene Transformation Strategies for Plant Genetic Engineering. *Plant J.* **2020**, *104*, 880–891. [CrossRef]
54. Dunbar, T.; Tsakirpaloglou, N.; Septiningsih, E.M.; Thomson, M.J. Carbon Nanotube-Mediated Plasmid DNA Delivery in Rice Leaves and Seeds. *Int. J. Mol. Sci.* **2022**, *23*, 4081. [CrossRef]
55. Wang, Z.-P.; Zhang, Z.-B.; Zheng, D.-Y.; Zhang, T.-T.; Li, X.-L.; Zhang, C.; Yu, R.; Wei, J.-H.; Wu, Z.-Y. Efficient and Genotype Independent Maize Transformation Using Pollen Transfected by DNA-Coated Magnetic Nanoparticles. *J. Integr. Plant Biol.* **2022**, *64*, 1145–1156. [CrossRef]
56. Zhao, X.; Meng, Z.; Wang, Y.; Chen, W.; Sun, C.; Cui, B.; Cui, J.; Yu, M.; Zeng, Z.; Guo, S.; et al. Pollen Magnetofection for Genetic Modification with Magnetic Nanoparticles as Gene Carriers. *Nat. Plants* **2017**, *3*, 956–964. [CrossRef]
57. Vejlupkova, Z.; Warman, C.; Sharma, R.; Scheller, H.V.; Mortimer, J.C.; Fowler, J.E. No Evidence for Transient Transformation via Pollen Magnetofection in Several Monocot Species. *Nat. Plants* **2020**, *6*, 1323–1324. [CrossRef]
58. Ma, X.; Zhang, X.; Liu, H.; Li, Z. Highly Efficient DNA-Free Plant Genome Editing Using Virally Delivered CRISPR-Cas9. *Nat. Plants* **2020**, *6*, 773–779. [CrossRef]
59. Li, T.; Hu, J.; Sun, Y.; Li, B.; Zhang, D.; Li, W.; Liu, J.; Li, D.; Gao, C.; Zhang, Y.; et al. Highly Efficient Heritable Genome Editing in Wheat Using an RNA Virus and Bypassing Tissue Culture. *Mol. Plant* **2021**, *14*, 1787–1798. [CrossRef]
60. Hu, J.; Li, S.; Li, Z.; Li, H.; Song, W.; Zhao, H.; Lai, J.; Xia, L.; Li, D.; Zhang, Y. A Barley Stripe Mosaic Virus-Based Guide RNA Delivery System for Targeted Mutagenesis in Wheat and Maize. *Mol. Plant Pathol.* **2019**, *20*, 1463–1474. [CrossRef]
61. Feng, Q.; Xiao, L.; He, Y.; Liu, M.; Wang, J.; Tian, S.; Zhang, X.; Yuan, L. Highly Efficient, Genotype-Independent Transformation and Gene Editing in Watermelon (*Citrullus lanatus*) Using a Chimeric ClGRF4-GIF1 Gene. *J. Integr. Plant Biol.* **2021**, *63*, 2038–2042. [CrossRef]
62. Liu, Y.; Zhang, L.; Li, C.; Yang, Y.; Duan, Y.; Yang, Y.; Sun, X. Establishment of Agrobacterium-Mediated Genetic Transformation and Application of CRISPR/Cas9 Genome-Editing System to Brassica Rapa Var. Rapa. *Plant Methods* **2022**, *18*, 98. [CrossRef]
63. Che, P.; Wu, E.; Simon, M.K.; Anand, A.; Lowe, K.; Gao, H.; Sigmund, A.L.; Yang, M.; Albertsen, M.C.; Gordon-Kamm, W.; et al. Wuschel2 Enables Highly Efficient CRISPR/Cas-Targeted Genome Editing during Rapid de Novo Shoot Regeneration in Sorghum. *Commun. Biol.* **2022**, *5*, 344. [CrossRef]
64. Aesaert, S.; Impens, L.; Coussens, G.; van Lerberge, E.; Vanderhaeghen, R.; Desmet, L.; Vanhevel, Y.; Bossuyt, S.; Wambua, A.N.; van Lijsebettens, M.; et al. Optimized Transformation and Gene Editing of the B104 Public Maize Inbred by Improved Tissue Culture and Use of Morphogenic Regulators. *Front. Plant Sci.* **2022**, *13*, 883847. [CrossRef]
65. Zhang, X.; Xu, G.; Cheng, C.; Lei, L.; Sun, J.; Xu, Y.; Deng, C.; Dai, Z.; Yang, Z.; Chen, X.; et al. Establishment of an Agrobacterium-Mediated Genetic Transformation and CRISPR/Cas9-Mediated Targeted Mutagenesis in Hemp (*Cannabis sativa* L.). *Plant Biotechnol. J.* **2021**, *19*, 1979–1987. [CrossRef]
66. Pan, C.; Li, G.; Malzahn, A.A.; Cheng, Y.; Leyson, B.; Sretenovic, S.; Gurel, F.; Coleman, G.D.; Qi, Y. Boosting Plant Genome Editing with a Versatile CRISPR-Combo System. *Nat. Plants* **2022**, *8*, 513–525. [CrossRef]
67. Maher, M.F.; Nasti, R.A.; Vollbrecht, M.; Starker, C.G.; Clark, M.D.; Voytas, D.F. Plant Gene Editing through de Novo Induction of Meristems. *Nat. Biotechnol.* **2020**, *38*, 84–89. [CrossRef]
68. Wang, N.; Ryan, L.; Sardesai, N.; Wu, E.; Lenderts, B.; Lowe, K.; Che, P.; Anand, A.; Worden, A.; van Dyk, D.; et al. Leaf Transformation for Efficient Random Integration and Targeted Genome Modification in Maize and Sorghum. *Nat. Plants* **2023**, *9*, 255–270. [CrossRef]
69. Belhaj, K.; Chaparro-Garcia, A.; Kamoun, S.; Nekrasov, V. Plant Genome Editing Made Easy: Targeted Mutagenesis in Model and Crop Plants Using the CRISPR/Cas System. *Plant Methods* **2013**, *9*, 39. [CrossRef]
70. Jiang, W.; Zhou, H.; Bi, H.; Fromm, M.; Yang, B.; Weeks, D.P. Demonstration of CRISPR/Cas9/SgRNA-Mediated Targeted Gene Modification in Arabidopsis, Tobacco, Sorghum and Rice. *Nucleic Acids Res.* **2013**, *41*, e188. [CrossRef]

71. Xie, K.; Yang, Y. RNA-Guided Genome Editing in Plants Using a CRISPR-Cas System. *Mol. Plant* **2013**, *6*, 1975–1983. [CrossRef]
72. Mao, Y.; Zhang, H.; Xu, N.; Zhang, B.; Gou, F.; Zhu, J.-K. Application of the CRISPR-Cas System for Efficient Genome Engineering in Plants. *Mol. Plant* **2013**, *6*, 2008–2011. [CrossRef]
73. Voytas, D.F. Plant Genome Engineering with Sequence-Specific Nucleases. *Annu. Rev. Plant Biol.* **2013**, *64*, 327–350. [CrossRef]
74. Bock, C.; Datlinger, P.; Chardon, F.; Coelho, M.A.; Dong, M.B.; Lawson, K.A.; Lu, T.; Maroc, L.; Norman, T.M.; Song, B.; et al. High-Content CRISPR Screening. *Nat. Rev. Methods Primers* **2022**, *2*, 8. [CrossRef]
75. Hanna, R.E.; Doench, J.G. Design and Analysis of CRISPR-Cas Experiments. *Nat. Biotechnol.* **2020**, *38*, 813–823. [CrossRef]
76. Conant, D.; Hsiau, T.; Rossi, N.; Oki, J.; Maures, T.; Waite, K.; Yang, J.; Joshi, S.; Kelso, R.; Holden, K.; et al. Inference of CRISPR Edits from Sanger Trace Data. *CRISPR J.* **2022**, *5*, 123–130. [CrossRef]
77. de Muinck, E.J.; Trosvik, P.; Gilfillan, G.D.; Hov, J.R.; Sundaram, A.Y.M. A Novel Ultra High-Throughput 16S RRNA Gene Amplicon Sequencing Library Preparation Method for the Illumina HiSeq Platform. *Microbiome* **2017**, *5*, 68. [CrossRef]
78. Illumina. Illumina 16S Metagenomic Sequencing Library Preparation (Illumina Technical Note 15044223). 2013. Available online: https://support.illumina.com/documents/documentation/chemistry_documentation/16s/16s-metagenomic-library-prep-guide-15044223-b.pdf (accessed on 5 August 2022).
79. Oyola, S.O.; Otto, T.D.; Gu, Y.; Maslen, G.; Manske, M.; Campino, S.; Turner, D.J.; Macinnis, B.; Kwiatkowski, D.P.; Swerdlow, H.P.; et al. Optimizing Illumina Next-Generation Sequencing Library Preparation for Extremely AT-Biased Genomes. *BMC Genom.* **2012**, *13*, 1. [CrossRef]
80. Quail, M.A.; Corton, C.; Uphill, J.; Keane, J.; Gu, Y. Identifying the Best PCR Enzyme for Library Amplification in NGS. *bioRxiv*, 2022; *preprint*. [CrossRef]
81. Park, S.H.; Cao, M.; Bao, G. Detection and Quantification of Unintended Large On-Target Gene Modifications Due to CRISPR/Cas9 Editing. *Curr. Opin. Biomed. Eng.* **2023**, *28*, 100478. [CrossRef]
82. Chu, P.; Agapito-Tenfen, S.Z. Unintended Genomic Outcomes in Current and Next Generation GM Techniques: A Systematic Review. *Plants* **2022**, *11*, 2997. [CrossRef]
83. Connelly, J.P.; Pruett-Miller, S.M. CRIS.Py: A Versatile and High-Throughput Analysis Program for CRISPR-Based Genome Editing. *Sci. Rep.* **2019**, *9*, 4194. [CrossRef]
84. Altpeter, F.; Springer, N.M.; Bartley, L.E.; Blechl, A.E.; Brutnell, T.P.; Citovsky, V.; Conrad, L.J.; Gelvin, S.B.; Jackson, D.P.; Kausch, A.P.; et al. Advancing Crop Transformation in the Era of Genome Editing. *Plant Cell* **2016**, *28*, 1510–1520. [CrossRef]

plants

MDPI

Article

Validation of Novel Reference Genes in Different Rice Plant Tissues through Mining RNA-Seq Datasets

Xin Liu [1], Yingbo Gao [1], Xinyi Zhao [1], Xiaoxiang Zhang [2], Linli Ben [1], Zongliang Li [1], Guichun Dong [1], Juan Zhou [1], Jianye Huang [1,*] and Youli Yao [1,*]

1 Jiangsu Key Laboratory of Crop Genetics and Physiology/Co-Innovation Center for Modern Production Technology of Grain Crops, Yangzhou University, Yangzhou 225009, China; liuxinkiq@163.com (X.L.); yingbogao_yzu@163.com (Y.G.); xyzhao_o@163.com (X.Z.); m18752788661@163.com (L.B.); lzl4737@163.com (Z.L.); gcdong@yzu.edu.cn (G.D.); juanzhou@yzu.edu.cn (J.Z.)

2 Lixiahe Agricultural Research Institute of Jiangsu Province, Yangzhou 225007, China; zhngyz@126.com

* Correspondence: jyhuang@yzu.edu.cn (J.H.); yaoyl@yzu.edu.cn (Y.Y.)

Abstract: Reverse transcription quantitative real-time PCR (RT-qPCR) is arguably the most prevalent and accurate quantitative gene expression analysis. However, selection of reliable reference genes for RT-qPCR in rice (*Oryza sativa*) is still limited, especially for a specific tissue type or growth condition. In this study, we took the advantage of our RNA-seq datasets encompassing data from five rice varieties with diverse treatment conditions, identified 12 novel candidate reference genes, and conducted rigorous evaluations of their suitability across typical rice tissues. Comprehensive analysis of the leaves, shoots, and roots of two rice seedlings subjected to salt (30 mmol/L NaCl) and drought (air-dry) stresses have revealed that *OsMED7*, *OsACT1*, and *OsOS-9* were the robust reference genes for leaf samples, while *OsACT1*, *OsZOS3-23*, and *OsGDCP* were recommended for shoots and *OsMED7*, *OsOS-9*, and *OsGDCP* were the most reliable reference genes for roots. Comparison results produced by different sets of reference genes revealed that all these newly recommended reference genes displayed less variation than previous commonly used references genes under the experiment conditions. Thus, selecting appropriate reference genes from RNA-seq datasets leads to identification of reference genes suitable for respective rice tissues under drought and salt stress. The findings offer valuable insights for refining the screening of candidate reference genes under diverse conditions through the RNA-seq database. This refinement serves to improve the accuracy of gene expression in rice under similar conditions.

Keywords: reference genes; RNA-seq datasets; *Oryza sativa*; RT-qPCR; gene expression

Citation: Liu, X.; Gao, Y.; Zhao, X.; Zhang, X.; Ben, L.; Li, Z.; Dong, G.; Zhou, J.; Huang, J.; Yao, Y. Validation of Novel Reference Genes in Different Rice Plant Tissues through Mining RNA-Seq Datasets. *Plants* **2023**, *12*, 3946. https://doi.org/10.3390/plants12233946

Academic Editors: Zanmin Hu, Han Xiao, Chengming Fan and Yi Ren

Received: 6 November 2023
Revised: 20 November 2023
Accepted: 22 November 2023
Published: 23 November 2023

1. Introduction

Gene expression analysis is a critical approach to exploring functions of target genes in molecular research. Since gene expression is fundamentally regulated at transcription level, studies often need to elaborate the expression level of mRNA of a target gene. Common techniques for measuring gene expression include: Northern blot, in situ hybridization, reverse transcription polymerase chain reaction (PCR), microarray, and RNA-sequence (RNA-seq). Among them, reverse transcription quantitative real-time PCR (RT-qPCR) is much more frequently used for quantifying the mRNA levels of specific genes thanks to its specificity, sensitivity, flexibility, scalability, and high-throughput [1,2]. Quantification of RT-qPCR products are generally indicated and calculated by the relative values of cycle numbers to certain chosen stably expressed genes reference genes [3]. Unless quite stably expressed genes are chosen as reference, the expression level in a specific sample is often overestimated or underestimated, and differential expressions are prone to errors. A specific gene of research interest is likely displaying a contrasting expression level across tissues, especially when different reference genes are used. For example, when *OsUBQ* [4] and *OsACTIN* [5] are separately used as a reference gene in the same laboratory, their same target gene (LOC_Os01g12890) displays a contrasting ratio of (1;0.6;1.4;0.6) vs (1;3;10;6) in roots, stems, leaves, and young panicles, respectively. Apparently, in addition to inconsistency in their probable growth condition, choosing a different reference gene likely contributes to the inconsistency in their results. Obviously, choosing an unsuitable reference gene in gene expression assays may lead to confounding results and confusing conclusions [6]. The validity of a reference gene is critical for generating reliable and accurate RT-qPCR results [7,8]. In a vast gene expression analysis, a variety of genes are generally considered to be reference genes in rice plants, such as UBIQUITIN (*UBQ*), ACTIN (*ACT*), TUBULIN (*TUB*), GLYCERALDEHYDE-3-PHOSPHATE DEHYDROGENASE (*GAPDH*), TYPE 2 C PROTEIN PHOSPHATASES (*PP2C*), ELONGATION FACTOR (*EF*), F-BOX DOMAIN PROTEIN (*F-box*) and NAC TRANSCRIPTION FACTOR (*NAC*) genes [9–13]. However, as studies accumulate, it has been demonstrated that the transcript levels of these genes are responsive to multiple common growth factors and vary substantially in development stages [9,14,15]. *UBQ* has traditionally been regarded as a reliable reference gene across various species. However, its expression stability in sugarcane under drought conditions has been proven to be inconsistent [16]. Similarly, *ACTIN* exhibits instability at different stages of grape development [17]. As for rice, *Osβ-TUB4* (LOC_Os01g59150) and *OsGAPDH* (LOC_Os04g40950) emerge as the least stable candidate reference genes for expression in various organs and different stages of rice plants [15]; *OsUBQ10* (LOC_Os06g46770) and *OsGAPDH* (LOC_Os04g40950) exhibit the lowest stability rankings in a 7-day-old rice seedlings treated with phytohormones and stress conditions [9]. These evidently manifest the need to refine the screening of reference genes and identify genes with higher expression stability [18,19].

RNA-seq data provide us a comprehensive and accurate global expression profile of every active gene in the genome. The use of RNA-seq data has been demonstrated to be a highly reliable method for selecting reference genes in rice. By digging into RNA-seq and microarray data of developing endosperms, *OsFb15*, *OseIF-4a*, and *OsUBQ5* are proved to be stable reference genes under normal or high temperature conditions [20]. Screening from publicly available transcriptomic datasets exposed to heavy metal stress identifies *OsOBP* and *OsCK1a.3* to be the most reliable reference genes under similar conditions [12]. Nonetheless, their selection of RNA-seq datasets was limited to specific treatment scenarios. To identify reference genes with a broader applicability, we conducted a screening of RNA-seq datasets encompassing diverse varieties, organs, and environmental conditions. Before our study, other researchers had employed public transcriptome databases to identify stable reference genes for rice [15,21]. However, when we employed our own database for screening, we found that all of these reference genes exceeded the range of 0.75 to 1.25 (FPKM/FPKM average) in one or more RNA-seq datasets, with a majority of them surpassing this range in the case of salt-stressed RNA-seq data. This underscores the notion that the reference genes identified in prior studies may not be universally applicable and indirectly validates the robustness of our screening method as well as the comprehensiveness of our RNA-seq dataset. Taking the advantage of our accumulated multiple RNA-seq datasets, we screened out 12 genes with relatively stable expression as candidate reference genes (*OsACT1*, *OsF-Box1*, *OsPP2C1*, *OsPP2C2*, *OsUBQ1*, *OsUBQ2*, G-PATCH DOMAIN CONTAINING PROTEIN (*OsGDCP*), HELICASE FAMILY PROTEIN (*OsHeFP*), COMPONENT OF THE MIDDLE MODULE OF THE MEDIATOR COMPLEX (*OsMED7*), HOMOLOGUES OF YEAST YOS-9P (OsOS-9), RETROTRANSPOSON PROTEIN (*OsReTP*), and ZOS3-23-C2H2 ZINC FINGER PROTEIN (*OsZOS3-23*)). Yet, to avoid inconsistency in gene expression of the target genes due to suboptimal performance of endogenous control, it is necessary to validate these candidate genes [22]. Different statistical algorithms and methods such as geNorm ver. 3.5 [23], NormFinder ver. 0.953 [24], BestKeeper ver. 1.0 [25], the comparative ΔCt [26], and RefFinder [27] have been developed to evaluate the best suited reference genes for normalization of RT-qPCR data in a given set of samples.

Moreover, RiceXPro (Rice Expression Profile Database; http://ricexpro.dna.affrc.go.jp/; accessed on 1 October 2022) databases provide access to curated microarray analysis data [28]. RiceXPro is a comprehensive platform that contains gene expression profiles under different growth stages and experimental conditions, which are grouped into three categories: field/development, plant hormone, and cell and tissue type. In order to identify reference genes with broader applicability across different scenarios, we commenced by analyzing the stability of reference genes that have been commonly used in previous research. Subsequently, we screened 12 candidate reference genes using six RNA-seq datasets, which encompassed a diverse range of rice varieties, tissues, and treatment conditions. To further assessing their stability, we utilized both the RNA-seq dataset and the RiceXPro database. Finally, to confirm their reliability, we used the candidate reference genes to validate the expression patterns of three target genes in the leaf, shoot, and root tissues under salt and drought stress conditions: *OsbZIP71*, which is responsive to salt and drought [29]; *OskTN80b*, which is non-responsive to salt and drought [30], and a nitrate transporter *OsNRT1.1B*, which is assumingly not closely related to these stress factors [31].

2. Results

2.1. Validation of Already Reported Common Reference Genes

Based on rice research articles from a public database, we summarized eight existing reference genes that are commonly used: OsUBQ5 (LOC_Os06g44080), OsGAPDH-1 (LOC_Os04g40950), Osβ-TUB (LOC_Os01g59150), OseIF4α (LOC_Os02g12840) [9], OsACT11 (LOC_Os03g50885) [13], UBIQUITIN-CONJUGATING ENZYME 32 (OsUBC32) (LOC_Os02g42314) [10], OsGAPDH-2 (LOC_Os04g51690) [11], and CASEIN KINASE 1a.3 (OsCK1a.3) (LOC_Os01g56580) [12]. These eight commonly adopted reference genes have been used in multiple rice tissues, such as those exposed to drought stress, diseases, iron toxicity, and heavy metal stress. To check their stability, we examined their FPKM values in our RNA-seq datasets. Among them, the FPKM values of OseIF4α was low, indicating limited abundance in expression; and OsACT11, OsGAPDH-1, and Osβ-TUB had high FPKM values, indicating rich abundance, yet, with a wide variation range (Figure 1a). Among them, only OsCK1a.3, OsGAPDH-2, and OsUBC32 have relatively stable expression abundances, although there are some outliers in the FPKM values of OsCK1a.3 and OsUBC32.

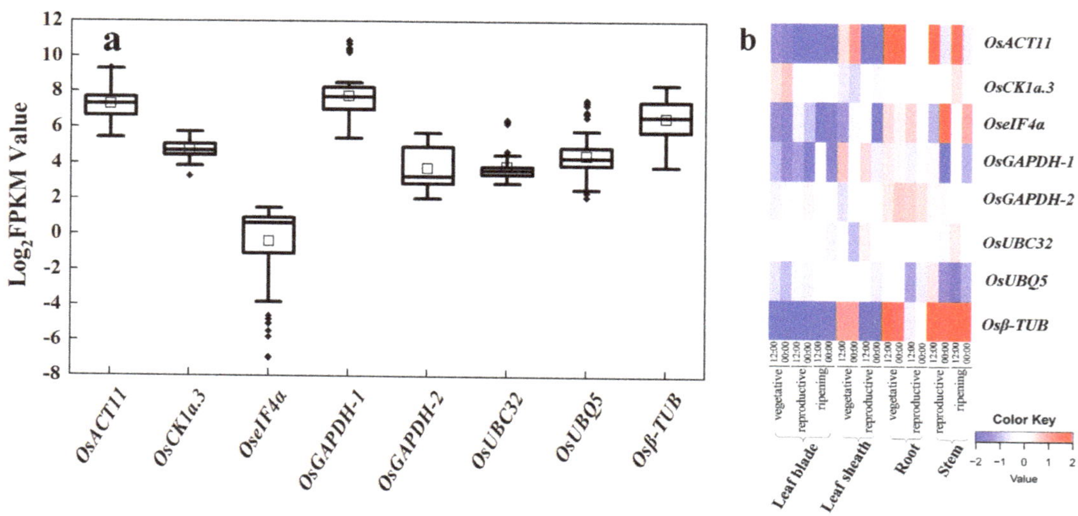

Figure 1. Expression profile of eight commonly used reference genes. (**a**) Expression displayed as Log_2(FPKM) values for eight reference genes (*OsACT11*, *OsCK1a-3*, *OseIF4α*, *OsGAPDH-1*, *OsGAPDH-2*,

OsUBC32, *OsUBQ5*, and *Osβ-TUB*) in the six RNA-seq datasets. The horizontal line in the box is the median line, the white box is the mean, and the black dot outside the box is the outlier. (**b**) Heatmap of expression of eight commonly used reference genes, created using data from RiceXPro (including different developmental stages and circadian time points). Blue indicates low transcript levels, and red indicates high transcript levels in log2 values.

RiceXPro (Rice Expression Profile Database; http://ricexpro.dna.affrc.go.jp/) provides access to curated microarray analysis data. We utilized this database to examine the stability of these eight existing reference genes in the various organs and tissues of rice in different growth stages and constructed a heatmap of these eight genes at different developmental stages (Figure 1b). Results from RiceXPro did corroborate the RNA-seq results. The transcript levels of OsACT11, OseIF4α, and Osβ-TUB showed higher fluctuation with log2 differences greater than four (transcript level at 16 times); the log2 differences of OsGAPDH-1 were greater than two (four times at transcript level). While there are outliers in the FPKM values of OsCK1a.3 and OsUBC32, their data in RiceXPro still exhibit relative stability. In conclusion, most of these existing reference genes were not as stable as they are assumed to be. Therefore, it appears necessary to find more suitable reference genes from the existing RNA-seq datasets.

2.2. Screening for Stably Expressed Reference Genes from RNA-Seq Datasets

In our quest for more universally applicable candidate reference genes, we screened our existing RNA-seq datasets, which comprise responses of two *japonica* rice varieties to salt stress, one *japonica* and one *indica* rice variety to three nitrogen application rate [32], one *japonica* and one *indica* rice variety to γ-irradiation dosage [33], and a *japonica* rice variety to carbon dioxide concentration and nitrogen rate [34] treatments. These datasets include samples from root, shoot, 7-day-young whole plants, filling grain at seven days post anthesis, and shoot tip growth points of rice, containing RNA-seq results from a total of 72 biological samples of various tissues. These RNA-seq datasets encompass a range of conditions, including non-stress normal growth conditions, abiotic stress, and extreme growth conditions. The goal was to identify candidate reference genes with broad applicability. After subjecting the FPKM values of genes in these datasets to an ANOVA test, 2776 genes were found showing non-significant change ($p > 0.05$) across different varieties and growth conditions. Genes that presented at least twice in the six databases were selected (totally 89 genes), and their expression data were collected in the Genevestigator. Pearson correlation analysis on the expression levels of these genes across various tissues identified 12 genes with the highest positive correlation coefficients ($R^2 > 0.95$, $p < 0.01$). Among them, six genes (*OsACT1*, *OsF-Box1*, *OsPP2C1*, *OsPP2C2*, *OsUBQ1* and *OsUBQ2*) belong to the commonly adopted reference gene families, but are not widely used, while the other six genes (*OsGDCP*, *OsHeFP*, *OsMED7*, *OsOS-9*, *OsReTP* and *OsZOS3-23*)) belong to new groups (Table 1). Following a comprehensive data screening process, we observed that the ratio of FPKM values to the average values of commonly used reference genes, *OsCK1a.3*, *OsGAPDH-2*, and *OsUBC32*, in the RNA-seq dataset of salt stress conditions fell outside the expected range of 0.75–1.25. Apparently, these three genes demonstrated limited stability and failed to be reference genes in a broader treatment condition.

To validate the applicability of these new candidate reference genes in a wet laboratory, we designed RT-qPCR primers by using the NCBI Primer-BLAST web-based tools (see methods section for details) and tested the amplification efficiency of these primers in actual RT-qPCR analysis (Table 1). The amplification efficiencies ranged from 91.08% to 108.05%, the linear regression correlation coefficients (R^2) for all 12 genes were ≥0.98, and all exhibited suitable ranges [35]. Therefore, these 12 genes were selected as novel candidate reference genes for further stability validation.

Table 1. Locus and primers info of 12 candidate references and three target genes.

Gene	Locus ID	Primer Sequence	Amplicon Product Tm (°C)	Amplicon Product Size (bp)	Efficiency (%)	R^2
OsACT1	LOC_Os08g28190	F: GTTCCCTTGTGTTGTTGGGC R: CCTCAAGTCAGCACAAGCCG	59	103	95.8	0.99
OsF-Box1	LOC_Os10g26990	F: TGTATATGATGGCAAGTG R: ATTGGATGATGGTAGGTA	56.3	82	94.6	0.99
OsPP2C1	LOC_Os05g51490	F: CATTGTTGTCCATCTTGTT R: CTCATCAGCACCTATCAG	59	79	100.7	0.99
OsPP2C2	LOC_Os06g43640	F: CAGGTTCATATCACTCAAG R: ATCATACTGGTGCTCATT	59	117	96.3	0.99
OsUBQ1	LOC_Os04g48770	F: ATTATTGAGAGGACTGTGAAGT R: TGATGGTTGCTGCGGATT	56.3	78	106.7	0.98
OsUBQ2	LOC_Os04g37950	F: ACTGCAACAGTGTCCTCCAG R: TCAAAAGGTTCTCCTCCGCA	56.3	111	95.1	0.98
OsGDCP	LOC_Os01g34190	F: CGAGTTCTGCTCCTCCGTAA R: CCGTCCCGTACTTGCTACAC	59	94	94.9	1.00
OsHeFP	LOC_Os03g38990	F: ACACACGATGTGGTTGCTCT R: CAAGGCGGAGAGGGCTATTT	56.3	151	103.7	0.99
OsMED7	LOC_Os04g56640	F: TTCATCTGCACCTGAACCTCC R: TCCTCTAAGCTTGGGAGCAC	61.2	96	103.4	0.99
OsOS-9	LOC_Os06g43710	F: CGGGAAAATCCGACAAGTCCA R: GCATCAGCATCATATTCACCCA	61.2	78	97.8	0.99
OsReTP	LOC_Os08g01070	F: TGGGATTAAGAGAACAAGGAG R: TGCAGTCGCAGCATCTACTT	56.3	133	105.1	0.99
OsZOS3-23	LOC_Os03g61640	F: TGAATGTGTCGTTAAGATCAGC R: ACACGCATAGACATGCCCAA	56.3	87	91.1	0.98
OsbZIP71	LOC_Os09g13570	F: TGTGTGCCCTAACTGACATCCTGA R: AAGTTATGGGTGGCTGGTTCCAT	59	133	108.0	0.99
OskTN80b	LOC_Os04g58130	F: TGATGCAGGGACAAAAACATGA R: CCAGGTTTTGCATCTTCCCGGGT	59	120	93.1	1.00
OsNRT1.1B	LOC_Os10g40600	F: GGCAGGCTCGACTACTTCTA R: AGGCGCTTCTCCTTGTAGAC	59	104	98.2	1.00

Note: The last three genes are target genes detected in the trial tests; R^2: The linear regression correlation coefficients in actual RT-qPCR analysis.

2.3. *Expression Profiling of the Novel Candidate Reference Genes*

In quantitative PCR, the Cq value is a scale indicating transcript abundance, with a smaller Cq value reflecting a higher template number (transcript level), and vice versa. The mean Cq values of the 12 candidate reference genes in trial samples varied from 24.19 to 39.99, with the least variations observed for *OsACT1* and *OsReTP* (Figure 2a).

In parallel, we compared the Log_2(FPKM) values from the RNA-seq datasets for these candidate reference genes (Figure 2b). Overall, the FPKM value indicated that their expression levels were all less variable than previous commonly used reference genes (Figure 2b vs. Figure 1a).

The expression profiles of these 12 candidates at different developmental stages from RiceXPro were also shown in a heatmap. All these novel candidates showed a relatively more stable transcript level (Figure 2c) compared with previous commonly used reference genes (Figure 1b). These indicate the novel candidate reference genes vary less in their expression levels.

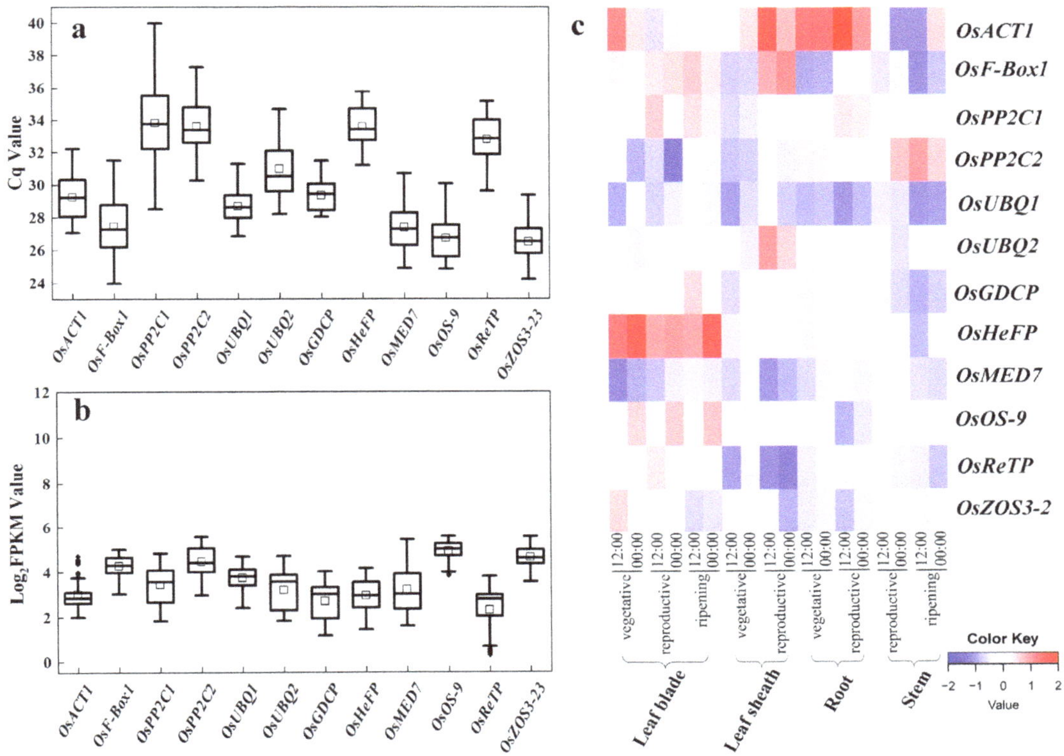

Figure 2. Expression profiling of the 12 candidate reference genes (*OsACT1*, *OsF-Box1*, *OsPP2C1*, *OsPP2C2*, *OsUBQ1*, *OsUBQ2*, *OsGDCP*, *OsHeFP*, *OsMED7*, *OsOS-9*, *OsReTP*, and *OsZOS3-23*). (**a**) The Cq values of the 12 candidate reference genes across the leaf, shoot, and root samples of rice varieties NPB and 9522 in drought and salt stress treatments. (**b**) The Log_2(FPKM) values for each candidate reference gene in the six RNA-seq datasets. The horizontal line in the box is the median line, the white box is the mean, and the black dot outside the box is the outlier. (**c**) Heatmap of the expression of candidate reference genes retrieved from RiceXPro data including various tissue/organs of different developmental stages and circadian time points. Blue indicates low transcript levels, and red indicates high transcript levels in log2 value.

2.4. Expression Stability of the Novel Candidate Reference Genes Based on Five Different Statistical Algorithms

To further compare their relative suitability as a reference gene, we used five common statistical algorithms to quantify their differences.

geNorm analysis ranks the candidate reference genes according to their average expression stability (*M* value) by using the Cq value of all samples. This evaluation is biased more on the abundance of a candidate. Genes with the lowest *M* values are considered the most stable ones. Based on the geNorm analysis, *M*-values were calculated for leaf, shoot, and root subjected to different treatments, respectively. The *M* ranking of the reference genes apparently differed in various tissues. The lowest *M*-values (most stable) were *OsMED7* and *OsACT1* in the leaf tissue, *OsZOS3-23* and *OsGDCP* in the shoots, and *OsMED7* and *OsOS-9* in the roots (Figure 3a–c). On the contrary, higher *M*-values (unstable) were observed for *OsReTP* and *OsF-Box1* in most leaf samples, *OsUBQ2* and *OsPP2C1* in shoot samples, and *OsPP2C2* and *OsPP2C1* in the roots. These suggest that it is better to choose the proper reference gene according to the tissue type of detection.

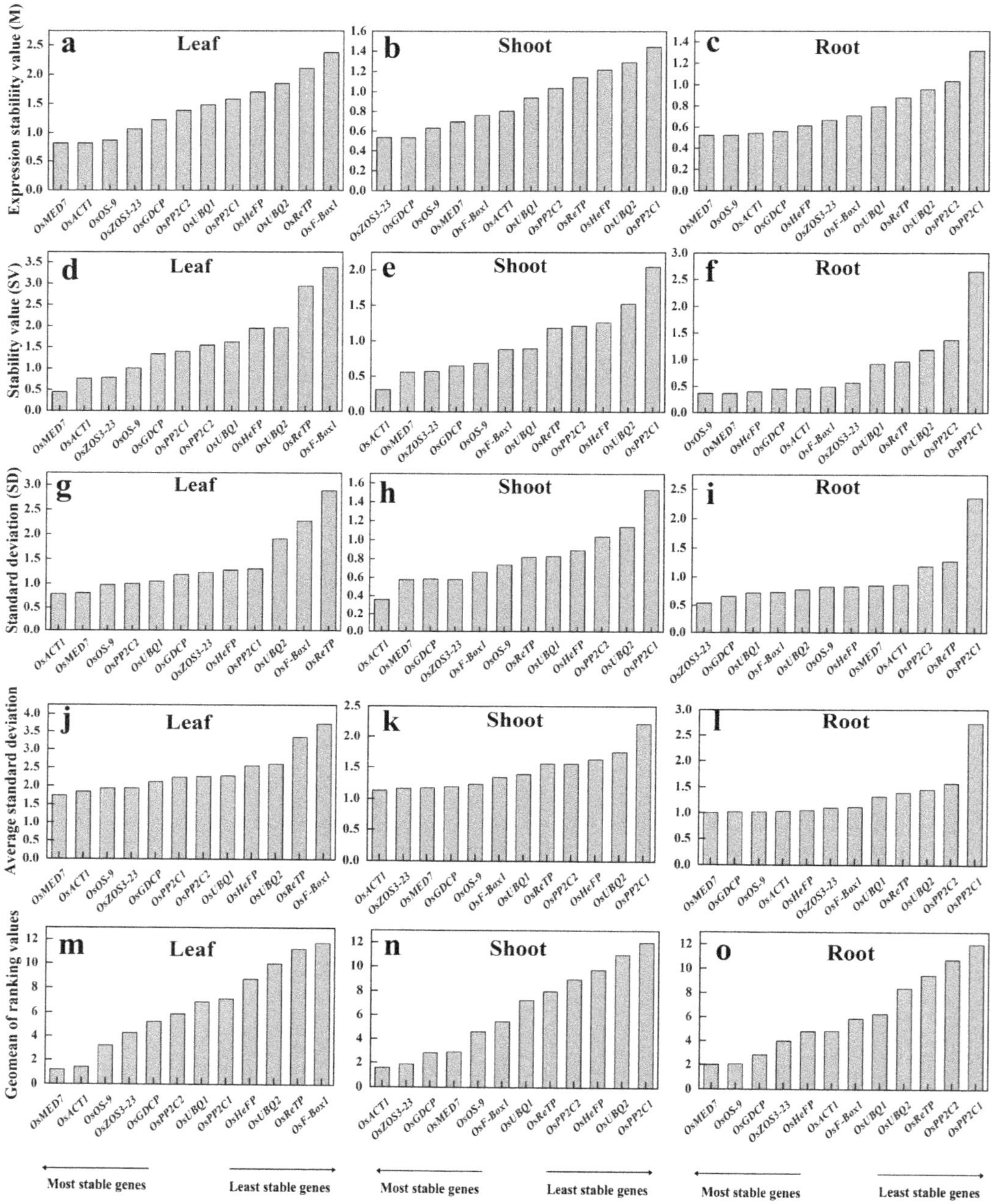

Figure 3. Expression stability of the 12 candidate reference genes (OsACT1, OsF-Box1, OsPP2C1, OsPP2C2, OsUBQ1, OsUBQ2, OsGDCP, OsHeFP, OsMED7, OsOS-9, OsReTP, and OsZOS3-23) by using algorithms geNorm (**a**–**c**), NormFinder (**d**–**f**), BestKeeper (**g**–**i**), ΔCt (**j**–**l**) and RefFinder (**m**–**o**).

NormFinder evaluates the stability of all candidate reference genes by their intra-group and inter-group variations, with lower values indicating more stability. In the trial samples of different tissues, *OsMED7* and *OsACT1* were shown to be the most stable references in both leaves and shoots, yet with a different order; whereas *OsOS-9* and *OsMED7* appeared the most stable in the roots (Figure 3d–f). The least stable ones were *OsReTP* and *OsF-Box1* in leaf samples, *OsUBQ2* and *OsPP2C1* in shoots, and *OsPP2C2* and *OsPP2C1* in root samples. The list of stable reference genes screened out in the leaf and root by NormFinder was basically consistent with those produced by using the geNorm algorithm but with a different one in shoot tissue.

BestKeeper determines the stability of a reference gene by the extent of SD, with a higher SD value referring to low stability of the housekeeping genes and vice versa. Results from the trial samples revealed that the most stable reference genes were *OsACT1* and *OsMED7* in both leaf and shoot samples and *OsZOS3-23* and *OsGDCP* in the roots (Figure 3g–i). The least stable reference genes were *OsF-Box1* and *OsReTP* in leaves, *OsUBQ2* and *OsPP2C1* in shoots, and *OsReTP* and *OsPP2C1* in the roots. The stable ones with this algorithm were the same as with the previous two methods in the leaf, with results being very different in the root or shoot.

The ΔCt method defines the stability of housekeeping genes by using comparative ΔCt based on standard deviation (SD). The higher the SD value is, the lower the stability of a reference gene, and vice versa. By using this approach, *OsMED7* and *OsACT1* were found to be the most stable genes for the leaf, *OsACT1* and *OsZOS3-23* were the most stable for the shoots, and *OsMED7* and *OsGDCP* were the most stable for the roots (Figure 3j–l). Apparently, the list of the most stable references found by using this method was consistent with the previous algorithms only for the leaf.

RefFinder calculates the comprehensive ranking of a candidate reference gene by combining multiple analysis programs (geNorm, NormFinder, BestKeeper). Results from RefFinder showed that the top three stable reference genes were *OsMED7*, *OsACT1*, and *OsOS-9* in leaf samples, *OsACT1*, *OsZOS3-23*, and *OsGDCP* in the shoot, and *OsMED7*, *OsOS-9*, and *OsGDCP* in the root (Figure 3m–o). The least stable reference genes were *OsReTP* and *OsF-Box1* for leaf samples, *OsUBQ2* and *OsPP2C1* for shoot samples, and *OsPP2C2* and *OsPP2C1* for root samples, respectively.

Different approaches generated inconsistent results regarding the stability of candidate reference genes, indicating the limitation of each algorithm, and suggesting flexible adoption of different reference genes in various tissues under specific growth conditions. Since RefFinder produces a comprehensive ranking of the candidate reference genes, and the top list from this method overlaps the most with the other four algorithms, we choose to use the three top-ranked genes to compare their performances in actual PCR quantification analysis. For comparison purposes, two candidate reference genes in the bottom rank of each tissue type were used as negative controls for gene expression normalization.

2.5. Expression Profiling of Target Genes by Using Different Novel Candidate Reference Genes

The three target genes tested in the trials are: *OsbZIP71*, which is sensitive to salt and drought [29], *OskTN80b*, which is non-responsive to salt and drought [30], and nitrate transporter *OsNRT1.1B* as a by-stander [31]. The results are shown in Figure 4.

It was evident that the leaf expression levels and patterns of *OsbZIP71*, *OskTN80b*, and *OsNRT1.1B* in response to drought and salt treatments were generally similar, either normalized singularly by *OsMED7*, *OsACT1*, or *OsOS-9* or by the three combined (*OsMED7* + *OsACT1* + *OsOS-9*). Yet, it was obviously different from those normalized by *OsReTP* or *OsF-Box1*. Similar consistency and differences were also noticeable in the shoots, and roots, respectively, by using our newly selected proper reference gene or by using the least preferred ones. This indicated that choosing one proper reference gene to calculate the expression level of a target gene did not compromise reflecting the response pattern to a treatment factor, compared with using three reference genes combined. Therefore, these results validated the reliability of the these newly selected candidate reference genes. It is interesting to note that even though *OskTN80b* is claimed to be non-responsive to drought and salt stress, its expression responded similarly with a confirmed responsive gene *OsbZIP7*, though at a much lower scale.

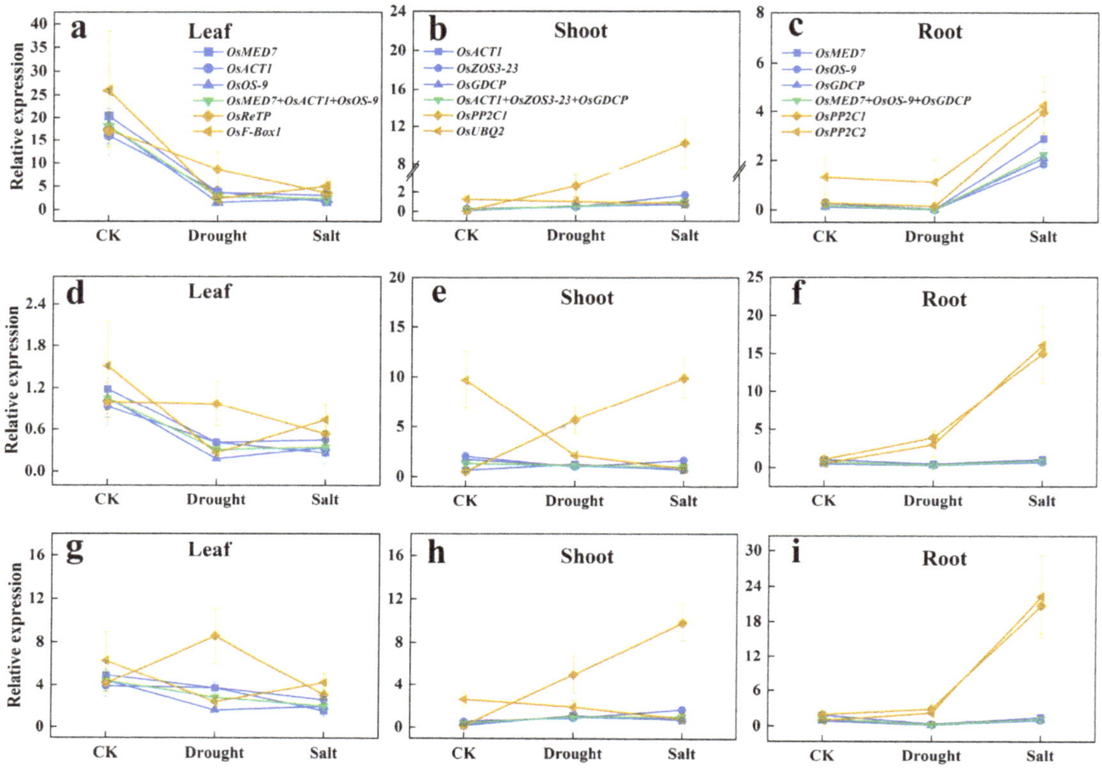

Figure 4. Relative expression levels of target genes in different sample sets normalized by the most or the least stable novel reference genes. (**a–c**) Relative expression levels of OsbZIP71 in different tissues. (**d–f**) Relative expression levels of OsKTN80b in different tissues. (**g–i**) Relative expression levels of OsNRT1.1B in different tissues. (The color legends used for the leaf, shoot, and root modules are consistent. Blue lines: normalized by each of the top three most stable candidate reference genes; Green line: normalized by combination of the top three most stable candidate reference genes; Yellow lines: normalized by each of the bottom two candidate reference genes; CK: mock treatment with water).

3. Discussion

Though RT-qPCR is a fast, reliable, and sensitive technique for quantifying the transcripts of an expressing gene of interest, it depends on the normalization procedures by using suitable reference genes to minimize the variation in sample preparation and reactions [36]. Ideally, a reference gene should be expressing constantly, with a minimal variation in transcript level, irrespective of tissue type or experimental treatment [37]. However, accumulated studies have revealed that the expression of most adopted reference genes undergoes significant changes in different organs [15,38] across various growth stages [39] and cultivation conditions, even in a specific variety of the same species [16]. Probably there are no perfectly stable expressing reference genes at all. Unless we fully adopt digital PCR [40] to obtain an absolute quantification of the templates, we must compromise by choosing proper reference genes in RT-qPCR.

On the other hand, rice is a vital crop with substantial socio-economic significance. It serves as the staple food for more than 50% of the world's population. Furthermore, it stands as a model crop for scientific research [41,42], and numerous studies have been conducted on the identification of reference genes in rice [9,10,12,13,15,20]. Nonetheless, these studies frequently exhibit limitations related to growth conditions, rice varieties, and rice organs.

Along with the sharp cost reduction in RNA-seq [43], the booming accumulation of transcriptomic data has generated accurate and comprehensive profiles of global gene expression in different species, various organs, across development phases, and growth conditions, which provide a great opportunity to revisit choosing a proper reference gene [14,36]. Here, by digging into our six existing RNA-seq datasets of 72 independent biological samples from rice plants, we screened out 12 new candidate reference genes with relatively more stable expression. Their FPKM values were greater than 70% of the expressing genes overall and showed significant correlation in expression level across a variety of rice cultivars, tissue types, growth stages, and treatment conditions. Quantification in test samples proved that they belong to the abundantly expressing group. Extraction data of these locus from RiceXPro database show that they vary less than most of those previous commonly adopted reference genes. Statistical analysis by using the established reference gene appraisal methods revealed that ranking of the newly selected reference genes was generally consistent for the top two to four candidates, but with a distinct variation among tissues. Quantification of evaluation indicated that for the leaf samples in rice, *OsMED7*, *OsACT1*, and *OsOS-9* were the most reliable reference genes, while for the shoot samples, *OsACT1*, *OsZOS3-23*, and *OsGDCP* were recommended; as for the root samples, *OsMED7*, *OsOS-9*, and *OsGDCP* were the best choices. Additionally, we harnessed the web-based edition of LinRegPCR (https://www.gear-genomics.com/rdml-tools/; accessed on 30 October 2023) for in-depth analysis of the RT-qPCR data, leading to quality assessments grounded in amplification and melting curve data [44]. Our findings confirmed that the individual PCR efficiency for all genes employed in quantification exceeded 1.7.

It has been proposed as a golden rule of RT-qPCR to include at least four reference genes to reduce the deviation by a single reference gene [6]. However, since including more reference genes brings in extra tedious lab work and a higher cost of chemicals and labor, most researchers prefer a singleton. Our results in validation quantification of target genes using novel reference genes (Figure 4) showed that with the choice of a singleton from these priority lists, the relative expression quantification of the target gene was literally consistent. A correct choice of reference gene will not only provide accurate assessment of the transcript responses of the gene of interest but also can save lab costs and labor time as well.

Nevertheless, despite the stability of the reference genes identified in this study across diverse parameters such as development, tissue, and various stresses in six RNA-seq datasets, it is essential to validate the stability of the respective choice of the selected internal reference genes [19]. As more transcriptomic data accumulates, it will be interesting to collect as many RNA-seq datasets as possible and provide a more comprehensive search for specific optimal reference genes by species/cultivar, tissue type, developmental stage, growth condition, and biotic/abiotic stress treatment. This may help researchers in swiftly tailoring their appropriate reference genes for their specific species, tissues, or experimental conditions.

4. Materials and Methods

4.1. Plant Material and Treatments

To further validate the stability of candidate reference genes, two *japonica* rice cultivars were chosen for conducting subsequent RT-qPCR experiments. Seeds were sterilized and imbibed in distilled water at a 4 °C refrigerator for three days before being subjected to germination at 32 °C in an incubator for three days under a 12/12 h of light/dark cycle. Seedlings with similar shoot length were transferred to grow hydroponically (in water) for seven days at 28 °C and the same light/dark cycle in a growth chamber at an illumination of 150 μmol $m^{-2} \cdot s^{-1}$. For salt stress treatment, the hydroponic solution was subsequently replaced with 30 mmol/L NaCl; for the drought treatment, the seedlings were subjected to air-drying; and mock treatment with a water served as a control. The leaves, shoots, and roots were collected thirty minutes after treatment, promptly frozen in liquid nitrogen, and stored at −80 °C before RNA extraction.

4.2. RNA Isolation and cDNA Synthesis

RNA was extracted using the RNA Easy Fast Plant Tissue Kit and any genomic DNA residue was removed using RNase-free DNase I per the manufacturer's instructions (Tiangen, Beijing, China). Concentration and purity of the RNA samples were measured with the spectrophotometer NanoDrop 2000 (Thermo Fisher Scientific, Wilmington, DE, USA). In accordance with the manufacturer's instructions, 1.0 μg of total RNA was reverse-transcribed using a 20 μL volume of a Fast Quant RT Kit (Tiangen, Beijing, China).

4.3. Selection of Candidate Reference Genes and Design of Primers

Candidate reference genes were screened and selected according to the bioinformatic analysis of six RNA-seq datasets from diverse organs of different rice varieties. These data were generated from varieties including Nipponbare (*japonica*), 9311 (Yangdao 6, *indica*), Wuyunjing 8 (*japonica*), Wuyunjing 30 (*japonica*), and Zhen23309 (*japonica*); tissue types including germinating shoot, germinating root, shoot tip with ~50 mm of the growth point, whole young shoot, young root, and filling grain at seven days post anthesis; and treatment factors including nitrogen rates, atmospheric CO_2 levels, γ-irradiation, and NaCl stresses. Selection of candidate reference genes was conducted by using the following criteria: (i) the gene must present at a high level in all tissues across all treatments and data sets included in the analysis [that is, at least one of its protein-coding transcripts must have an expression level higher than 70% of the whole sample's average fragments per kilobase million (FPKM)]; (ii) the ratio between the individual treatments' FPKM and the global average FPKM of the transcript in the specific data set [(individual FPKM)/(average FPKM)] must be in the range 0.75–1.25; (iii) the variability of transcript expression should be low within all tissues and across all treatments as evidenced by the p value of ANOVA >0.05. After the screening, we selected genes that were often represented in many datasets, examined their expressions in various rice tissues using Genevestigator (Immunai/NEBION AG, Zurich, Switzerland), and selected candidate reference genes based on their ranking by correlation coefficient (R^2) using correlation analysis (SPSS25, IBM, Chicago, IL, USA).

The primers for the twelve candidate reference genes and three target genes were generated using the NCBI Primer-BLAST web-based tools (https://blast.ncbi.nlm.nih.gov/Blast.cgi; accessed on 1 October 2022) with the following parameters: (1) a GC content of 40–60%; (2) an amplicon size of 78–133 base pairs (bp); and (3) a melting temperature (Tm) of 56.3–61.2 °C. Table 1 lists the primer sequences, amplicon length, and Tm of twelve candidate reference genes and three target genes. Primers were purchased from Qingke Company Ltd. (Nanjing, China).

4.4. Quantitative PCR Conditions

RT-qPCR was performed in 96-well plates using Bio-Rad CFX Real-Time PCR equipment with a SYBR green-based PCR system (Bio-Rad, Hercules, CA, USA). Each reaction had a final volume of 12 μL and contained the following components: 1 μL cDNA, 6 μL SYBR Green (Bio-Rad, Hercules, CA, USA), 0.5 μL forward primer (10 μmol/L), 0.5 μL reverse primer (10 μmol/L), and 4 μL ddH2O. The reaction was conducted at 98 °C for 1.5 min, followed by 44 cycles of denaturation at 95 °C for 5 s, annealing at 56.3–61.2 °C for 5 s, and extension at 72 °C for 10 s. After 44 cycles, the dissociation curve profiles were examined to determine the specificity of the amplicons (melting curve). Each RT-qPCR involved three biological replicates and three technical replicates.

4.5. Data Analysis

Bio-Rad CFX Manager software (Bio-Rad, Hercules, CA, USA) was used for the calculation of the Cq values of each amplification. The expression stability of each candidate reference gene was assessed using five distinct methodologies: geNorm ver. 3.5 [19],

NormFinder ver. 0.953 [20], BestKeeper ver. 1.0 [21], the comparative Ct [22], and RefFinder [23] (https://blooge.cn/RefFinder/; accessed on 1 October 2022) and the method therein. geNorm computes the *M* value to indicate the stability of a reference gene; the greater the stability of a reference gene, the smaller the *M* value is. NormFinder uses an ANOVA-based model to check the expression stability of a reference gene by evaluating inter- and intra-group variation; the gene with the lowest value has the most stable expression. BestKeeper estimates the stability of gene expression by using standard deviation (SD), a lower SD indicates greater stability of the reference gene. The Ct method compares the relative expression of all pairwise combinations of candidate reference genes to determine which gene has the most stable expression by calculating the standard deviation of the pair's relative expression; the lower the average standard deviation, the more stable the expression of a candidate reference gene. RefFinder combines multiple algorithms to identify the most appropriate reference gene.

5. Conclusions

Twelve novel candidate reference genes from multiple RNA-seq datasets were mined out, and their applicability was evaluated in different tissues of the rice plant (*Oryza sativa*). Our findings indicate that *OsMED7*, *OsACT1*, and *OsOS-9* emerged as the most robust reference genes for leaf samples, while *OsACT1*, *OsZOS3-23*, and *OsGDCP* were the most suitable for shoots and *OsMED7*, *OsOS-9*, and *OsGDCP* were the most reliable reference genes for roots. A singleton of these reference genes can be utilized in subsequent experiments in different rice tissues under drought or salt stress conditions to ensure accurately assessing expression of a target gene. Meanwhile, our research suggests that the RNA-seq analysis method holds great potential for the discovery of new, stable reference genes, which will contribute to assessing gene expression levels not only in rice but also in other species.

Author Contributions: Conceptualization, X.L. and Y.Y.; methodology, X.L. and Y.Y.; validation, X.Z. (Xinyi Zhao) and X.Z. (Xiaoxiang Zhang); formal analysis, X.L. and Y.G.; investigation, Y.G., X.Z. (Xinyi Zhao), and X.Z. (Xiaoxiang Zhang); resources, L.B., Z.L. and J.Z.; data curation, Y.G., X.Z. (Xinyi Zhao), L.B., and Z.L.; writing—original draft preparation, X.L.; writing—review and editing, Y.Y.; visualization, G.D.; supervision, J.H. and Y.Y.; project administration, J.H.; funding acquisition, X.Z. (Xiaoxiang Zhang) and Y.Y. All authors have read and agreed to the published version of the manuscript.

Funding: This research was funded by the Natural Science Foundation of Jiangsu (BK20201219), the National Natural Science Foundation of China (31571608), and Natural Science Fund for Colleges and Universities in Jiangsu Province (15KJA210003), Key R&D Program of Jiangsu Province Modern Agriculture (BE2022335), Yangzhou Modern Agriculture Project (YZ2022043), Jiangsu Province Agricultural Major Technology Collaborative Promotion Plan (2023-ZYXT-03-2).

Data Availability Statement: The data presented in this study are available upon request from the corresponding author.

Acknowledgments: Authors would like to thank Junpeng Wang for his helpful advice regarding experimental details.

Conflicts of Interest: The authors declare no conflict of interest.

References

1. Bustin, S.A.; Benes, V.; Nolan, T.; Pfaffl, M.W. Quantitative Real-Time RT-PCR—A Perspective. *J. Mol. Endocrinol.* **2005**, *34*, 597–601. [CrossRef] [PubMed]
2. Huggett, J.; Dheda, K.; Bustin, S.; Zumla, A. Real-Time RT-PCR Normalisation; Strategies and Considerations. *Genes Immun.* **2005**, *6*, 279–284. [CrossRef] [PubMed]
3. Huis, R.; Hawkins, S.; Neutelings, G. Selection of Reference Genes for Quantitative Gene Expression Normalization in Flax (*Linum usitatissimum* L.). *BMC Plant Biol.* **2010**, *10*, 71. [CrossRef] [PubMed]

4. Zheng, M.; Wang, Y.; Wang, Y.; Wang, C.; Ren, Y.; Lv, J.; Peng, C.; Wu, T.; Liu, K.; Zhao, S.; et al. *DEFORMED FLORAL ORGAN1 (DFO1)* Regulates Floral Organ Identity by Epigenetically Repressing the Expression of *OsMADS58* in Rice (*Oryza Sativa*). *New Phytol.* **2015**, *206*, 1476–1490. [CrossRef] [PubMed]
5. Liu, X.; Yang, C.Y.; Miao, R.; Zhou, C.L.; Cao, P.H.; Lan, J.; Zhu, X.J.; Mou, C.L.; Huang, Y.S.; Liu, S.J.; et al. DS1/OsEMF1 Interacts with OsARF11 to Control Rice Architecture by Regulation of Brassinosteroid Signaling. *Rice* **2018**, *11*, 46. [CrossRef] [PubMed]
6. Udvardi, M.K.; Czechowski, T.; Scheible, W.-R. Eleven Golden Rules of Quantitative RT-PCR. *Plant Cell* **2008**, *20*, 1736–1737. [CrossRef] [PubMed]
7. Nolan, T.; Hands, R.E.; Bustin, S.A. Quantification of mRNA Using Real-Time RT-PCR. *Nat. Protoc.* **2006**, *1*, 1559–1582. [CrossRef]
8. Bustin, S.A. Why the Need for qPCR Publication Guidelines?—The Case for MIQE. *Methods* **2010**, *50*, 217–226. [CrossRef]
9. Jain, M.; Nijhawan, A.; Tyagi, A.K.; Khurana, J.P. Validation of Housekeeping Genes as Internal Control for Studying Gene Expression in Rice by Quantitative Real-Time PCR. *Biochem. Biophys. Res. Commun.* **2006**, *345*, 646–651. [CrossRef]
10. Fang, P.; Lu, R.; Sun, F.; Lan, Y.; Shen, W.; Du, L.; Zhou, Y.; Zhou, T. Assessment of Reference Gene Stability in Rice Stripe Virus and Rice Black Streaked Dwarf Virus Infection Rice by Quantitative Real-Time PCR. *Virol. J.* **2015**, *12*, 175. [CrossRef]
11. Santos, F.I.D.C.D.; Marini, N.; Santos, R.S.D.; Hoffman, B.S.F.; Alves-Ferreira, M.; De Oliveira, A.C. Selection and Testing of Reference Genes for Accurate RT-qPCR in Rice Seedlings under Iron Toxicity. *PLoS ONE* **2018**, *13*, e0193418. [CrossRef] [PubMed]
12. Soni, P.; Shivhare, R.; Kaur, A.; Bansal, S.; Sonah, H.; Deshmukh, R.; Giri, J.; Lata, C.; Ram, H. Reference Gene Identification for Gene Expression Analysis in Rice under Different Metal Stress. *J. Biotechnol.* **2021**, *332*, 83–93. [CrossRef] [PubMed]
13. Wang, Y.; Wan, L.; Zhang, L.; Zhang, Z.; Zhang, H.; Quan, R.; Zhou, S.; Huang, R. An Ethylene Response Factor OsWR1 Responsive to Drought Stress Transcriptionally Activates Wax Synthesis Related Genes and Increases Wax Production in Rice. *Plant Mol. Biol.* **2012**, *78*, 275–288. [CrossRef] [PubMed]
14. Kumar, D.; Das, P.K.; Sarmah, B.K. Reference Gene Validation for Normalization of RT-qPCR Assay Associated with Germination and Survival of Rice under Hypoxic Condition. *J. Appl. Genet.* **2018**, *59*, 419–430. [CrossRef] [PubMed]
15. Zhao, Z.; Zhang, Z.; Ding, Z.; Meng, H.; Shen, R.; Tang, H.; Liu, Y.-G.; Chen, L. Public-Transcriptome-Database-Assisted Selection and Validation of Reliable Reference Genes for qRT-PCR in Rice. *Sci. China Life Sci.* **2020**, *63*, 92–101. [CrossRef] [PubMed]
16. De Andrade, L.M.; Dos Santos Brito, M.; Fávero Peixoto Junior, R.; Marchiori, P.E.R.; Nóbile, P.M.; Martins, A.P.B.; Ribeiro, R.V.; Creste, S. Reference Genes for Normalization of qPCR Assays in Sugarcane Plants under Water Deficit. *Plant Methods* **2017**, *13*, 28. [CrossRef] [PubMed]
17. Song, Y.; Hanner, R.H.; Meng, B. Genome-Wide Screening of Novel RT-qPCR Reference Genes for Study of GLRaV-3 Infection in Wine Grapes and Refinement of an RNA Isolation Protocol for Grape Berries. *Plant Methods* **2021**, *17*, 110. [CrossRef]
18. Czechowski, T.; Stitt, M.; Altmann, T.; Udvardi, M.K.; Scheible, W.-R. Genome-Wide Identification and Testing of Superior Reference Genes for Transcript Normalization in Arabidopsis. *Plant Physiol.* **2005**, *139*, 5–17. [CrossRef]
19. Gutierrez, L.; Mauriat, M.; Gunin, S.; Pelloux, J.; Lefebvre, J.-F.; Louvet, R.; Rusterucci, C.; Moritz, T.; Guerineau, F.; Bellini, C.; et al. The Lack of a Systematic Validation of Reference Genes: A Serious Pitfall Undervalued in Reverse Transcription-Polymerase Chain Reaction (RT-PCR) Analysis in Plants. *Plant Biotechnol. J.* **2008**, *6*, 609–618. [CrossRef]
20. Xu, H.; Bao, J.-D.; Dai, J.-S.; Li, Y.; Zhu, Y. Genome-Wide Identification of New Reference Genes for qRT-PCR Normalization under High Temperature Stress in Rice Endosperm. *PLoS ONE* **2015**, *10*, e0142015. [CrossRef]
21. Kudo, T.; Sasaki, Y.; Terashima, S.; Matsuda-Imai, N.; Takano, T.; Saito, M.; Kanno, M.; Ozaki, S.; Suwabe, K.; Suzuki, G.; et al. Identification of Reference Genes for Quantitative Expression Analysis Using Large-Scale RNA-Seq Data of Arabidopsis Thaliana and Model Crop Plants. *Genes Genet. Syst.* **2016**, *91*, 111–125. [CrossRef] [PubMed]
22. Xu, M.; Zhang, B.; Su, X.; Zhang, S.; Huang, M. Reference Gene Selection for Quantitative Real-Time Polymerase Chain Reaction in Populus. *Anal. Biochem.* **2011**, *408*, 337–339. [CrossRef] [PubMed]
23. Vandesompele, J.; De Preter, K.; Pattyn, F.; Poppe, B.; Van Roy, N.; De Paepe, A.; Speleman, F. Accurate Normalization of Real-Time Quantitative RT-PCR Data by Geometric Averaging of Multiple Internal Control Genes. *Genome Biol.* **2002**, *3*, research0034. [CrossRef] [PubMed]
24. Andersen, C.L.; Jensen, J.L.; Ørntoft, T.F. Normalization of Real-Time Quantitative Reverse Transcription-PCR Data: A Model-Based Variance Estimation Approach to Identify Genes Suited for Normalization, Applied to Bladder and Colon Cancer Data Sets. *Cancer Res.* **2004**, *64*, 5245–5250. [CrossRef] [PubMed]
25. Pfaffl, M.W.; Tichopad, A.; Prgomet, C.; Neuvians, T.P. Determination of Stable Housekeeping Genes, Differentially Regulated Target Genes and Sample Integrity: BestKeeper—Excel-Based Tool Using Pair-Wise Correlations. *Biotechnol. Lett.* **2004**, *26*, 509–515. [CrossRef]
26. Silver, N.; Best, S.; Jiang, J.; Thein, S.L. Selection of Housekeeping Genes for Gene Expression Studies in Human Reticulocytes Using Real-Time PCR. *BMC Mol. Biol.* **2006**, *7*, 33. [CrossRef]
27. Xie, F.; Xiao, P.; Chen, D.; Xu, L.; Zhang, B. miRDeepFinder: A miRNA Analysis Tool for Deep Sequencing of Plant Small RNAs. *Plant Mol. Biol.* **2012**, *80*, 75–84. [CrossRef]
28. Sato, Y.; Antonio, B.A.; Namiki, N.; Takehisa, H.; Minami, H.; Kamatsuki, K.; Sugimoto, K.; Shimizu, Y.; Hirochika, H.; Nagamura, Y. RiceXPro: A Platform for Monitoring Gene Expression in Japonica Rice Grown under Natural Field Conditions. *Nucleic Acids Res.* **2011**, *39*, D1141–D1148. [CrossRef]
29. Liu, C.; Mao, B.; Ou, S.; Wang, W.; Liu, L.; Wu, Y.; Chu, C.; Wang, X. OsbZIP71, a bZIP Transcription Factor, Confers Salinity and Drought Tolerance in Rice. *Plant Mol. Biol.* **2014**, *84*, 19–36. [CrossRef]

30. Wan, L.; Wang, X.; Li, S.; Hu, J.; Huang, W.; Zhu, Y. Overexpression of OsKTN80a, a Katanin P80 Ortholog, Caused the Repressed Cell Elongation and Stalled Cell Division Mediated by Microtubule Apparatus Defects in Primary Root in *Oryza Sativa*: A Katanin P80 Ortholog in Root Growth in Rice. *J. Integr. Plant Biol.* **2014**, *56*, 622–634. [CrossRef]
31. Zhang, J.; Liu, Y.-X.; Zhang, N.; Hu, B.; Jin, T.; Xu, H.; Qin, Y.; Yan, P.; Zhang, X.; Guo, X.; et al. NRT1.1B Is Associated with Root Microbiota Composition and Nitrogen Use in Field-Grown Rice. *Nat. Biotechnol.* **2019**, *37*, 676–684. [CrossRef] [PubMed]
32. Zhang, X.; Zhou, J.; Huang, N.; Mo, L.; Lv, M.; Gao, Y.; Chen, C.; Yin, S.; Ju, J.; Dong, G.; et al. Transcriptomic and Co-Expression Network Profiling of Shoot Apical Meristem Reveal Contrasting Response to Nitrogen Rate between Indica and Japonica Rice Subspecies. *Int. J. Mol. Sci.* **2019**, *20*, 5922. [CrossRef] [PubMed]
33. Zhang, X.; Huang, N.; Mo, L.; Lv, M.; Gao, Y.; Wang, J.; Liu, C.; Yin, S.; Zhou, J.; Xiao, N.; et al. Global Transcriptome and Co-Expression Network Analysis Reveal Contrasting Response of Japonica and Indica Rice Cultivar to γ Radiation. *Int. J. Mol. Sci.* **2019**, *20*, 4358. [CrossRef] [PubMed]
34. Zhou, J.; Gao, Y.; Wang, J.; Liu, C.; Wang, Z.; Lv, M.; Zhang, X.; Zhou, Y.; Dong, G.; Wang, Y.; et al. Elevated Atmospheric CO_2 Concentration Triggers Redistribution of Nitrogen to Promote Tillering in Rice. *Plant-Environ. Interact.* **2021**, *2*, 125–136. [CrossRef] [PubMed]
35. Ruijter, J.M.; Ramakers, C.; Hoogaars, W.M.H.; Karlen, Y.; Bakker, O.; Van Den Hoff, M.J.B.; Moorman, A.F.M. Amplification Efficiency: Linking Baseline and Bias in the Analysis of Quantitative PCR Data. *Nucleic Acids Res.* **2009**, *37*, e45. [CrossRef] [PubMed]
36. Nicot, N.; Hausman, J.-F.; Hoffmann, L.; Evers, D. Housekeeping Gene Selection for Real-Time RT-PCR Normalization in Potato during Biotic and Abiotic Stress. *J. Exp. Bot.* **2005**, *56*, 2907–2914. [CrossRef] [PubMed]
37. Bustin, S. Quantification of mRNA Using Real-Time Reverse Transcription PCR (RT-PCR): Trends and Problems. *J. Mol. Endocrinol.* **2002**, *29*, 23–39. [CrossRef]
38. Yang, Z.; Wang, K.; Aziz, U.; Zhao, C.; Zhang, M. Evaluation of Duplicated Reference Genes for Quantitative Real-Time PCR Analysis in Genome Unknown Hexaploid Oat (*Avena sativa* L.). *Plant Methods* **2020**, *16*, 138. [CrossRef]
39. Nguyen, D.Q.; Eamens, A.L.; Grof, C.P.L. Reference Gene Identification for Reliable Normalisation of Quantitative RT-PCR Data in Setaria Viridis. *Plant Methods* **2018**, *14*, 24. [CrossRef]
40. Quan, P.-L.; Sauzade, M.; Brouzes, E. dPCR: A Technology Review. *Sensors* **2018**, *18*, 1271. [CrossRef]
41. Yu, J.; Hu, S.; Wang, J.; Wong, G.K.-S.; Li, S.; Liu, B.; Deng, Y.; Dai, L.; Zhou, Y.; Zhang, X.; et al. A Draft Sequence of the Rice Genome (*Oryza sativa* L. Ssp. Indica). *Science* **2002**, *296*, 79–92. [CrossRef] [PubMed]
42. Ouyang, S.; Zhu, W.; Hamilton, J.; Lin, H.; Campbell, M.; Childs, K.; Thibaud-Nissen, F.; Malek, R.L.; Lee, Y.; Zheng, L.; et al. The TIGR Rice Genome Annotation Resource: Improvements and New Features. *Nucleic Acids Res.* **2007**, *35*, D883–D887. [CrossRef] [PubMed]
43. Wang, Z.; Gerstein, M.; Snyder, M. RNA-Seq: A Revolutionary Tool for Transcriptomics. *Nat. Rev. Genet.* **2009**, *10*, 57–63. [CrossRef] [PubMed]
44. Untergasser, A.; Ruijter, J.M.; Benes, V.; van den Hoff, M.J.B. Web-Based LinRegPCR: Application for the Visualization and Analysis of (RT)-qPCR Amplification and Melting Data. *BMC Bioinform.* **2021**, *22*, 398. [CrossRef]

plants

MDPI

Technical Note

Construction of an Efficient Genetic Transformation System for Watercress (*Nasturtium officinale* W. T. Aiton)

Jiajun Ran †, Qiang Ding †, Yunlou Shen, Zhanyuan Gao, Guangpeng Wang, Yue Gao, Xiaoqing Ma and Xilin Hou *

State Key Laboratory of Crop Genetics & Germplasm Enhancement, Key Laboratory of Biology and Genetic Improvement of Horticultural Crops (East China), Ministry of Agriculture and Rural Affairs of China, Engineering Research Center of Germplasm Enhancement and Utilization of Horticultural Crops, Ministry of Education of China, Nanjing Agricultural University, Nanjing 210095, China; 2020104055@stu.njau.edu.cn (J.R.); 2019204024@njau.edu.cn (Q.D.); 2020204027@njau.edu.cn (Y.S.); 2021204025@stu.njau.edu.cn (Z.G.); 2022204055@stu.njau.edu.cn (G.W.); 2021804158@njau.edu.cn (Y.G.); 2018204021@njau.cn (X.M.)

* Correspondence: hxl@njau.edu.cn; Tel./Fax: +86-025-8439-5917

† These authors contributed equally to this work.

Abstract: Based on the established efficient regeneration system for watercress in our laboratory, we optimized the processes of pretreatment, co-culture, and differentiation culture. Through GFP fluorescence and PCR identification, we successfully obtained transgenic watercress with the *DR5* gene, which allowed us to investigate the distribution details of auxin in the growth process of watercress. Our findings provide an effective method for gene function research and lay the foundation for innovative utilization of germplasm resources of watercress.

Keywords: *DR5* gene; transgenic watercress; auxin; efficient genetic transformation system

Citation: Ran, J.; Ding, Q.; Shen, Y.; Gao, Z.; Wang, G.; Gao, Y.; Ma, X.; Hou, X. Construction of an Efficient Genetic Transformation System for Watercress (*Nasturtium officinale* W. T. Aiton). *Plants* **2023**, *12*, 4149. https://doi.org/10.3390/plants12244149

Academic Editor: Iyyakkannu Sivanesan

Received: 5 November 2023
Revised: 7 December 2023
Accepted: 7 December 2023
Published: 13 December 2023

1. Introduction

Watercress (*Nasturtium officinale* W. T. Aiton.) is a widely cultivated aquatic vegetable that thrives in freshwater streams and mountain springs. Its natural growth is characterized by a robust root system and the ability to rapidly colonize available space through creeping vines and lateral branching. Watercress is rich in vitamins, minerals, flavonoids, and glucosinolates and has high nutritional value. Compounds extracted from this plant can be used for diuresis, hypoglycemia, neuroprotection, and anti-neuritis [1,2]. In addition, it is found that watercress has a certain ability to absorb heavy metal cadmium, which can alleviate cadmium pollution in the environment [3]. However, due to its aggressive growth, watercress has been classified as an invasive plant by certain countries [4]. Despite its innate vigor, watercress faces challenges in achieving efficient breeding and genetic manipulation. The absence of an effective regeneration system hinders the application of modern biotechnological approaches such as genetic transformation, asexual variation, and cell engineering in watercress breeding. Consequently, the identification and validation of desirable genes in watercress becomes a daunting task. Establishing a reliable and efficient regeneration system is therefore of paramount importance for watercress to achieve germplasm resource innovation. Moreover, it serves as a crucial pathway to expand market opportunities and achieve large-scale production of this valuable aquatic vegetable.

In this study, we aimed to investigate the root hair development mechanism in watercress and explore whether it is similar to that in Arabidopsis and other plants. We also aimed to develop watercress varieties with well-developed root systems that can adapt to land cultivation. Root hairs are crucial for water purification and nutrient absorption in watercress [5–8]. They develop from the differentiation of the root epidermis into H cells and are responsible for absorbing soil water and mineral elements and interacting with rhizosphere microorganisms [9]. Under unfavorable conditions such as drought or phosphorus deficiency, plants increase the production of root hairs to enhance water and phosphorus absorption and alleviate these conditions [10]. Auxin is known to be a direct influencing factor in plant root development, including root hair formation [11–13]. It regulates the development of the endodermis, root elongation, lateral root formation, and root hair formation. However, the mechanism of root hair formation in watercress and its similarity to other plants, such as Arabidopsis, is unknown. Watercress is commonly cultivated using hydroponics, but land cultivation is also possible. Expanding land cultivation of watercress is important for increasing the production scale [14]. Root activity is very important to plant growth and development. As an important part of the root system, root hair can not only increase the absorption area of the root system but also enhance the ability of the root system to absorb water and nutrients. They also interact with rhizosphere microbial communities to regulate the root environment [6]. The root is an important organ for watercress to adapt to dry land environments, purify water bodies, and increase yield. Watercress is mainly cultivated in water, which greatly limits the production and promotion of watercress. Watercress varieties with more developed roots can be obtained with genetic transformation, which is of great significance for watercress cultivation on land and is beneficial for watercress to gradually get rid of its dependence on hydroponics.

In conclusion, this study aimed to investigate the root hair development mechanism in watercress and develop watercress varieties suitable for land cultivation. Understanding the root hair formation process in watercress and its similarity to other plants will contribute to the development of improved varieties and increase production scale.

2. Materials and Methods

2.1. Materials

Watercress (serial number: DBC-001) came from the seed bank of the Chinese cabbage system biology laboratory of Nanjing Agricultural University, Nanjing, China.

Agrobacterium tumefaciens and related plasmids were provided by Nanjing Agricultural University.

2.2. Chemical Reagents

6-BA: 6-Benzylaminopurine (from Nanjing Zhengyang Biotechnology Co., Ltd., Nanjing, China); TDZ: Thidiazuron (from Nanjing Zhengyang Biotechnology Co., Ltd.); 2,4-D: 2,4-Dichlorophenoxyacetic acid (from Nanjing Zhengyang Biotechnology Co., Ltd.); NAA: Naphthaleneacetic acid (from Nanjing Zhengyang Biotechnology Co., Ltd.); KT: 6-Furfurylaminopurine (from Nanjing Zhengyang Biotechnology Co., Ltd.); ZT: Zeatin (from Nanjing Zhengyang Biotechnology Co., Ltd.); Cef: Cefalexin (from Nanjing Zhengyang Biotechnology Co., Ltd.); TMT: Timentin (from Nanjing Zhengyang Biotechnology Co., Ltd.); Kan: kanamycin (from Nanjing Zhengyang Biotechnology Co., Ltd.); AS: 3′,5′-Dimethoxy-4′-hydroxyacetophenone (from Nanjing Zhengyang Biotechnology Co., Ltd.).

2.3. Construction of *DR5::EGFP* Vector

To better understand the distribution of auxin in the growing points of watercress, we used the *DR5::EGFP* plasmid that had been constructed previously (Figure 1). The *DR5* promoter was cloned and inserted in front of the green fluorescent protein (GFP), and the expression of the GFP gene was induced by the *DR5* promoter.

Figure 1. *DR5::EGFP* vector. *DR5:* an artificial DNA sequence, auxin response element. The *DR5* promoter contains many binding sites of ARF transcription factors which can respond to auxin signals to induce the expression of neighboring genes. The activity of *DR5* depends not only on the distribution of endogenous auxin but also on the signal of auxin. The plant response to auxin can be visualized by expressing appropriate reporter genes under the control of *DR5*.

2.4. PCR Positive Detection

PCR detection was performed using primers designed for EGFP with SnapGene 2.8.3. See Table S1 for details of the primers and procedures.

2.5. Observation of GFP Fluorescence

The regenerated watercress was placed under a stereomicroscope equipped with a GFP filter. The γ value was adjusted to 2, and the exposure time was set to 6 s. The samples were observed and photographed under a bright field and GFP fluorescence (see Figure S1 for details).

2.6. Efficient Regeneration of Watercress

The efficient regeneration system and hormone ratios for watercress are described in Table S2.

2.7. Mannitol Concentration and Treatment of Negative Pressure Time

The experimental results of the mannitol concentration and negative pressure time required for the pretreatment of the watercress genetic transformation system are detailed in Table S5.

2.8. Transformation Steps

(1) Select healthy watercress with consistent growth, remove the leaves, and cut the stem into sections of about 0.5 cm.
(2) Place the stem in the Murashige and Skoog (MS) medium (pH 5.2, supplemented with 0.6 M mannitol + 4 mg/L 6-Benzylaminopurine (6-BA)) for 24 h of pretreatment.
(3) Prepare the bacterial solution in lysogeny broth (LB) medium with antibiotics and activate it 24 h in advance. Two hours before infection, resuspend the bacteria in MS liquid medium (pH 5.2, containing 50 mg/L 3′,5′-Dimethoxy-4′-hydroxyacetophenone (AS), and remove phosphates). Then, activate the bacterial solution at 24 °C and adjust the optical density (OD) to 0.2~0.3 during infection.
(4) Wash the pretreated stem with sterile water, place it in a syringe, add the bacterial solution, and seal it for 3 min of vacuum infiltration.
(5) Inoculate the stem into the MS medium (pH 5.2, containing 4 mg/L 6-BA + 1.5 mg/L Thidiazuron (TDZ) + 1.5 mg/L 2,4-Dichlorophenoxyacetic acid (2,4-D) + 50 mg/L inositol + 10 mg/L AS), and cultivate in the dark at 24 °C for 3 days.
(6) Wash the stem with sterile water (containing 200 mg/L cefotaxime sodium (Cef) and 100 mg/L Timentin (TMT)), dry it, and place it in the MS medium (pH 5.2, containing 4 mg/L 6-BA + 1.5 mg/L TDZ + 1.5 mg/L 2,4-D + 50 mg/L inositol + 200 mg/L Cef + 100 mg/L TMT). Induce callus formation and observe GFP fluorescence by cultivating under light for 16 h a day and at 24 °C for 15 days.
(7) Place the callus on MS medium (pH 5.2, supplemented with 3 mg/L 6-BA + 3 mg/L TDZ + 200 mg/L Kanamycin (Kan)), cultivate under light for 16 h a day at 24 °C for 15 days, and induce the differentiation of large amounts of resistant buds.
(8) Inoculate the resistant buds into the MS medium (pH 5.7, containing 200 mg/L Kan), cultivate under light for 16 h a day at 24 °C for 7 days, induce root formation, and observe. Before transplanting the seedlings, the cap of the tissue culture bottle was half-opened and half-closed 3 days in advance, and the regenerated plants of watercress were domesticated in the light culture box, and then, the seedlings were removed from the tissue culture bottle.
(9) The regenerated plants were adapted to tissue culture in vitro (it was cultured in water with Hogan's solution and cultured at 24 °C for 16 h), and PCR detection was performed.

2.9. Statistical Analysis

All of the data are presented as the means and standard deviation (SD) of at least three independent replicates for each experiment. Student's two-tailed t-test was conducted using GraphPad Prism version 8.0 (GraphPad Software, San Diego, CA, USA).

3. Results

3.1. Effects of Common Hormones on Stem Segments of Watercress

6-BA is one of the commonly used cytokinins which can induce callus formation of watercress (Figure 2A). When the concentration of 6-BA was 0.5 mg/L, the stem segment and callus were green, and the callus was smaller; when the concentration of 6-BA increased gradually, the stem segment gradually ruptured and the callus became larger; when 6-BA $\geq$ 5 mg/L, the stem segment browned, and the callus volume became smaller. Therefore, the suitable concentration range of 6-BA is 2~5 mg/L. When the concentration of TDZ is 0.5 mg/L, the stem segment bends, the callus is small, and the color is white; when the concentration of TDZ is 1 mg/L, the stem segment browns, and the callus is still small; when the TDZ concentration is 2 mg/L, the stem segment breaks, browns, and the callus is small and green; when the TDZ concentration $\geq$ 5 mg/L, the stem segment color turns white and does not form calli (Figure 2B). Therefore, the suitable concentration range of TDZ is 1~2 mg/L. KT is one of the commonly used cytokinins. Figure 2C shows that KT has a toxic effect on the stem segment of watercress, resulting in serious browning, which is not conducive to callus formation. The effect of ZT on watercress is similar to that of KT (Figure 2D) but not suitable for tissue culture of watercress. 2,4-D is one of the commonly used cell auxins which can induce the callus of watercress (Figure 2E). With the increase in the concentration of 2,4-D, the stem segment browns and the callus becomes larger; when 2,4-D $\geq$ 5 mg/L, the stem segment browns and does not form calli. Therefore, the suitable concentration range of 2,4-D is 1~2 mg/L. NAA is one of the plant auxins which can induce the rooting of watercress (Figure 2F). When the concentration of NAA was 0.5~2 mg/L, the stem segment was curved and yellowish green, and a large number of roots sprouted; when the concentration of NAA was greater than or equal to 5 mg/L, the stem segment was yellow, sprouting a little or not sprouting roots. Therefore, the NAA of 0.5~1 mg/L is suitable for tissue culture of watercress.

3.2. Establishment and Optimization of Regeneration System

When the stem segments of watercress were inoculated in the MS medium for 15 days and when the concentrations of 6-BA and TDZ were constant, the callus rate decreased gradually with the increase in 2,4-D concentration. The callus rate increased with the increase in TDZ when the concentrations of 6-BA and 2,4-D were constant. The callus rate increased with the increase in 6-BA concentration when the concentrations of TDZ and 2,4-D were constant. Finally, we found that 6-BA at the concentration of 4 mg/L, TDZ of 1.5 mg/L, and 2,4-D of 1.5 mg/L had the highest efficiency of inducing calli from stem segments (Table 1).

Table 1. The proportion of hormones needed to induce calli in stems of watercress.

Number	6-BA mg/L	TDZ mg/L	2,4-D mg/L	Number of Inoculated Stem Segments	Callus Rate (%)
CK	-	-	-	300	0.00 ± 0.00 [a]
1-1	2	1	1.5	300	80.32 ± 1.53 [b]
1-2	2	1	2	300	73.67 ± 2.08 [c]
1-3	2	1	2.5	300	64.67 ± 1.53 [d]
2-1	2	1.5	1.5	300	84.33 ± 3.05 [b]
2-2	2	1.5	2	300	79.67 ± 1.53 [c]
2-3	2	1.5	2.5	300	73.68 ± 2.08 [d]

Table 1. *Cont.*

Number	6-BA mg/L	TDZ mg/L	2,4-D mg/L	Number of Inoculated Stem Segments	Callus Rate (%)
3-1	3	1	1.5	300	92.67 ± 1.46 [b]
3-2	3	1	2	300	84.67 ± 1.45 [c]
3-3	3	1	2.5	300	76.33 ± 2.52 [d]
4-1	3	1.5	1.5	300	94.34 ± 1.53 [b]
4-2	3	1.5	2	300	89.00 ± 2.65 [c]
4-3	3	1.5	2.5	300	85.67 ± 3.07 [d]
5-1	4	1	1.5	300	91.67 ± 0.58 [b]
5-2	4	1	2	300	86.01 ± 1.00 [c]
5-3	4	1	2.5	300	80.33 ± 1.53 [d]
6-1	4	1.5	1.5	300	98.00 ± 2.00 [b]
6-2	4	1.5	2	300	95.01 ± 1.00 [c]
6-3	4	1.5	2.5	300	92.67 ± 2.08 [d]

Note: "-" means not to add. Values indicate mean ± SD. Letters indicate significant difference analysis ($p < 0.05$, Student's t-test).

When the callus was cultured on MS medium for 10 days and when the concentration of 6-BA was 2~3 mg/L, the number of adventitious buds increased with the increase in TDZ concentration, and when the concentration of 6-BA was 5 mg/L, the number of adventitious buds decreased with the increase in TDZ concentration. We finally decided to use the concentration of 3 mg/L 6-BA and TDZ to induce watercress calli to differentiate adventitious buds, and this combination had the highest efficiency in inducing calli to produce adventitious buds (Table 2).

Table 2. Hormone ratios for inducing adventitious buds from calli of watercress.

Number	6-BA mg/L	TDZ mg/L	2,4-D mg/L	NAA mg/L	Number of Inoculated Calli	Number of Buds
CK	-	-	-	-	300	0 [a]
1	2	1	-	-	300	1422 [b]
2	2	2	-	-	300	2267 [c]
3	2	3	-	-	300	3697 [d]
4	3	1	-	-	300	2788 [e]
5	3	2	-	-	300	4059 [f]
6	3	3	-	-	300	5111 [g]
7	5	1	-	-	300	2453 [h]
8	5	2	-	-	300	1930 [i]
9	5	3	-	-	300	1151 [j]

Note: "-" means not to add. Letters indicate significant difference analysis ($p < 0.05$, Student's t-test).

When the adventitious buds were inoculated into the MS medium for 10 days, we found that the adventitious buds of watercress could take root even without NAA in the medium. When the concentration of NAA in the medium was 0.3~0.7 mg/L, adventitious buds could be induced to take root, but the root length would be shorter (Table 3). Therefore, we suggest that MS medium be directly used to induce the rooting of adventitious buds of watercress.

Table 3. NAA concentration for inducing rooting of adventitious buds of watercress.

Number	NAA mg/L	Rooting Rate	Rooting Number (Root/Plant)	Root Length (cm)
CK	-	100%	21.67 ± 0.58 [a]	6.30 ± 0.1 [a]
1	0.3	100%	35.67 ± 0.58 [b]	5.33 ± 0.25 [b]
2	0.5	100%	30.00 ± 1 [c]	3.90 ± 0.36 [c]
3	0.7	100%	27.00 ± 1 [d]	1.63 ± 0.15 [d]

Note: "-" means not to add. Values indicate mean ± SD. Letters indicate significant difference analysis ($p < 0.05$, Student's t-test).

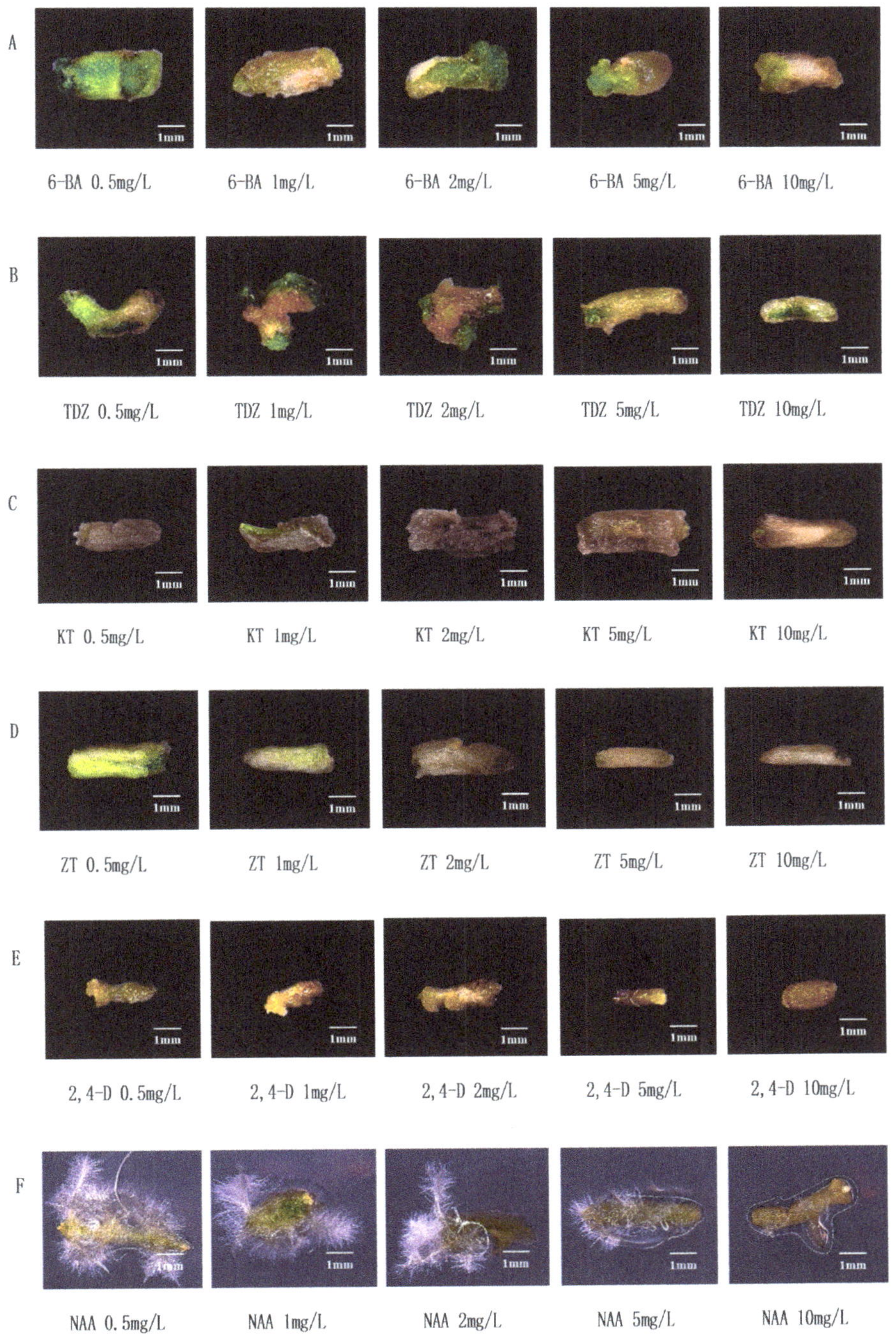

Figure 2. Effect of common plant hormones on stem segments of watercress. (**A**) Effect of 6-BA on stem segments of watercress. (**B**) Effect of TDZ on stem segments of watercress. (**C**) Effect of KT on stem segments of watercress. (**D**) Effect of ZT on stem segments of watercress. (**E**) Effect of 2,4-D on stem segments of watercress. (**F**) Effect of NAA on stem segments of watercress.

Inositol and culture medium pH had obvious effects on plant tissue dedifferentiation and callus formation. We took the regeneration system as control and observed the stem segment of watercress by adding different inositol concentrations and setting different pH. Figure 3 shows that the addition of 50 mg/L inositol to the original regeneration system can promote callus expansion. When the pH was close to 5.2, the number of calli formed at the wound of the stem segment was significantly higher. Therefore, adding 50 mg/L inositol to the regeneration system and adjusting pH to 5.2 can form a more efficient regeneration system.

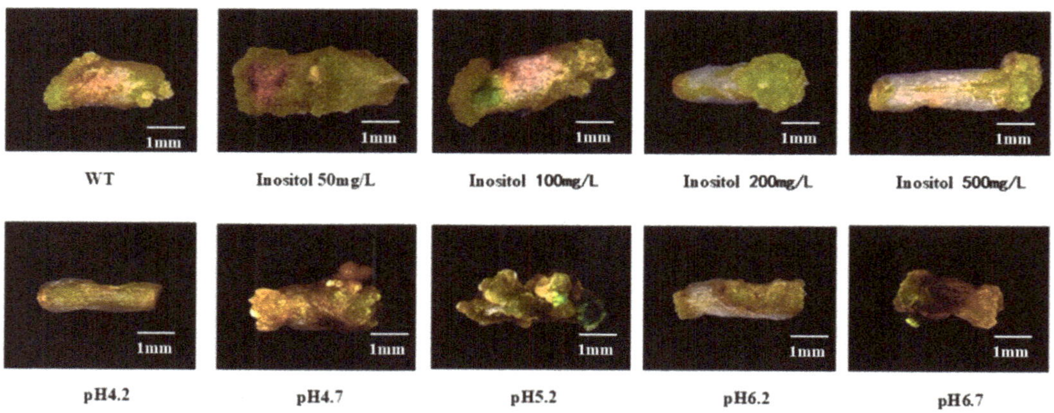

Figure 3. Effects of different inositol concentrations and pH on callus formation of watercress.

In summary, the stem segment of watercress was used as material to induce calli by adding 6-BA 4 mg/L + TDZ 1.5 mg/L + 2,4-D 1.5 mg/L + inositol 50 mg/L to the MS medium of pH 5.2, 6-BA 3 mg/L + TDZ 3 mg/L was added to the MS medium of pH 5.2 to induce callus differentiation and bud differentiation, and MS medium was used to induce adventitious bud rooting. Finally, an efficient regeneration system of watercress with a period of about 50 days was constructed (Figure 4).

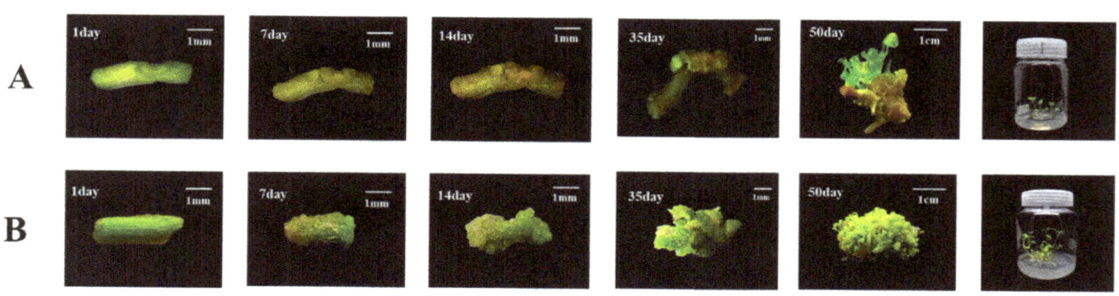

Figure 4. Comparison of watercress regeneration system before (**A**) and after (**B**) optimization.

3.3. Genetic Transformation Steps

As shown in Figure 5, the first step is to prepare *A. tumefaciens* and stem segments of the watercress. The second step is to pretreat the stem segments, and the third step is to infect the stem segments with an injection syringe for 3 min. The fourth step involves co-cultivation of the stem segments with *A. tumefaciens* in darkness. The fifth step is to induce callus formation by washing the stem segments and placing them in a specific medium. The sixth step is to induce the callus to form resistance buds. The seventh step involves the induction of resistant buds to form roots, and the eighth step involves the cultivation of plants with resistance through domestication and planting.

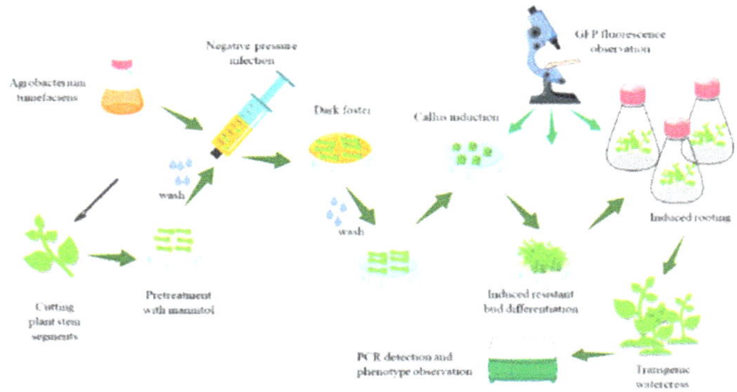

Figure 5. Steps of the genetic transformation system of watercress.

The specific operations are as follows: Cut off the bud primordia at the stem–leaf axil according to the ab dashed line in Figure 6A. After the stem is cut, preculture it in the MS medium (containing 0.6 M mannitol) for 24 h in the dark. Wash the pretreated stem segment, and place it together with the bacterial solution in the syringe. Draw in twice the volume of the stem segment of the bacterial solution, exclude the excess gas in the syringe, remove the needle, seal it with flames to form a closed environment, and pull the piston to form a negative pressure environment. This process lasts for about 3 min, shaking constantly to remove gas from the stem. If the time is too long, the stem will become transparent. If the time is too short, the bacterial solution cannot fully be in contact with the wound. After the infection, suck up the residual solution and place the stem in the co-culture medium, and co-culture in the dark at 24 °C for 3 days. Rinse the stem after the co-culture with sterile water for 10–20 min. The rinsed stem is then inoculated into the callus induction medium and incubated at 24 °C under 16 h of light for 15 days. During this process, the callus will gradually form and enlarge. The enlarged callus is then inoculated into the shoot differentiation medium and incubated at 24 °C under 16 h of light for 15 days. The callus will form resistant buds after Kan screening. Cut the appropriate size of resistant buds and inoculate them into the MS culture medium. Incubate at 24 °C under 16 h of light for 7 days, and the buds will spontaneously form roots. After the roots grow into the MS culture medium for 10 days, the regenerated seedlings are domesticated and transplanted.

3.4. Identification of Transgenic Watercress

We conducted a tracking study of the genetic transformation system of watercress (as shown in Figure 7A,B). During the 50-day regeneration process, we recorded the development of the control group and *DR5::EGFP* transgenic watercress and conducted GFP fluorescence and PCR detection. The transformation efficiency could reach up to 10.6% (Table S4). The color of the watercress stem segment changed from green to yellow after pretreatment. After 7–14 days of induction of calli, the control group formed a large number of yellow-green calli at the wound site, while the *DR5::EGFP* transgenic watercress formed yellow calli with stronger granules. After 21–35 days of growth on the shoot differentiation medium, the control group's callus continued to grow for a period of time before differentiating into a large number of green shoot buds. The *DR5::EGFP* transgenic watercress, on the other hand, differentiated into a small number of green-resistant shoot buds within a short period of time, and the buds continued to grow. At about 50 days, the control group of watercress formed a large number of shoots, while the number of regenerated shoots in *DR5::EGFP* transgenic watercress was rare.

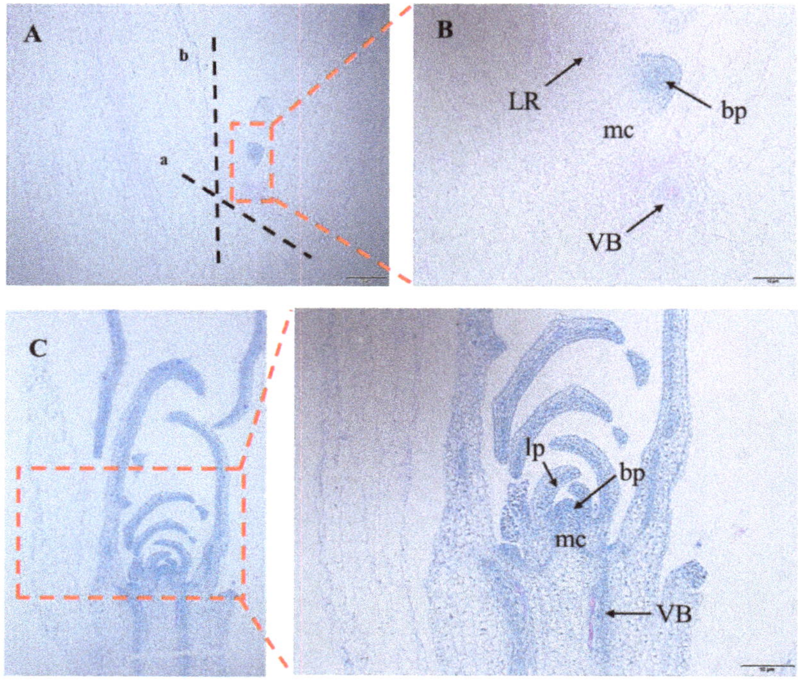

Figure 6. Meristem tissue slice of watercress. (**A**) Watercress leaf axil slice. (**B**) Enlarged view of leaf axil slice. (**C**) Slice of growth points of watercress. ab—cut point of watercress stem section. LR—root primordium. mc—meristematic cell. bp—bud primordium. lp—leaf primordium. VB—vascular bundle.

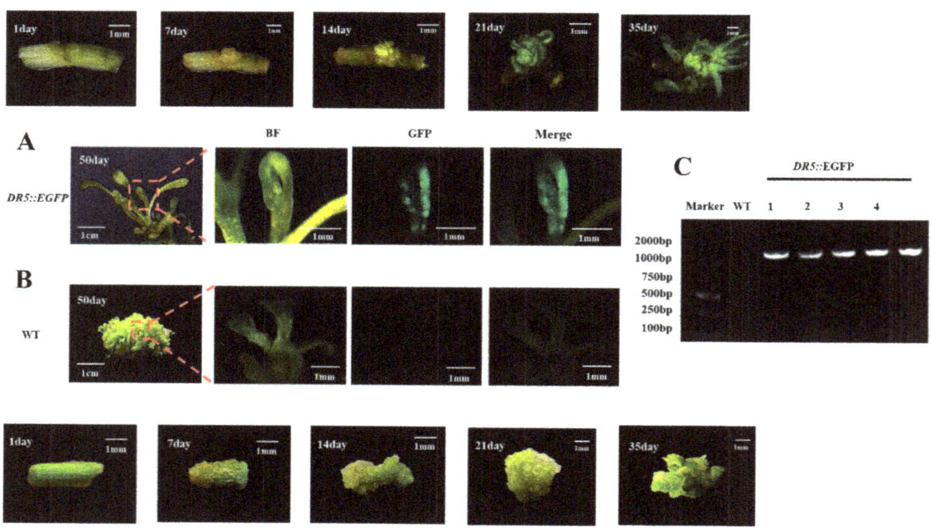

Figure 7. Regeneration process and PCR and GFP detection of the watercress. (**A**) Regeneration and GFP detection results of *DR5::EGFP* watercress. (**B**) Regeneration and GFP detection results of WT watercress. (**C**) PCR identification results of *DR5::EGFP* watercress.

After observing the GFP fluorescence of regenerated seedlings of the control group and *DR5::EGFP* transgenic watercress, it was found that the GFP fluorescence signal was mainly concentrated near the growing point of the regenerated seedlings. Therefore, the growing point of the watercress was made into paraffin sections (Figure 5). Combined with the GFP fluorescence distribution of *DR5::EGFP* transgenic watercress and the structure of the growing point, it was found that the main site of auxin in the growing point of watercress was in the bud primordium, leaf primordium, and meristematic zone, while there was no obvious GFP fluorescence signal in other areas. This indirectly indicates that during the growth process of watercress, the lateral branches of watercress can germinate, and the polarity distribution of auxin near the growing point is closely related.

4. Discussion

Watercress, a valuable source of nutrients with high nutritional and medicinal value, has garnered significant research attention in areas such as pharmacology, cultivation methods, and water purification [13,15]. However, studies focusing on the molecular mechanism and cell engineering breeding of watercress remain scarce. Building upon our exploration of the effects of common cytokinins and auxins on stem segments, we conducted gradient experiments to establish a more efficient regeneration system with a cycle of 50 days. Further optimization was achieved through inositol and pH gradients, ultimately leading to the development of a highly efficient regeneration system. Our work successfully bridges the gap in efficient watercress regeneration systems, both domestically and internationally.

Transcriptome sequencing analysis of watercress has revealed numerous candidate genes related to metabolism, such as phenylpropanoid biosynthesis and glucosinolate metabolism [16–18]. Through sequence alignment, many genes related to phosphorus starvation response, roots, and root hair development were identified in watercress [19]. However, the lack of an effective genetic transformation system has hindered the functional validation of these candidate genes. The construction of a genetic transformation system faces many unknown challenges. Different plant hormones have varying effects on watercress regeneration. After conducting gradient experiments with a single hormone concentration, we selected the plant hormone that had the least harm to stem segments and the most significant effect to establish the watercress regeneration system (Tables 1–3). Inositol is often added to plant tissue culture to regulate cell division, which has a significant effect on callus development. The level of endogenous inositol in soybean decreased and cell division was inhibited. On the contrary, in the process of tissue culture, the concentration of inositol was on the high side, and the inhibition of cell division was not conducive to callus growth [20,21]. The addition of 50 mg/L of inositol to the watercress callus induction medium significantly increased the callus volume (Figure 3). Tissue sections of the watercress showed that the cell walls at the leaf axils were relatively thick compared to other parts (Figure 5), which hindered sufficient contact between the Agrobacterium and meristematic cells. Therefore, we used different concentrations of mannitol to determine that 0.6 M mannitol could dehydrate the protoplasts of watercress while keeping them alive (Table S5). Watercress is rich in glucosinolates and can inhibit the activity of fungi, which is one of the reasons why genetic transformation is difficult to succeed [22]. According to the report, the tissue of grape and other plants will isolate the bacterial liquid, and the negative pressure infection is beneficial to the direct contact between the bacterial liquid and the material and increases the probability of infection [23,24]. Thus, stem segments were pretreated with 0.6 M mannitol + 4 mg/L 6-BA before infection, allowing a large number of meristematic cells to undergo protoplast wall separation while remaining active. The purpose of using negative pressure infiltration in the syringe was to ensure sufficient contact between the bacterial solution and watercress cells without affecting the viability of the stem segments. During the optimization process of the regeneration system, it was found that a pH 5.2 culture environment could induce abundant callus formation. We believed that this was due to the effect of the lower pH on the form of plant hormones such as auxin (Tables 1–3). Moreover, an acidic environment is also beneficial for activating the vir genes. Therefore, the pH of the environment was maintained at 5.2 from infection to the differentiation of resistant buds. From the success rate of our genetic transformation, it can be seen that mannitol pretreatment is the key to the genetic transformation system, while negative pressure infection can only increase the transformation success rate.

Currently, there are very few reports on genetic transformation methods in watercress, mainly through root-inducing Agrobacterium-mediated transformation to obtain transgenic watercress roots. Nam Il Park constructed a root-inducing Agrobacterium-mediated transformation method [25] and successfully obtained transgenic watercress roots with the reporter gene GUS [26]. However, there have been no reports on the genetic transformation of watercress shoots or the regeneration of whole transgenic watercress plants [27]. In our study, we successfully established a genetic transformation system for watercress shoots using Agrobacterium-mediated transformation. The key steps of this system include mannitol pretreatment of stem segments, negative pressure infiltration of the bacterial solution, and maintaining a pH 5.2 culture environment. By optimizing these steps, we were able to achieve high transformation efficiency and obtain transgenic watercress shoots. The transgenic watercress shoots obtained in our study were confirmed using PCR analysis and showed stable expression of the introduced gene. This provides a foundation for further functional studies and genetic improvement of watercress. In conclusion, we have developed a successful genetic transformation system for watercress shoots using Agrobacterium-mediated transformation. This system can be used for the introduction of desirable traits into watercress and for the study of gene functions in this important vegetable crop. Further research is needed to optimize the regeneration process and to explore the potential applications of this system in watercress breeding and genetic improvement.

Supplementary Materials: The following supporting information can be downloaded at: https://www.mdpi.com/article/10.3390//s1, Figure S1. *DR5: GFP* fluorescence of EGFP transgenic watercress; Table S1. PCR detection parameters; Table S2. Efficient regeneration parameters of watercress; Table S3. Formula of genetic transformation system of watercress; Table S4. Conversion rate of genetic transformation system of watercress; Table S5. Effect of different concentrations of mannitol on protoplast morphology of watercress.

Author Contributions: Conceptualization, X.H. and J.R. J.R. and Q.D. conducted the experiments, analyzed the data, and wrote the manuscript. Y.S. and Z.G. revised the manuscript. G.W., Y.G. and X.M. helped prepare the samples. Funding acquisition, X.H. J.R. and Q.D. contribute equally to the article. All authors have read and agreed to the published version of the manuscript.

Funding: This study was supported by the National Natural Science Foundation of China (32072575), the National Vegetable Industry Technology System (CARS-23-A16), Jiangsu province's seed industry revitalization project (JBGS [2021]015), and the Priority Academic Program Development of Jiangsu Higher Education Institutions (PAPD).

Data Availability Statement: Data are available within the article and Supplementary Materials.

Conflicts of Interest: The authors declare no conflict of interest.

References

1. Clemente, M.; Miguel, M.D.; Felipe, K.B.; Gribner, C.; Moura, P.F.; Rigoni, A.G.R.; Fernandes, L.C.; Carvalho, J.L.S.; Hartmann, I.; Piltz, M.T.; et al. Acute and sub-acute oral toxicity studies of standardized extract of *Nasturtium officinale* in Wistar rats. *Regul. Toxicol. Pharmacol.* **2019**, *108*, 104443. [CrossRef]
2. Lee, T.H.; Khan, Z.; Subedi, L.; Kim, S.Y.; Lee, K.R. New bis-thioglycosyl-1, 1′-disulfides from Nasturtium officinale R. Br. and their anti-neuroinflammatory effect. *Bioorganic Chem.* **2019**, *86*, 501–506. [CrossRef]
3. Li, K.; Lin, L.; Wang, J.; Xia, H.; Liang, D.; Wang, X.; Liao, M.A.; Wang, L.; Liu, L.; Chen, C.; et al. Hyperaccumulator straw improves the cadmium phytoextraction efficiency of emergent plant Nasturtium officinale. *Environ. Monit. Assess.* **2017**, *189*, 374. [CrossRef] [PubMed]
4. Ogita, S.; Usui, M.; Shibutani, N.; Kato, Y. A simple shoot multiplication procedure using internode explants, and its application for particle bombardment and Agrobacterium-mediated transformation in watercress. *J. Plant Res.* **2009**, *122*, 455–463. [CrossRef] [PubMed]

5. Vissenberg, K.; Claeijs, N.; Balcerowicz, D.; Schoenaers, S. Hormonal regulation of root hair growth and responses to the environment in Arabidopsis. *J. Exp. Bot.* **2020**, *71*, 2412–2427. [CrossRef] [PubMed]
6. Xue, C.; Li, W.; Shen, R.; Lan, P. Impacts of iron on phosphate starvation-induced root hair growth in Arabidopsis. *Plant Cell Environ.* **2023**, *46*, 215–238. [CrossRef]
7. Tanaka, H.; Dhonukshe, P.; Brewer, P.B.; Friml, J. Spatiotemporal asymmetric auxin distribution: A means to coordinate plant development. *Cell. Mol. Life Sci. CMLS* **2006**, *63*, 2738–2754. [CrossRef]
8. Tanaka, N.; Kato, M.; Tomioka, R.; Kurata, R.; Fukao, Y.; Aoyama, T.; Maeshima, M. Characteristics of a root hair-less line of *Arabidopsis thaliana* under physiological stresses. *J. Exp. Bot.* **2014**, *65*, 1497–1512. [CrossRef]
9. Kleine-Vehn, J.; Wabnik, K.; Martinière, A.; Łangowski, Ł.; Willig, K.; Naramoto, S.; Friml, J. Recycling, clustering, and endocytosis jointly maintain PIN auxin carrier polarity at the plasma membrane. *Mol. Syst. Biol.* **2011**, *7*, 540. [CrossRef]
10. Zhang, Y.; Xu, F.; Ding, Y.; Du, H.; Zhang, Q.; Dang, X.; Cao, Y.; Dodd, I.C.; Xu, W. Abscisic acid mediates barley rhizosheath formation under mild soil drying by promoting root hair growth and auxin response. *Plant Cell Environ.* **2021**, *44*, 1935–1945. [CrossRef]
11. Tabesh, M.; Sh, M.E.; Etemadi, M.; Naddaf, F.; Heidari, F. The antibacterial activity of *Nasturtium officinale* extract on common oral pathogenic bacteria. *Niger. J. Clin. Pract.* **2022**, *25*, 1466–1475. [PubMed]
12. He, F.; Nair, G.R.; Soto, C.S.; Chang, Y.; Hsu, L.; Ronzone, E.; DeGrado, W.F.; Binns, A.N. Molecular basis of ChvE function in sugar binding, sugar utilization, and virulence in Agrobacterium tumefaciens. *J. Bacteriol.* **2009**, *191*, 5802–5813. [CrossRef] [PubMed]
13. Sadeghi, H.; Azarmehr, N.; Razmkhah, F. The hydroalcoholic extract of watercress attenuates protein oxidation, oxidative stress, and liver damage after bile duct ligation in rats. *J. Cell. Biochem.* **2019**, *120*, 14875–14884. [CrossRef] [PubMed]
14. El-Bayoumy, K.; Stoner, G. Use of Freeze-dried Watercress for Detoxification of Carcinogens and Toxicants in Smokers: Implications of the Findings and Potential Opportunities. *Cancer Prev. Res.* **2022**, *15*, 139–141. [CrossRef]
15. Schulze, H.; Hornbacher, J.; Wasserfurth, P.; Reichel, T.; Günther, T.; Krings, U.; Krüger, K.; Hahn, A.; Papenbrock, J.; Schuchardt, J.P. Immunomodulating Effect of the Consumption of Watercress (*Nasturtium officinale*) on Exercise-Induced Inflammation in Humans. *Foods* **2021**, *10*, 1774. [CrossRef] [PubMed]
16. Bong, S.J.; Jeon, J.; Park, Y.J.; Kim, J.K.; Park, S.U. Identification and analysis of phenylpropanoid biosynthetic genes and phenylpropanoid accumulation in watercress (*Nasturtium officinale* R. Br.). *Biotech* **2020**, *10*, 260. [CrossRef] [PubMed]
17. Wielanek, M.; Królicka, A.; Bergier, K.; Gajewska, E.; Skłodowska, M. Transformation of *Nasturtium officinale*, Barbarea verna and Arabis caucasica for hairy roots and glucosinolate-myrosinase system production. *Biotechnol. Lett.* **2009**, *31*, 917–921. [CrossRef]
18. Jeon, J.; Bong, S.J.; Park, J.S.; Park, Y.-K.; Arasu, M.V.; Al-Dhabi, N.A.; Park, S.U. De novo transcriptome analysis and glucosinolate profiling in watercress (*Nasturtium officinale* R. Br.). *BMC Genom.* **2020**, *18*, 401. [CrossRef]
19. Hibbert, L.; Taylor, G. Improving phosphate use efficiency in the aquatic crop watercress (*Nasturtium officinale*). *Hortic. Res.* **2022**, *9*, uhac011. [CrossRef]
20. Staudt, G. The Effect of myo-Inositol on the Growth of Callus Tissue in Vitis. *J. Plant Physiol.* **1984**, *116*, 161–166. [CrossRef]
21. Biffen, M.; Hanke, D.E. Reduction in the level of intracellular myo-inositol in cultured soybean (Glycine max) cells inhibits cell division. *Biochem. J.* **1990**, *265*, 809–814. [CrossRef] [PubMed]
22. Freitas, E.; Aires, A.; de Santos Rosa, E.A.; Saavedra, M.J. Antibacterial activity and synergistic effect between watercress extracts, 2-phenylethyl isothiocyanate and antibiotics against 11 isolates of Escherichia coli from clinical and animal source. *Lett. Appl. Microbiol.* **2013**, *57*, 266–273. [CrossRef] [PubMed]
23. Simmons, C.W.; VanderGheynst, J.S.; Upadhyaya, S.K. A model of Agrobacterium tumefaciens vacuum infiltration into harvested leaf tissue and subsequent in planta transgene transient expression. *Biotechnol. Bioeng.* **2008**, *102*, 965–970. [CrossRef] [PubMed]
24. Tague, B.W.; Mantis, J. In planta Agrobacterium-mediated transformation by vacuum infiltration. *Arab. Protoc.* **2006**, *323*, 215–223.
25. Voutsina, N.; Payne, A.C.; Hancock, R.D.; Clarkson, G.J.J.; Rothwell, S.D.; Chapman, M.A.; Taylor, G. Characterization of the watercress (*Nasturtium officinale* R. Br.; *Brassicaceae*) transcriptome using RNASeq and identification of candidate genes for important phytonutrient traits linked to human health. *BMC Genom.* **2016**, *17*, 378.
26. Martin, R.E.; Marzol, E.; Estevez, J.M.; Muday, G.K. Ethylene signaling increases reactive oxygen species accumulation to drive root hair initiation in *Arabidopsis*. *Development* **2022**, *149*, dev200487. [CrossRef]
27. Park, N.I.; Kim, J.K.; Park, W.T.; Cho, J.W.; Lim, Y.P.; Park, S.U. An efficient protocol for genetic transformation of watercress (*Nasturtium officinale*) using Agrobacterium rhizogenes. *Mol. Biol. Rep.* **2011**, *38*, 4947–4953. [CrossRef]

MDPI
St. Alban-Anlage 66
4052 Basel
Switzerland
www.mdpi.com

Plants Editorial Office
E-mail: plants@mdpi.com
www.mdpi.com/journal/plants

www.ingramcontent.com/pod-product-compliance
Lightning Source LLC
Chambersburg PA
CBHW061501210926
51850CB00008B/1782
9783725810352